T0215284

Philosophy of Mathematics

PHILOSOPHY OF MATHEMATICS

AN INTRODUCTION

DAVID BOSTOCK

A John Wiley & Sons, Ltd., Publication

This edition first published 2009
© 2009 by David Bostock

Blackwell Publishing was acquired by John Wiley & Sons in February 2007. Blackwell's publishing program has been merged with Wiley's global Scientific, Technical, and Medical business to form Wiley-Blackwell.

Registered Office
John Wiley & Sons Ltd, The Atrium, Southern Gate, Chichester, West Sussex, PO19 8SQ, United Kingdom

Editorial Offices
350 Main Street, Malden, MA 02148-5020, USA
9600 Garsington Road, Oxford, OX4 2DQ, UK
The Atrium, Southern Gate, Chichester, West Sussex, PO19 8SQ, UK

For details of our global editorial offices, for customer services, and for information about how to apply for permission to reuse the copyright material in this book please see our website at www.wiley.com/wiley-blackwell.

The right of David Bostock to be identified as the author of this work has been asserted in accordance with the Copyright, Designs and Patents Act 1988.

Library of Congress Cataloging-in-Publication Data

Bostock, David.
  Philosophy of mathematics : an introduction / by David Bostock.
    p. cm.
  Includes bibliographical references and index.
  ISBN 978-1-4051-8992-7 (hardcover : alk. paper) — ISBN 978-1-4051-8991-0 (pbk. : alk. paper)
1. Mathematics—Philosophy. I. Title.
  QA8.4.B675 2009
  510.1—dc22

                                                                              2008026377

A catalogue record for this book is available from the British Library.

Set in 10.5/13pt Minion by Graphicraft Limited, Hong Kong

1 2009

Contents

Contents   vii

Preface

This book is intended for those who have already encountered a little philosophy, and who wish to find out something about one of its more specialized areas. I hope that it will also be of interest to those studying mathematics who would like an outsider's view of their subject. I have had to presuppose an elementary knowledge of modern logic, for without this the subject cannot be understood. But it is a background which I expect that both budding philosophers and budding mathematicians will have acquired already. I do not presuppose any mathematics beyond what can be learnt at school. While there are indeed some interesting philosophical questions to be asked about much more advanced areas of mathematics, they are not pursued here, for an introduction should confine attention to the problems that one meets at an elementary level.

The book is organized historically. The first three chapters concern philosophical contributions to the subject from the beginning up to about 1900, and the fourth notes some developments in mathematics during the same period. Particularly in the first two chapters, but occasionally elsewhere, it has been necessary to discuss some problems of interpretation, i.e. problems in establishing what a philosopher's views actually were. But the emphasis throughout is not on this, but on becoming clear about the merits and demerits of those views. While the first three chapters aim to say something of all the major philosophical contributions before 1900, the fourth is much more selective. It is concerned only with those developments in mathematics itself which have influenced the philosophers' views of it. I have not kept rigidly to the date 1900, but it does roughly correspond to an important distinction. With some exceptions one can say that developments before 1900, in both philosophy and mathematics, were very little influenced by what we now call modern logic. But after 1900 that influence was fundamental. One obvious exception is Frege, whose positive views I have treated as post-1900, though in fact his main work was done before then. This is because it was Frege who created modern logic, and he was the first to emphasize its importance for our subject. In the other direction I have made an exception of the Vienna Circle, and its English representative A.J. Ayer, by treating their views as pre-1900. In fact their work belongs to the 1920s and 1930s, but they were consciously responding to earlier problems.

The next four chapters are devoted to four main schools of thought that one may think of as opening the twentieth century, namely logicism, formalism, intuitionism, and what I have called predicativism. Again, this date is only rough. Logicism begins with Frege, and I have already said that I have therefore treated him as post-1900. Formalism also had its adherents well before 1900, but I have passed over them in silence, in order to begin with Hilbert, for he first gave the subject an interesting and a detailed exposition. The tradition has been to treat logicism and formalism and intuitionism as 'the big three' of this period,[1] and to give no separate treatment of predicative theories. This is understandable, since the predicative line of thought was in effect introduced by Russell, and he thought of it as forming part of his logicist programme. But in my view this was a mistake on Russell's part, and the two should be separated.

The final chapter takes up some issues belonging to the later part of the twentieth century, and here my own views have played a larger part in the discussion than they do in earlier chapters. It will be obvious that these views of mine are controversial, and by no means orthodox. Elsewhere in the book I have of course pointed to difficulties in the views of others, but my own ideas have remained rather in the background.

I have been interested in the philosophy of mathematics ever since I first began teaching and lecturing on philosophy 45 years ago. Naturally my views have developed over time, in response to various criticisms from pupils and colleagues, but I am surprised to find how many of my earlier opinions I still retain. The first book that I published was on this subject, namely my *Logic and Arithmetic*, and although I now see well enough that much of the first volume (1974) was mistaken,[2] I still think that the second

---

[1] For example Shapiro (2000) entitles his discussion of these schools 'The Big Three', and this attitude is common.

[2] I still endorse the main points made in its first and last chapters, but much of the rest of the book was taken up with a mistaken approach to type-neutrality. I have tried to put that right in my (1980).

volume (1979) contains a lot of truth. So in chapter 9 of this book I have found it natural on several occasions to refer back to that one, for a fuller and more detailed account of the approach to ontology that I still endorse.

I end this preface by extending a general thank you to all who have contributed to my thoughts, for they are too many to list individually. But I add a special thank you to an anonymous reader for Blackwell Publishing, who gave many detailed and helpful comments on all parts of the book. I have often profited from his or her advice.

David Bostock
Merton College, Oxford

Chapter 1

Plato versus Aristotle

A. Plato

1. The Socratic background[1]

Plato's impetus to philosophize came from his association with Socrates, and Socrates was preoccupied with questions of ethics, so this was where Plato began.

A point which had impressed Socrates was that we all used the notions of goodness and beauty and virtue – and again of the particular virtues such as courage, wisdom, justice, piety, and so on – but could not *explain* them. Faced with a question such as 'what really is goodness?', or 'what really is justice?', we soon found ourselves unable to answer. Most ordinary people would begin by seeing no real problem, and would quite confidently offer a first-off response, but Socrates would then argue very convincingly that this response could not be right. So they would then try various other answers, but again Socrates would show that they too proved unsatisfactory when properly examined. So he was led to conclude that actually we did not know what we were talking about. As he said, according to the speech that Plato gives him in his *Apology* (21b–23b), if there is any way in which he is wiser than other people, it is just that others are not aware of their lack of knowledge. On the questions that concern him he is alone in knowing that he does *not* know what the answers are.

It may be disputed whether and to what extent this is a fair portrait of the actual Socrates, but we need not enter that dispute. At any rate it is

---

[1] This section involves several conjectures on my part. I have attempted to justify them in my (1986, pp. 94–101).

Plato's view of Socrates, as is clear from Plato's early writings, and that is what matters to us.

Now Socrates was not the only one to have noticed that there were such problems with the traditional Greek ethics, and other thinkers at the time had gone on to offer their own solutions, which were often of a subjectivist – or one might even say nihilist – tendency. For example, it was claimed that justice is simply a matter of obeying the law, and since laws are different from one city to another, so too justice is different from one city to another. Moreover, laws are simply human inventions, so justice too is simply a human invention; it exists 'by convention' and not 'by nature'. The same applies to all of morality. And the more cynical went on: because morality is merely a human convention, there is no reason to take it seriously, and the simple truth of the matter is just that 'might is right'.[2]

Socrates did not agree at all. He was firmly convinced that this explanation did not work, and that morality must be in some way 'objective'. Plato concurred, and on this point he never changed his mind. He always held that when we talk about goodness and justice and so on then *there is* something that we are talking about, and it exists quite independently of any human conventions. So those who say that might is right are simply mistaken, for that is not the truth about what rightness is. But there is a truth, and our problem is to find it. However, this at once gives rise to a further question: why does it seem to be so very difficult to reach a satisfying view on such questions? And I believe that Plato's first step towards a solution was the thought that the difficulty arises because there are no clear *examples* available to us in this world.

Normally the meaning of a word is given (at least partly) by examples, for surely you could not know what the word 'red' means if you had never seen any red objects. But Plato came to think that the meaning of *these* words cannot be explained in the same way. One problem is that what is the right thing to do in one context may also be the wrong thing to do in another. Another problem is that we *dispute* about alleged examples of rightness, in a way in which we never seriously dispute whether certain things are red or one foot long, or whatever. In such cases we have procedures which are generally accepted and which will settle any disputed questions, but there are no such generally accepted procedures for

---

[2] For some examples of such opinions see, e.g., the views attributed to Callicles in Plato's *Gorgias* (482c–486c), and to Thrasymachus in Plato's *Republic* (336b–339a). See also the theory proposed by Antiphon in his fragment 44A (D/K).

determining whether something is right or wrong. (To mention some modern examples, consider abortion, euthanasia and capital punishment.) Plato also gives other reasons – different reasons at different places in his early writings – for holding that, in the cases which concerned him, there are no unambiguous examples available to us. I shall not elaborate on these, but simply observe that this situation evidently leads to a puzzle: how *do* we understand what these words mean, and how could our understanding be improved?

It was this question, I believe, that first led Plato to a serious interest in mathematics, for it seemed to provide a hopeful analogy. In mathematics too there are no clear examples, available to our perception, of the objects being considered (e.g. the numbers), and yet mathematical knowledge is clearly possible. So perhaps the same might apply to ethics?

2.   The theory of recollection

(i) Meno *80b–86d*

In the first half of Plato's dialogue *Meno*, the topic has been 'what is virtue?' The respondent Meno has offered a number of answers to this question, and Socrates has (apparently) shown that each of them is inadequate. Naturally, Meno is frustrated, and he asks how one could set about to seek for an answer to such a question, and how – even if one did happen to stumble upon the answer – one could recognize it as the right answer. In broad terms his question is: how is it possible ever to enquire into a topic such as this? In answer Socrates turns to an example from mathematics, which he hopes will show that one should not despair. He puts forward the theory that all (genuine) knowledge is really 'recollection', and offers to demonstrate this by a simple lesson in geometry.

He summons one of Meno's slave-boys, who has had no education in geometry, and poses this question: if we begin with a square of a given area (in this case of four square feet), how shall we find a square that has double that area? As is usual with a Socratic enquiry, the slave first offers an answer that is obviously wrong, and then thinks a little and proposes another which is also quite clearly wrong. He is then stumped. So (as *we* would say) Socrates then takes him through a simple proof to show that, if we start with any square, then the square on its diagonal is twice the area of the one that we began with. To prove this, Socrates himself draws a crucial figure: start with the given square, then add to it three more squares that are equal to it, to give a larger square that is four times the given area. Then draw in the diagonals of the squares as shown:

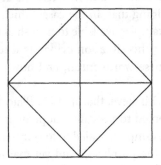

It is now easy to argue that the central square, formed by the four diagonals, is twice the area of the one that we began with. As Plato describes the lesson, at each step of the proof Socrates merely *asks* the boy some question, and the boy provides the correct answer, as if this at least is a point that he already knows. (For example: does the diagonal of a square bisect that square into two equal halves? – Yes. And what do we get if we add four of those halves together? And so on.) The moral that we are invited to draw is that the boy already possessed the knowledge of each individual step in the argument, and so Socrates' questioning has merely brought back to his mind some result that he really knew all along. Here Plato is evidently exaggerating his case, for although our slave-boy might have known beforehand all the premises to the proof, there is obviously no reason to think that, before Socrates questioned him, he had ever put those premises together in the right way to see what conclusion followed from them. But this criticism is of no real importance to the example.

We may press home the force of Plato's illustration in several ways that Plato himself fails to emphasize. First, it is obvious to all readers that *Socrates* must know the answer to the problem before he starts, and that that is how he knows what figure to draw and what questions to ask about it. But this cannot be essential, for it is obvious that whoever first discovered this geometrical theorem did not have such help from someone else who already knew it. As Plato might have said, one can ask *oneself* the right questions, for mathematical discovery is certainly possible. Second, although Plato insists that the slave-boy was just 'drawing knowledge out of himself', he does not insist as he might have done that this knowledge cannot be explained as due simply to his perception of the diagram that Socrates has drawn in the sand. At least three reasons might be given for this: (i) the diagram was no doubt somewhat inaccurate, as diagrams always are, but that does not prevent us from grasping the proof; (ii) since it is indeed a *proof* that we grasp, we can see on reflection that the result will hold also for all other squares, whatever their size, and not just for the one drawn here; (iii) moreover, we see that this is a *necessary* truth, and that there *could not* be any exceptions to it, but no one diagram could reveal that.

So one moral that Plato certainly wishes us to draw is that mathematics uses *proof*, and that a proof is seen to be correct *a priori*, by using (as he might say) the 'eye of the mind', and not that of the body. Another moral that he probably wishes us to draw is that in this example, as in other cases of mathematical proof, the premises are also known *a priori* (e.g. one knows *a priori* that the diagonal of a square bisects it into two equal halves). If so, then mathematics is a wholly *a priori* study, nowhere relying on our perceptions. (But we shall see that if he did think this when he wrote the *Meno* – and I guess that he did – then later in the *Republic* he will change his mind.) A final moral, which he certainly draws explicitly to our attention, is that since we 'draw this knowledge out of ourselves' (i.e. not relying on perception), we must have been born with it. It is, in later language, 'innate'. From this he infers further that the soul (or mind, i.e. *psychē*) must have existed before we were born into this world, and so is immortal. But the *Meno* is itself rather evasive on how a previous existence might explain this supposedly innate knowledge, and for this I move on to my next passage.

(ii) Phaedo 72e–77d

The theme throughout Plato's dialogue *Phaedo* is the immortality of the soul. Several arguments for this are proposed, and one of them begins by referring back to the *Meno*'s recollection theory. But it then goes on to offer a rather different argument for this theory. Whereas the *Meno* had invoked recollection to explain how we can come to see the *truths* of mathematics (i.e. by means of proof), in the *Phaedo* its role is to explain how we grasp the *concepts* involved. Although the example is taken from mathematics in each case, still the application that is ultimately desired is to ethics, i.e. to such concepts as goodness and justice and so on.

The example chosen here is the concept of equality, and the overall structure of the argument is completely clear. The claim is that we do understand this concept, but that our understanding cannot be explained as due to the examples of equality perceived in this world, for there are no unambiguous examples. That is, whatever in this world is correctly called 'equal' is *also* correctly called 'unequal'. But of course the words 'equal' and 'unequal' do not mean the same as one another, so their difference in meaning must be explained in another way. The suggestion is that it should be explained by invoking our 'experience', in 'another world', of genuine and unambiguous examples. This we somehow bring with us when we are born into the world of perceptible things, and what we perceive here can trigger our recollection of it, but cannot by itself provide the understanding. The same is supposed to apply to all those other concepts that Plato is finding problematic.

Unfortunately Plato's discussion in the *Phaedo* does not make it clear just what 'defect' infects all perceptible examples of equality, though he does claim that we all recognize that there always is such a 'defect'. On one interpretation his point is just that no two perceptible things, e.g. sticks or stones, are ever *perfectly* equal, say in length or in weight or whatever. I think myself that this interpretation is highly improbable, for why might Plato have believed such a thing? Two sticks – e.g. matchsticks – can certainly *look* perfectly equal in length, and though *we* might expect that a microscope would reveal to us that the equality was not exact, we must recall that there were no microscopes in Plato's day. (In any case, would not the example of two things that *look* perfectly equal, e.g. in length, be enough to provide our understanding of the concept of perfect equality, at least in length?) A different interpretation of Plato's thought, which I find very much more plausible, is that although two perceptible things may be (or appear) perfectly equal in *one* respect, still they will also be unequal in *another*. One might apply such an idea in this way: even if two sticks do seem to be exactly the same length, the same shape, the same weight, the same colour, and so on, still they will not be in the same place as one another, and in that respect they are bound to be unequal. The intended contrast will then be with the objects of pure mathematics, which simply do not have places. For example, the two 'ones' which are mentioned in the equation '$2 = 1 + 1$' are really equal to one another in absolutely *every* way.[3] I am inclined to think myself that Plato's real thought was even more surprising than this suggests, but I do not need to explore that suggestion here.[4] At any rate, the main point is clearly this: the only examples of equality

---

[3]   Plato did think that this equation mentions two 'ones', or as we might say 'two units' which together compose the number 2, as I show in a moment.

[4]   I have defended my preferred interpretation in my (1986, chapter 4).

in this world are 'defective', because they are also examples of inequality, so they cannot explain what we do in fact understand.

It is clear that we are expected to generalize: there are no satisfactory examples in this world of *any* of the things that mathematics is really about. For example, arithmetic is about pluralities of units, units which really are equal to one another in all ways, and are in no way divisible into parts, but there are no such things in this world (*Republic*, 525d–526a; cf. *Philebus* 56d–e). Or again, geometry is about *perfect* squares, circles, and so on, which are bounded by lines of no thickness at all, and the things that we can perceive in this world are at best rough approximations (*Republic*, 510d–511a). Mathematics, then, is not about this world at all, but about what can metaphorically be called 'another world'. So our understanding of it can be explained only by positing something like an 'experience' of that 'other world', which on this theory will be something that happens *before* our birth into this world.

I should add one more detail to this theory. The recollection that ordinary people are supposed to have of that 'other world' is in most cases only a *dim* recollection. That is why most of us cannot *say* what goodness is, or what justice is, or even what equality is. We do have some understanding of these concepts, for we can use them well enough in our ordinary thought and talk, but it is not the full understanding that would enable us to 'give an account', i.e. to frame explicit definitions of them. So the philosopher's task is to turn his back on sense-perception, and to search within himself, trying to bring out clearly the knowledge that is in some sense latent within him. For that is the only way in which real understanding is to be gained. We know, from the case of mathematics, that this *can* be done. This gives us reason to hope that it can also be done for ethics too, for – as Plato sees them – the cases are essentially similar: there are no unambiguous examples in this world, but we do have some (inarticulate) understanding, and only recollection could explain that.

3. Platonism in mathematics

Henceforth I set aside Plato's views on ethics. What is nowadays regarded as 'Platonism' in the philosophy of mathematics has two main claims. The first is ontological: mathematics is about real objects, which must be regarded as genuinely existing, even though (in the metaphor) they do not exist 'in this world'. This metaphor of 'two worlds' need not be taken too seriously. An alternative way of drawing the distinction, which is also present in Plato's own presentation, is that the objects of mathematics are not 'perceptible objects', but 'intelligible objects'. This need not be taken to mean that they exist 'in a different place', but perhaps that they exist 'in a different way'. The main claim is just that they do exist, but are not objects that we perceive by sight or by touch or by any other such sense. The second claim is epistemological: we do know quite a lot about these objects, for mathematical knowledge genuinely is *knowledge*, but this knowledge is not based upon perception. In our jargon, it is *a priori* knowledge. This second claim about epistemology is quite naturally thought of as a consequence of the first claim about ontology, but – as I shall explain at the end of this chapter – there is no real entailment here. Similarly the first claim about ontology may quite naturally be thought of as a consequence of the second, but again there is no real entailment. However, what is traditionally called 'Platonism' embraces both of them.

Platonism is still with us today, and its central problem is always to see how the two claims just stated can be reconciled with one another. For if mathematics concerns objects which exist not here but 'in another world', there is surely a difficulty in seeing how it can be that we know so much about them, and are continually discovering more. To say simply that this knowledge is '*a priori*' is merely to give it a name, but not to explain how it can happen. As we have seen, in the *Meno* and the *Phaedo* Plato does have an account of how this knowledge arises: it is due to 'recollection' of what we once upon a time experienced, when we ourselves were in that 'other world'. This was never a convincing theory, partly because it takes very literally the metaphor of 'two worlds', but also because the explanation proposed quickly evaporates. If we in our present embodied state cannot even conceive of what it would be like to 'experience' (say) the number 2 itself, or the number 200 itself, how can we credit the idea that we did once have such an 'experience', when in a previous disembodied state, and now recollect it? Other philosophers have held views which have some similarity to Plato's theory of recollection, e.g. Descartes' insistence upon some ideas being innate, but I do not think that anyone else has ever endorsed his theory.

Indeed, it seems that quite soon after the *Meno* and the *Phaedo* Plato himself came to abandon the theory. At any rate, he does not mention it in the account of genuine knowledge that occupies much of the central books of his *Republic*, which was written quite soon after. Nor does it recur in his later discussion of knowledge in the *Theaetetus*.[5] Moreover, if he did quite soon abandon it, that would explain why Aristotle never mentions it in any of his numerous criticisms of Plato's theories. But, so far as one can see, Plato never proposed an alternative theory of how mathematical knowledge is possible, and this was left as a problem for his successors.

4. Retractions: the Divided Line in *Republic* VI (509d–511e)

On many subjects Plato's views changed as time went by, and this certainly applies to his views on the nature of mathematics. I give just one example, namely the simile of the Divided Line that we find in book VI of the *Republic*. Unfortunately it is not entirely clear just how this simile is to be interpreted, so I shall merely sketch the two main lines of interpretation. But I remark at the outset that my preference is for the second.

At this stage in the *Republic* the topic is what we have come to call Plato's 'theory of forms'.[6] The theory has been introduced in the *Phaedo*, where it is these so-called 'forms' that we are supposed to have encountered in a previous existence, and which we now dimly recollect. I think it is clear that in the *Phaedo* Plato's view of these supposed forms is muddled. On the one hand they are regarded as properties, common to the many perceptible things that are said to participate in them, and on the other hand they are also taken to be perfect examples of those properties, which perceptible things imperfectly resemble. Thus there is a single form of beauty, which all beautiful things participate in, but it is itself an object that is supremely beautiful, and which is imitated by other things that are beautiful, but are always less beautiful than it is. Similarly there is a form of justice that is itself perfectly just, a form of largeness that is itself perfectly large, and so on. This is the theory that is already familiar to us before we come to the *Republic*.

Our simile is introduced as a contrast between things that are visible and things that are intelligible, and the initial idea is the familiar point that the former are images or copies of the latter. But then Plato adds that this relation also holds *within* each realm, e.g. as some visible things are

---

[5]　The theory recurs at *Phaedrus* 249b–c, but I am inclined to think that it is there regarded as one of the 'poetical embellishments' that Socrates later apologizes for at *Phaedrus* 257a, i.e. as something that is not to be taken too seriously.

[6]　The word 'form' in this context is a more or less conventional translation of Plato's words '*idea*' and '*eidos*'.

Philosophy of Mathematics

images (shadows or reflections) of others. To represent this we are to consider a line, divided into two unequal parts, with each of those parts then subdivided in the same ratio. The one part represents the visible, and the other the intelligible, and the subdivisions are apparently described like this:

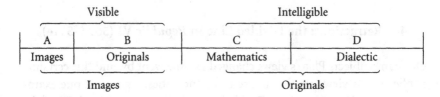

The stipulation is that A is to B (in length) as C is to D, and also as A + B is to C + D. (It follows that B is the same length as C, though Plato does not draw attention to that point.) The main problem of interpretation is obvious from the labels that I have attached to this diagram. Plato certainly introduces A + B and C + D as representing *objects* that are either visible or intelligible. He also seems to describe the sections A and B as each representing *objects* (namely: in A there are shadows and reflections, e.g. in water or in polished metal; in B there are the material objects which cause such images, e.g. animals and plants and furniture and so on). But, when he comes to describe the relationship between C and D, his contrast seems to be between different *methods of enquiry*, i.e. the method used in mathematics and the method that he calls dialectic. The first line of interpretation supposes that what Plato really has in mind all along is distinct kinds of *objects*, so (despite initial appearances) we must supply different kinds of object for sections C and D. The second supposes that Plato is really thinking throughout of different *methods*, so (despite initial appearances) we must supply different methods for sections A and B. I begin with the first.

There are different versions of this line of interpretation, but the best seems to me to be one that draws on information provided by Aristotle, though the point is not clearly stated anywhere in Plato's own writings.[7] Aristotle tells us that Plato distinguished between the forms proper and the objects of mathematics. Both are taken to be intelligible objects rather than perceptible ones, but the difference is that there is only one of each

---

[7]   As recent and distinguished proponents of this interpretation I mention Wedberg (1955, appendix) and Burnyeat (2000).

proper form (e.g. the form of circularity) whereas mathematics demands many perfect examples of each (e.g. many perfect circles). So the objects of mathematics are to be viewed as 'intermediate' between forms and perceptible things: they are like forms in being eternal, unchangeable, and objects of thought rather than perception; but they are like perceptible things insofar as there are many of each kind (Aristotle, *Metaphysics* A, 987^{b}14–18). If we grant this doctrine, it will be entirely reasonable to suppose that section C of the divided line represents these 'intermediate' objects of mathematics, while section D represents the genuine forms. But the problem with this interpretation is whether the doctrine should be granted.

One must accept that there is good reason for holding such a view, and that Aristotle's claim that Plato held it cannot seriously be questioned.[8] But one certainly doubts whether Plato had already reached this view at the time when he was writing the *Republic*, and I myself think that this is very unlikely. For, if he had done, why should he never state it, or even hint at it in any recognizable way?[9] Why should he tell us that what a geometer is really concerned with is 'the square itself' or 'the diagonal itself', when this is his standard vocabulary for speaking of *forms* (such as 'the beautiful itself', 'the just itself')? If he did at this stage hold the theory of intermediates, would you not expect him here to use plural expressions, such as 'squares themselves'? There is also a more general point in the background here. The theory which Aristotle reports surely shows the need to *distinguish* between a form, as a universal property, and any (perfect) instances of that property that there may be. But I do not believe that Plato had seen this need at the time when he was writing the *Republic*, for the confusion is clearly present at 597c–d of that work.[10] For these reasons I am sceptical of this first line of interpretation. Let us come to the second.

This second approach is to see the simile as really concerned throughout with methods of enquiry, and it is easy to see how to apply this line of thought. The sections of the line should be taken to represent:

[8]   The claim is repeated in chapter 1 of his *Metaphysics*, book M, where he also describes how other members of Plato's Academy reacted to this idea. He cannot just be making it up.

[9]   *Republic* 525d–526a can certainly be seen as implying that in mathematics there are *many* 'number ones'. But I doubt whether Plato had absorbed this implication.

[10]   It seems probable that Plato *later* recognized this as a confusion, for the dialogue *Parmenides* constructs several arguments which rely upon it and which quite obviously have unacceptable conclusions. (The best known is the argument that, since Aristotle, has been called the argument of 'The Third Man'. This is given at *Parmenides* 132a, cf. 132d–e.)

A: The indirect study of ordinary visible objects, *via* their images (e.g. shadows or reflections).

B: The direct study of such objects in the usual way.

C: The indirect study of intelligible forms, *via* their visible images (e.g. geometrical diagrams). This is the method of mathematics.

D: The direct study of such objects, using (let us say) pure reason and nothing else. This is the method of dialectic (i.e. of philosophy).

On this interpretation we do not have to suppose that Plato intends any ontological distinction between the objects of mathematics and those of dialectic. Visible images are more obviously relevant in the one case than in the other, but it is open to us to hold that the method of dialectic *could* be applicable to mathematical forms, and that the method of mathematics could be used in any study of forms. Moreover, there are some hints that this is what Plato intends. But before I come to this I must be more specific on what Plato takes the method of mathematics to be.

He proposes *two* features as characteristic of this method. The first is the one that we have mentioned already, namely that it uses visible diagrams. The second is that it proceeds by deduction from 'hypotheses', which are taken to be evident to all, and for which no justifications are given. Moreover, he apparently sees a close connection between these two features, saying that the one necessitates the other.[11] The usual explanation of this point is that the reason why these 'hypotheses' are thought to be evident is just that they seem obviously to fit the visible diagrams. (Plato's text never quite says this, but 511a comes close to it, and no better connection suggests itself.) Just what these 'hypotheses' are, in the case of mathematics, is not clearly explained,[12] but their overall role is obvious: they are the premises from which mathematical proofs start, and Plato has now recognized that there must be such premises.

Unfortunately we do not know enough about the state of mathematics in Plato's day to be able to say what he must have been thinking of. This is because Euclid's well-known *Elements*, which was written between 50 and

---

[11]   At *Republic* 510b the thought appears to be that the method is forced to use hypotheses because it employs visible images, but at 511a the connection appears to be the other way round. The latter suggestion is I think the better.

[12]   510c gestures towards some examples, but they appear to be examples of the *subject-matter* of such hypotheses, and we are left to guess at what *propositions* about these subjects are intended. (The text is: 'hypothesizing the odd and the even, and the [geometrical] figures, and three kinds of angle'.) See next note.

100 years later, was so clearly an improvement on what had come before it that it eclipsed all earlier work. We do know that there had been earlier 'Elements', and presumably they were known to Plato, but they have been lost, and we cannot say how closely their style resembled what we now find in Euclid. (Euclid distinguishes the premises into definitions, common notions, and postulates. Both of the last two we would class as axioms.) I think myself that it is quite a possible conjecture that earlier 'Elements' did not admit to any starting points that we would call axioms, but only to what we would classify as definitions. So I think it is quite possible that the 'hypotheses' which Plato is thinking of were mainly, and perhaps entirely, what we would call definitions.[13] But in any case what he has now come to see is that mathematical proofs do have starting points, and that these are not justified within mathematics itself, but simply assumed. For that reason he now says that they do not count as *known* (in the proper sense of the word), and hence what is deduced from them is not known either. In the *Meno* mathematics had certainly been viewed as an example of knowledge, but now in the *Republic* it is denied that status. This is a notable change of view.

It may not be quite such a clear-cut change as at first appears, for at 511d there is a strong hint that one *could* apply the dialectical method to the hypotheses of mathematics, thereby removing their merely hypothetical status. If so, then the implication is that mathematics *could* become proper knowledge, even though as presently pursued (i.e. at the time when Plato was writing the *Republic*) it is not. In the other direction I remark that something like the method of mathematics, i.e. a method which (for the time being) simply accepts certain hypotheses without further justification, can evidently be employed in many areas, including an enquiry into the nature of the moral forms. In fact *Meno* 86e–87c attempts to do just that in its enquiry into what virtue is (though the attempt does not succeed (*Meno* 86e–99c)); and *Republic* 437a invokes a 'hypothesis', which is left without further justification, and which plays an important role in its analysis of what justice is. In broad terms, Plato thinks of the method of mathematics as one that starts by assuming some hypotheses and then goes 'downwards' from them (i.e. by deduction), whereas the method of philosophy (i.e. dialectic) is to go 'upwards' from the initial

---

[13] This would explain why the elucidation offered at 510c (previous note) simply mentions certain concepts – i.e. concepts to be defined? – and does not give any propositions about them.

hypotheses, finding reasons for them (when they are true), until eventually they are shown to follow from an 'unhypothetical first principle'. While the 'downward' method is something which we have no difficulty in understanding, it is not easy to say quite how the 'upward' method is supposed to work, but I cannot here pursue that problem.[14]

At any rate, the upshot is this. Plato himself did not always remain quite the 'Platonist' about mathematics that I described in the previous section. This is because he came to think that mathematics (as presently pursued) begins from unjustified assumptions – or from assumptions that are 'justified' only in the wrong way, i.e. by appeal to visible diagrams – and that this means that it is not after all an example of the best kind of knowledge. Perhaps he also thought that this defect could in principle be remedied, but at any rate he has certainly pointed to a problem with the usual 'Platonic' epistemology: proofs start from premises, and it is not clear how we know that those premises are true. As for the ontology, he remains always a 'Platonist' in that respect. Either in the *Republic*, or (as I think more probable) at a later date, he became clearer about just what the objects of mathematics are, namely not the forms themselves but perfect examples thereof. But in any case they remain distinct from the ordinary perceptible objects of this world, accessible only to thought rather than perception, for they have a 'perfection' which is not to be found in this world. That is the main reason why, in the *Republic*, he lays down for the aspiring philosopher a lengthy and arduous preliminary training in mathematics: it is because this subject directs our mental gaze away from ordinary material things and towards what is 'higher'. (It may be that he *also* thought that the philosopher's 'upward path' towards an 'unhypothetical first principle' would start from reflection on the hypotheses of mathematics. But that is a mere speculation.[15])

In brief, the *Republic* indicates at least a hesitation over the epistemology, but no serious shift in the ontology. By way of contrast, let us now turn to Aristotle, whose views on both these issues were very different.

[14] A classic discussion of Plato's 'method of hypothesis', which assembles all the relevant evidence, is Robinson (1953, chapters 6–13). I have offered a few observations myself in my (1986, chapter 8).
[15] Book VII outlines five areas of mathematics, which are all to be studied, until their 'kinship' with one another is seen (531d). Is that perhaps because such a study will allow one to begin on the project of explaining the several initial hypotheses, by seeing how each may be viewed as an instance of some more general truth that explains them all?

B. Aristotle[16]

5. The overall position

In several places Aristotle gives us an outline sketch of his position on the nature of mathematics, and it is very clear that he rejects the Platonic account.[17]

First, he thinks that Plato was quite wrong to 'separate' the objects of knowledge from the ordinary objects of this world. He will agree with Plato that there are things which can be called 'forms', and that knowledge (properly speaking) is always knowledge of such forms, but he claims against Plato that these forms have no existence apart from their instances in this world. For example, there is a form of man, and there is a form of circle, but these forms exist only *in* actual men and actual circular objects (such as the top of a round table). He firmly denies the Platonic idea that there are, as it were, 'two worlds', one containing perceptible objects and the other imperceptible but intelligible objects. There is only the one world, and it is that world that mathematics is about.

It may *look* as if in geometry we are concerned with a special kind of object, as we say what properties 'the square' or 'the circle' possess, but actually we are speaking in very general and abstract terms of the properties which all ordinary square or circular objects have in common, simply in virtue of being square or circular. When engaged in a geometrical investigation, one does not think of the geometrical figures being studied as things which have weight or temperature or mobility and so on. But that is because it is only the geometrical properties of an object that are here in question. Of course, any ordinary object will have many other properties too, and the truths of geometry are truths about ordinary objects, but their other properties are here ignored as irrelevant. Aristotle presumably intends a similar account to apply to arithmetic. It may *look* as if we are concerned with some rather special objects called 'numbers', which have none of the properties of ordinary perceptible things. But actually we are just speaking at a very general and abstract level of what one might call 'embodied numbers', e.g. the number of cows in the field, or the number of coins on the table, and so on. These are things

---

[16] Much of this discussion is taken from the fuller treatment in my (2009b).

[17] The principal passages are *Metaphysics* M.3 and *Physics* II.2, 193^{b}22–194^{a}12. But see also *De Anima* III.7, 431^{b}12–17 and *Metaphysics* K.3, 1061^{a}28–^{b}4.

that we can perceive. That is the broad outline: the truths of mathematics are truths about perfectly ordinary objects, but truths at a high level of generality. Compared with Plato, it seems like a breath of fresh air.

On the topic of epistemology Aristotle similarly claims that our knowledge of such truths is again perfectly ordinary empirical knowledge, based upon perception in much the same way as all other scientific knowledge is based upon perception. He quite naturally supposes that if mathematics need not be understood as concerned with a special and 'separate' kind of object, then equally our mathematical knowledge need not be credited to a special and rather peculiar faculty for 'a priori' knowledge. So both the Platonic ontology and the Platonic epistemology are to be rejected.

That is the broad outline of his position. Unfortunately we do not find very much by way of argument for it. Certainly, Aristotle very frequently states objections to Plato's general theory of forms, construed as objects which enjoy a separate existence of their own. But that is not the end of the argument. For, as we have already noted (pp. 10–11), Plato eventually came to distinguish between the forms themselves and the objects studied in mathematics, regarding these latter as 'intermediate' between forms proper and perceptible things. Granted this distinction, one might for the sake of argument concede to Aristotle that he has good reason for rejecting the Platonic view of forms, but insist that the question of the separate existence of the objects of mathematics is not thereby settled. For these objects are not supposed to be forms, but to be perfect examples. And mathematics might still need perfect examples, which exist 'separately' from all the imperfect examples in this world, even if the same does not apply to the forms themselves.

There is only one place where Aristotle seriously addresses this question, namely in chapter 2 of book M of the *Metaphysics*, and his arguments there are less than compelling. I here pass over all the details, noting only this one general point. The two main arguments that Aristotle gives, at 1076^b11–39 and 1076^b39–1077^a14, aim to show that if we must assume the existence of those intermediates that Plato desires, then we must also assume the existence of many other intermediates too, which will lead to an incredible and quite needless duplication of entities. But he never tells us *why* Plato thought that these intermediates were in fact needed, nor how his reasons should be countered. As I have said, the usual explanation is this: Plato held that mathematics was about *perfect* examples, and so – since mathematics is true – there must *be* perfect examples. But no examples that we can perceive are perfect, so there must somewhere be imperceptible examples, available to the intellect but not to perception. Let us assume that this is indeed Plato's argument for supposing that his 'intermediates' are *needed*. Then the chief weakness in the counter-arguments that Aristotle presents in *Metaphysics* M.2 is that they have nothing to say about what is wrong with this Platonic argument. They simply do not address the opposition's case. Nor is there anywhere else in his writings where this line of thought is explicitly considered. So I now turn to consider what we might say on Aristotle's behalf.

6. Idealizations

It is a vexing feature of Aristotle's discussion that we cannot even be sure of whether he himself did or did not accept the Platonic premise that there are no perfect examples in this world. His main discussions are quite silent on this point, and although there are a couple of asides elsewhere they are not to be trusted.[18] What he *should* have done is to accept the premise for geometry but deny it for arithmetic, so let us take each of these separately.

Arithmetic
We, who have been taught by Frege, can clearly see that Plato was mistaken when he claimed that this subject introduces idealizations. The source of his error is that he takes it for granted that, when numbers are applied to ordinary perceptible objects, they are applied 'directly'; i.e. that it is the object itself that is said to have this or that number. But Frege made it quite clear that this is not so. In his language, a 'statement of number' makes an assertion about a *concept*, not an object, i.e. it says how many objects fall under that concept. (An alternative view, which for present purposes we need not distinguish, is that numbers apply not to physical objects but to sets of those objects, and they tell us how many members the set has.) To illustrate, one may ask (say) how many cows there are

---

[18] In the preliminary discussion of problems in *Metaphysics* B, we find the claim that perceptible lines are never perfectly straight or perfectly circular (997^{b}35–998^{a}6). But in the context there is no reason to suppose that Aristotle is himself endorsing this claim, rather than mentioning it as a point that might appeal to his Platonist opponent. On the other side, a stray passage in *De Anima* I.1 apparently claims that a material straight edge really does touch a material sphere at just one point (403^{a}12–16). But I am very suspicious of this passage, for as it stands it makes no sensible contribution to its context. I have discussed the passage in an appendix to my (2009b).

in this field, and then one is asking of the *concept* 'a cow in this field' how many objects fall under it. The answer will (in most cases) be entirely unambiguous, say sixteen. There is nothing 'imperfect' in this application of the number. One may say that here we take as our 'units' the cows in the field, but there is no implication that a cow is an indivisible object, or that the cows are 'equal' to one another in any respect beyond all being cows in this field. Nor is it implied that the matter in question could not be counted under some other concept – i.e. taking something else as the 'unit' – say pairs of cows or kilograms of cow.

Aristotle has grasped this point. He frequently compares counting to measuring, with the idea that in each case one *chooses* something as the 'unit', which is then *treated* for that purpose as indivisible. (E.g.: 'The measure must always be something that is the same for all [the things measured], for example if the measure is a horse then horses [are being measured], and if a man then men' (*Metaphysics* N.1, 1088^a8-9).[19] Notice that my supplement 'measured' would in each case be very naturally replaced by 'counted'.) What is somewhat surprising is that he never presents this as a criticism of Plato. He certainly argues against the conclusions that Plato was led to, and he points to a number of difficulties in the view that a number is 'really' a plurality of 'perfect units' which enjoy a 'separate' existence. (This is the theme of most of chapters 6–8 of book M, to 1083^b23. The arguments are often cogent, but I shall not discuss them here.[20]) But he does not seem to have asked just what it was that led Plato astray, so we get no diagnosis of the opponent's errors. Worse, there are hints in the positive account which he does give – and which I come to shortly – which leave us wondering just what he himself is proposing as an alternative to Plato's picture.

Geometry

It is fair to say that geometry 'idealizes', in that it concerns what has to be true of *perfect* squares, circles, and so on. But the first thing to say is that

---

[19]   I translate the mss. reading. Ross prefers to emend to '. . . if horses [are being measured] then the measure is a horse, and if men then a man'. But in either case the main idea is the same. Other passages of the *Metaphysics* which clearly show a good understanding of how numbers are applied in practice are: Δ.6, 1016^b17-24; I.1, 1052^b15-17, 1053^a24-30, 1054^a4-9; M.7, 1082^b16-19; N.1, $1087^b33-1088^a14$. Cf. also $1052^b31-1053^a2$; 1092^b19-20; *Physics* IV.12, 220^a19-22.

[20]   I shall also leave undiscussed Aristotle's arguments, which occupy most of book N, against the Platonic idea that the numbers (and other things) are somehow 'generated' from 'the one' and 'the indefinite dyad'.

what geometry claims about perfect circles may very well be true even if there are no perfect circles at all, for the claims may be construed hypothetically: *if* there are any perfect circles, then such-and-such will be true of them (e.g. they can touch a perfectly straight line at just one point and no more). One might ask how geometry can be so useful in practice if there are no such entities as it speaks of, but (a) this is a question for the Platonist too (since 'in practice' means 'in this perceptible world'), and anyway (b) the question is quite easy to answer.

We are nowadays familiar with a wide range of scientific theories which may be said to 'idealize'. Consider, for example, the theory of how an 'ideal gas' would behave – e.g. it would obey Boyle's law precisely[21] – and this theory of 'ideal' gases is extremely helpful in understanding the behaviour of actual gases, even though no actual gas is an ideal gas. This is because the ideal theory simplifies the actual situation by ignoring certain features which make only a small difference in practice. (In this case the ideal theory ignores the actual size of the molecules of the gas, and any attractive or repulsive force that those molecules exert upon one another.) But no one nowadays could suppose that because this theory is helpful in practice there must really *be* 'ideal gases' somewhere, if not in this world then in another; that reaction would plainly be absurd. Something similar may be said of the idealizations in geometry. For example, a carpenter who wishes to make a square table will use the geometric theory of perfect squares in order to work out how to proceed. He will know that in practice he cannot actually produce a *perfectly* straight edge, but he can produce one that is very nearly straight, and that is good enough. It obviously explains why the geometric theory of perfect squares is in practice a very effective guide for him. We may infer that geometry may perfectly well be viewed as a study of the spatial features – shape, size, relative position, and so on – of ordinary perceptible things. As ordinarily pursued, especially at an elementary level, it does no doubt involve some idealization of these features, but that is no good reason for saying that it is not really concerned with this kind of thing at all, but with some other 'ideal objects' that are not even in principle accessible to perception. All this, however, is on the assumption that geometry may be construed hypothetically: it tells us that *if* there are perfect squares, perfect circles, and so on, then they must have such-and-such

---

[21] For a body of gas maintained at the same temperature, where 'P' stands for the pressure that it exerts on its container, and 'V' for its volume, Boyle's law states that $PV = k$, for some constant k.

properties. That is helpful, because it implies that the approximate squares and circles which we perceive will have those properties approximately. But, one may ask, does not geometry (as ordinarily pursued) assert outright that *there are* perfect circles? That would distinguish it from the theory of ideal gases, and is a question which I must come back to.

7. Complications

Within the overall sketch of his position that Aristotle gives us in *Metaphysics* M.3, there are two brief remarks which certainly introduce a complication. At 1078ª2–5 he says, somewhat unexpectedly, that mathematics is not a study of what is perceptible, even if what it studies does happen to be perceptible. More important is 1078ª17–25, where he says that the mathematician *posits* something as separate, though it is not really separate. He adds that this leads to no falsehood, apparently because the mathematician does not take the separateness as one of his premises. A similar theme is elaborated at greater length in the outline given in *Physics* II.2, at 193ᵇ31–5. There again we hear that the mathematician, since he is not concerned with features accidental to his study, does *separate* what he is concerned with, for it can be separated 'in thought', even if not in fact. And again we are told that this leads to no falsehood. On the contrary, Aristotle seems to hold that such a fictional 'separation' is distinctly helpful, both in mathematics and in other subjects too (1078ª21–31).

Our texts do not tell us what kind of thing a mathematical object is conceived as being, when it is conceived as 'separate'. I think myself that the most likely answer is that it is conceived as the Platonist would conceive it, i.e. as existing in its own 'separate world', intelligible and not perceptible. Further, if – as seems probable – Aristotle concedes that perceptible objects do not *perfectly* exemplify the properties treated in elementary geometry, then it will presumably be this mental 'separation' that smoothes out the actual imperfections. But it must be admitted that this is pure speculation, and cannot be supported from anything in our texts.[22]

---

[22] I believe that Aristotle holds that when a geometrical figure is conceived as separate, it is also conceived as made of what he calls 'intelligible matter' (*Metaphysics* Z.10, 1036ª1–12; Z.11, 1036ᵇ32–1037ª5; H.6, 1045ª33–6). This is what allows us to think of a plurality of separate figures – e.g. circles – that are all exactly similar to one another. For what distinguishes them is that each is made of a different 'intelligible matter'.

I note further that in *Metaphysics* M.3, and for the most part in his programmatic discussions elsewhere, Aristotle is mainly thinking of geometry. But he clearly believes that something very similar applies to arithmetic too, so presumably we are to take it that the mathematician also conceives of the numbers as separate entities, though that too is a fiction. And if in the geometrical case it is the separation in thought that also allows for idealization, then we should perhaps think that when numbers are conceived as separate then they too are conceived in an idealized way. One supposes that this will again be a Platonic way, in which a number is conceived as made up of units which each have their own separate existence, which are perfectly equal to one another in every way, and which are in no way divisible. As I have said, much of *Metaphysics* M.6–8 argues (very successfully) that numbers cannot *really* be like this, but perhaps Aristotle means to concede that that is how mathematicians do in practice think of them.

That is one way in which the outline sketch that we began with becomes complicated. Another is that the objects of mathematics are said to exist *potentially* rather than actually.[23] One presumes that his thought here is that these objects, when considered as existing separately, can be said to exist potentially because it is possible for them to exist actually, i.e. to exist in actual physical objects. Thus a circle exists actually in a circular table-top, and the number 7 exists actually wherever there are (say) 7 cows. Then the idea will be that some rather complex geometrical figures, e.g. a regular eicosahedron, may not actually exist anywhere in the physical world, but this figure still has a potential existence because it could do so. The same would apply to a very large number, too large to be exemplified. But there may be a further complication to be added here, namely the idea that a mathematical object is brought into actual existence not only by being physically exemplified but also just by being thought of. At any rate, at *Metaphysics* Θ.9, 1051ª21–3, Aristotle notes that geometers will often prove some result by 'constructing' lines additional to those already given, and he comments that this construction makes actual what was previously only potential. Moreover, he must be taking it to be the construction *in thought* that matters, for he adds 'and the explanation is that thinking is actuality'. In any case, the main point is that Aristotle is conceding that, in

---

[23] *Metaphysics* M.3, 1078ª28–31 says that they exist not actually but 'in the way of matter'. Aristotle does constantly think of matter as existing potentially and not actually, and I do not believe that he intends the comparison with matter to extend any further than this.

a sense, 'there are' many more mathematical objects than are either actually embodied or actually thought of. He wishes to say that 'there are' such things in the sense that they have a potential existence if not an actual one. I shall consider in my next section whether this is an adequate solution to what must, for Aristotle, be a serious problem.

Meanwhile, let us sum up the position so far. Aristotle holds that mathematics is the study of certain properties which perfectly ordinary perceptible objects possess, and in that way its objects are just ordinary perceptible objects. But the study proceeds at a high level of generality, paying no heed to all the non-mathematical properties of these objects. It is therefore a convenient fiction to suppose that it concerns some special and peculiar objects which have *no* properties other than the mathematical ones. There are not really any such objects, but it does no harm to imagine that there are, and at the same time both to 'smoothe out' the small geometrical irregularities which actual physical objects are likely to display, and to expand the account by including geometrical and arithmetical properties that may not be exemplified at all. That is permissible because we are still concerned with objects that have at least a potential existence, if not an actual one. But still, the foundation of the subject must be the actual physical bodies and their actual geometrical and arithmetical properties. For that is where our understanding must begin.

This outline sketch leaves many questions unanswered, and one can only guess at the answers that Aristotle might have given. For example, I would *expect* him to say that a simple statement of pure arithmetic, such as '7 + 5 = 12', should not be interpreted as referring to some puzzling entities called 'the numbers themselves', but as generalizing over ordinary things in some such way as this: if there are 7 cows in one field, and 5 in another, then there are 12 in both fields taken together; and the same holds not only for cows and fields but also for everything else too. In fact he does not actually say this, or anything like it; he is completely silent on the meaning of arithmetical equations. Again, I would *expect* him to say that we find out that 7 + 5 = 12 by the ordinary procedure of counting cows in fields, and other such familiar objects. But in fact he never does explicitly address the question of how we come to know such truths of simple arithmetic, and he never does respond to the Platonic claim that the knowledge must be *a priori*.

Here too all that we have are some very general and programmatic pronouncements. In the well-known final chapter of the *Posterior Analytics* (i.e. II.19) he claims that *all* knowledge stems from experience. Perception is of particulars, but memory allows one to retain many particular cases in one's mind, and this gives one understanding of universals. This is put forward as an account of how one grasps 'by induction' the first principles of any science, and the similar account in *Metaphysics* A.1 makes it clear that mathematics is not an exception (981^a1–3, $^b20$–5). But it is quite clear that this says far too little. Indeed, Aristotle claims that we must somehow come to see that these first principles are *necessary* truths, but has no explanation of how we could ever do this. Elsewhere we find the different idea that what Aristotle calls 'dialectic' also has a part to play in the discovery of first principles, but again the discussion stays at a very superficial level, and Aristotle really has nothing useful to say about how it could do so. One can only conclude that he must think that our knowledge of mathematics (like our knowledge of everything else) is empirical and not *a priori*, but he has not addressed the problem in any detail.[24]

As we shall see more fully in later chapters, there are many objections which this kind of empiricism has to meet. But here I shall mention only one, because it raises a problem that Aristotle himself did see and did discuss, namely the use that is made in mathematics of the notion of infinity. Even the elementary arithmetic and geometry that Aristotle was familiar with often invoke infinities. But how could this be, if they are based upon perception? For surely we do not perceive infinities?

8. Problems with infinity

Aristotle's treatment of infinity is in chapters 4–8 of *Physics* III. After introducing the subject in chapter 4, the first positive claim for which he argues (in chapter 5) is that there is not and cannot be any body that is infinitely large. This is because he actually believes something stronger, namely that the universe is a finite sphere, which (he assumes) cannot either expand or contract over time, so the size of the universe is a maximum size that cannot ever be exceeded. In his view, there is absolutely nothing outside this universe, not even empty space, so there is a definite limit even to the possible sizes of things. I shall not rehearse his arguments, which – unsurprisingly – carry no conviction for one who has been brought up to

---

[24] Does he really think that absolutely *all* knowledge is based upon perception, e.g. including our knowledge of what follows from what (as first codified in his own system of syllogisms)? All that one can say is that he never draws attention to any exception.

believe in the Newtonian infinity of space. I merely note that this is his view.

In consequence he must deny one of the usual postulates of ordinary Euclidean geometry, namely that a straight line can be extended in either direction to any desired distance (Euclid, postulate 2). For in his view there could not be any straight line that is longer than the diameter of the universe. It follows that he cannot accept the Euclidean definition of parallel lines (Euclid, definition 23), as lines in the same plane which will never meet, however far extended. But parallelism can easily be defined in other ways, and of course one can apply Euclidean geometry to a finite space, as in effect Aristotle says himself. At 207^{b}27–34 he claims that his position 'does not deprive the mathematicians of their study', since they do not really need an infinite length, nor even the permission always to extend a finite length. His idea is that whatever may be proved on this assumption could instead be proved by considering a smaller but similar figure, and then arguing that what holds for the smaller figure (which *is* small enough to be extended as desired) must also hold for the larger original, if the two are exactly similar.[25] So his denial of an infinite length is indeed harmless from the mathematician's point of view, but his other claims are less straightforward.

At the start of chapter 6, which opens his positive account of infinity, Aristotle mentions three serious reasons for supposing that there is such a thing:

> If there is, unqualifiedly, no infinite, it is clear that many impossible things result. For there will be a beginning and an end of time, and magnitudes will not be divisible into magnitudes, and number will not be infinite. (206^{a}9–12)

His position is that there must be *some* sense in which these things can be said to be infinite, even if it is not 'unqualifiedly'. I must here set aside his views on time, with the excuse that this is a question in physics or metaphysics, rather than in mathematics, but the infinite divisibility of

---

[25] As was in effect discovered by the English mathematician John Wallis (1616–1703), and known to Gerolamo Saccheri in his book *Euclides ab omni naevo vindicatus* (1733), the assumption that, for any figure, there is a similar but smaller figure of any size you please, is characteristic of a Euclidean space, and could replace Euclid's parallel postulate. So as it happens Aristotle's response is relying on Euclidean geometry. (I take the information from Heath, 1925, pp. 210–12.)

geometrical magnitudes, and the infinity of the numbers, are central to our concerns.

Aristotle quite frequently says that his main claim is that all infinity is always potential, which apparently implies that it cannot ever be actual. But in fact this misdescribes his real position, and further elucidation is certainly required. The central assumption that he makes, which is an assumption that he never argues for, and never even states in a clear way, is that an infinite totality could exist only as the result of an infinite process being completed. Moreover he (quite understandably) believes that an infinite process – i.e. a process that has no end – cannot ever be completed. So there can never be a (completed) infinite totality, though there may perfectly well be an (ongoing) infinite process. No doubt *most* processes will stop at some time, though they may be said to be potentially infinite because they *could* always have been continued further. That is the usual case. There are some processes which never will stop – in Aristotle's view the process of one day succeeding another is an example – and these are processes which are actually infinite. But there cannot be a process which both stops and is unending, which is to say that no infinite process can ever be completed, and hence that there cannot be a time at which there exists an infinite totality.

Aristotle applies this general view to the supposed infinite divisibility of a geometrical object, such as a line. There could (in theory) be an unending process of dividing a finite line into parts. To cite Zeno's well known example, one may take half of a line, and then half of what remains, and then half of what still remains, and so on for ever. But Aristotle holds that these parts, and the points that would divide them from one another, do not *actually* exist until the divisions are actually made. This is because, if they do exist, then one who moves over a finite distance must have *completed* an infinite series of smaller movements, each half as long as its predecessor, which he regards as impossible. So his idea is that one who simply moves in a continuous way over a certain distance does not count as 'actualizing' any point on that distance. To do so he would have to pause at the point, or stick in a marker, or just to count the point as he passes it. The general idea, I think, is that to 'actualize' a point one must *do* something, at or to that point, which singles it out from all neighbouring points. And the (rather plausible) thought is that no infinite series of such *doings* could be completed in a finite time.[26] Consequently a finite line never will

---

[26]   I have argued in my (1972/3) that although this thought is 'plausible' still it is false.

contain infinitely many actual points (or actual parts), but we can still say that its divisibility is 'potentially infinite', on the ground that, however many divisions have been made so far, another is always possible.

This position is compatible with the basic assumptions of Greek geometry, because in Greek geometrical practice points, lines, planes, and solids were all taken as equally basic entities. Besides, a common view was that the most basic kind of entity is the solid, since planes may be regarded as the surfaces of solids, lines as the boundaries of planes, and points as the limits of lines. On this view points are the *least* basic of geometrical entities, and we do not have to suppose that infinitely many of them are needed simply in order to 'construct' lines, planes, and so on, from them. (Indeed, Aristotle opens book VI of his *Physics* with an argument which aims to show that lines *cannot* be made up of nothing but points.)

Where one might expect a tension is over the existential postulates of geometry, for do not these assume that *there are* points, lines, planes, and so on, even when there is nothing that has marked them out? But, on reflection, this is not obvious. At any rate, it is not one of Euclid's demands. As it happens, Euclid makes no explicit claim about the existence of points (though he should have done), but he is quite definite about lines. His first three postulates are:[27]

Let the following be postulated:
1   to draw a straight line from any point to any point;
2   to produce a finite straight line continuously in a straight line;
3   to describe a circle with any centre and distance.

I say no more about the second, for I have earlier remarked that it is not essential, but what of the first and third? Are they not straightforward claims to existence? Well, that is certainly not how Euclid himself presents them. He may readily be interpreted as claiming not that these lines do (already) exist, but that they can (if we wish) be brought into existence by being drawn. (His proofs often require them to *be* drawn.) It is often said of Greek geometry that it is 'constructive' in the sense that it does not assume the existence of any figure that cannot be constructed (with ruler and compasses). In that case Aristotle need have no quarrel with it.[28] For such a construction

[27]   I quote Heath's translation, in his (1925, pp. 195–9).
[28]   Plato does have a quarrel. He complains that lines do not have to be drawn in order to exist (*Republic* VII, 527a–b).

must be describable, and Aristotle would seem to be entitled to assume that it does not exist until it has been described (in thought, or on paper). On that view, there will never be a time when there are more than finitely many geometrical figures in existence, so his account of infinity may still stand. But although it is commonly held that Greek geometry was 'constructive', in the sense roughly indicated here, I do not think that the same is ever said about arithmetic. This is a more serious problem for him.

As I have already noted, he has earlier acknowledged that everyone believes that there are infinitely many numbers. He has suggested that this is because the numbers do not give out 'in our thought' (203^{b}22–5), but apparently he is committed to saying that they do give out in fact. For if a number exists only when there is a collection of physical objects that has that number, then there cannot (according to him) be infinitely many of them. This is because he holds that the universe is finite in extent, and that physical objects never are infinitely divided, from which it must follow that every actual collection is finite. The same conclusion holds if we add, as Aristotle might desire, that simply thinking of a particular number is enough to 'actualize' it. For still at any one time there will be only finitely many numbers that have actually been thought of. There is no doubt a potential infinity, but Aristotle seems to be committed to the view that, even in this case, there is no actual infinity.

To confirm this point we may note a line of argument that Aristotle sometimes uses against his Platonist opponent. The Platonist thinks that all the numbers have an actual existence, in separation from the perceptible things of this world. Aristotle assumes that in this case the Platonist must accept that there is such a thing as the number of all the numbers. If so, then this number must (on plausible assumptions) be an infinite number. But Aristotle thinks that it is impossible for there to be such a thing as an infinite number, (a) because, since a number can always be reached by counting, this would mean that one could count to infinity, and (b) because every number must be either odd or even, but an infinite number would not be either.[29] Now Aristotle presents these thoughts as creating a difficulty for the Platonic conception of numbers as existing separately, but we may certainly ask whether they would also apply to numbers construed in Aristotle's way, i.e. as existing only *in* (collections of) independently existing items. So far as one can see, the points made would apply equally in either case, and hence there equally cannot be infinitely many Aristotelian

---

[29] For (a) see *Physics* III.5, 207^{b}7–10; for (b) see *Metaphysics* M.8, 1083^{b}37–1084^{a}4.

numbers. So there is here no exception to his overall position. The series of numbers is *potentially* infinite, but only potentially. That is to say that at no time will there *actually* be more than finitely many of them.

One cannot be content with this conclusion. One asks, for example, how many numbers there are *today*, and surely no answer is possible. The conception is that more may be realized tomorrow, but that there must be some definite (and finite) answer to how many there are now, and every such answer is ridiculous. It is of course true that the numbers do not give out 'in our thought', but it is very difficult to take seriously the idea that they do give out 'in fact'.

Could Aristotle evade this criticism? Well, perhaps he could, but only by applying to arithmetic a line of thought that he must accept for geometry, although he never candidly admits it. Geometry is an idealizing subject. It *assumes* the existence of all kinds of perfect figures, and surely most of them are not actually exemplified in the physical world. In that way it is a fiction. But obviously it is a very useful fiction, and its practical applications are immensely valuable. I think that Aristotle should say the same about arithmetic. It too 'idealizes' by assuming that every number really does have a successor. That is its way of 'smoothing out' the imperfections of the world that we actually inhabit. But at the same time he might hope to argue that this 'fictional', 'idealized', 'smoothed out' theory is something which, in practice, we cannot do without. We shall have more on this theme in what follows.

I conclude with a question. In opposition to Plato, Aristotle is aiming for an empirical theory of mathematics, which does not accept the real existence of any objects that are not empirically available. There are many problems for such theories, and we shall find more as we go on. But one of the most central, and the most difficult, is that which Aristotle himself noticed when he first put forward an empirical theory. How can it explain infinity? If we cannot answer this, then must we be forced back to Platonism?

C.   Prospects

The basic feature of a Platonic ontology is just that it posits a special kind of objects, nowadays usually described as abstract objects, to be what mathematics is about. These objects are thought of as existing both independently of any physical instances or exemplifications, and independently of all human thought. The basic feature of a Platonic epistemology is that it insists that our knowledge of mathematics is a special kind of knowledge, called *a priori* knowledge, i.e. knowledge which is (in principle) independent of what we can learn from our perceptual experience of the physical world. In each case, this is a very brief and very general characterization of an overall approach which can take different forms in different philosophers. I have here given some description of Plato's own version, partly because that is of some historical interest, and partly because one finds even in Plato himself a hesitation over both the ontology and the epistemology.

So far as the ontology is concerned, Plato began with the idea that mathematics was about what he called forms, but then changed to the idea that it was about 'intermediates', i.e. perfect examples of those forms. Subsequent philosophers have said instead that it is about universals, e.g. such things as properties and relations and functions and so on, all Platonically construed. A more common version these days is that it is about sets, but (in its purest version) about sets which nowhere involve ordinary perceptible objects in their composition. (These are the so-called 'pure sets'; I shall introduce them in chapter 5, section 4.) All of these are different versions of 'the Platonic ontology'.

As for the epistemology, we have seen that Plato himself began with a theory to explain *a priori* knowledge, namely that it is a matter of 'recollecting' a previous existence that was not on this earth, but apparently he quite soon abandoned this theory. Since Plato there have been other theories aiming to explain how *a priori* knowledge is possible, which we shall come to in due course. So far as Plato himself is concerned, we have seen that he at least saw a difficulty with this claim to *a priori* knowledge. For although it is true that mathematics is full of proofs, and that our ability to construct and follow a proof is often held to be *a priori*, still Plato noticed that mathematical proofs do begin from axioms. So there is still the problem of how (if at all) we know these axioms to be true, and later Platonizing philosophers have given very different accounts of this.

The basic feature of an Aristotelian ontology is essentially negative: we do not need to posit such special and peculiar objects as the Platonist desires, so we should not do so. The general idea is that mathematics can perfectly well be construed as a study of entirely ordinary objects, even if the initial appearance suggests otherwise. To make a convincing case for this, one must of course say just how the basic propositions of mathematics can be construed as propositions concerning ordinary objects, and this is a task which Aristotle himself does not really address. I imagine that he took it to be quite obvious in the case of geometry, and that he assumed without much further consideration that the case of arithmetic would be similar. In more modern developments geometry plays no important role (for reasons that I give in chapter 3, section 1), but arithmetic has become central. How are we to explain, in broadly Aristotelian terms, the apparent reference to such things as numbers? I would say that the first worthwhile attempt at this question was by Bertrand Russell (which I discuss in chapter 5, section 2), and this attempt certainly showed that the problem was by no means simple.

As for the Aristotelian epistemology, one may say that its basic idea is just that the method of mathematics is no different in principle from the methods of what we call 'the natural sciences'. Everyone will agree that knowledge gained in this way counts as empirical knowledge, though there will be disagreement on just what entitles it to be knowledge, and on how in detail it is achieved. Nowadays one would say that there are basically two sides to it, one being the observation of those happenings that we take to be observable, and the other being the construction of theories to explain and predict these observations. It is fair to say that Aristotle himself had no good account of the second, and that his successors have had to pay more attention to this. But the empirical approach that one associates with him is certainly still with us today.

I remark that it is quite *natural* for the Platonic ontology and epistemology to go together with one another, and similarly the Aristotelian ontology and epistemology. To put it crudely: if the objects of mathematics are really just ordinary perceivable objects, then it is natural to suppose that it is perception that tells us about them, and vice versa. On the other hand, if mathematical objects are objects of a special kind, available to thought but not to perception, then it is natural to suppose that the thought in question must be a special kind of thought. However, these links are not inevitable. For example, it is possible to combine the Aristotelian ontology with a Platonic epistemology, and Bertrand Russell's approach might be described in this way. For he aims to offer a reductive analysis of mathematical statements, which reveals them as generalizations about ordinary objects rather than singular statements about a special kind of object called a number. But at the same time he holds that these generalizations are not ordinary empirical generalizations but truths of logic, and that logic is known *a priori*. That combination of views is quite easy to understand (though – as Russell discovered – not too easy to maintain. I shall discuss it in chapter 5, section 2.). In the other direction it is possible to combine the Aristotelian claim that all knowledge is empirical with the Platonic claim that mathematics is about objects of a very special kind, namely abstract objects which are not the sort of thing that could have a location in space. The idea here is that such objects are assumed by our scientific theories, and that if we do not accept this assumption then we shall be deprived of all the usual scientific explanations of familiar physical phenomena. (This line of thought is mainly due to Quine and Putnam. I shall discuss it in chapter 9, section 3.)

To sum up: there are many different versions of Platonism, and of the empiricism that Aristotle proposed. There are also approaches which combine some aspects of the one with some aspects of the other. We shall find all these themes constantly recurring in the thinking of the twentieth century. At the same time we shall also find a quite different theme which has not yet emerged at all, and which I introduce in the next chapter. In broad outline this is the idea that the objects which mathematics is about exist only in minds and nowhere else.

Suggestions for further reading

Plato

The passages cited from Plato's *Meno* and *Phaedo* are relevant, and comparatively straightforward. There are many translations of Plato's dialogues that are easily available, and any of them will do perfectly well. The problems with what the *Republic* has to say about mathematics have been much debated by Plato scholars, and for present purposes this debate might reasonably be bypassed, since it is of little relevance to contemporary philosophy. But for those who wish to pursue the topic I recommend reading more of the *Republic*, say from 506b to 534d, and beginning with the following commentaries: Robinson (1953, chapters 10–11); Malcolm (1962); Cross and Woozley (1964, chapters 9–10). As I have noted, Wedberg (1955, Appendix), is a classic discussion opposed to these. For something more recent I suggest Mueller (1992) and Burnyeat (2000).

Aristotle

For chapters 2–3 of his *Metaphysics*, book M, I recommend Annas (1976), which contains a convenient translation (with notes), and an introductory essay on the philosophy of mathematics in both Plato and Aristotle. The several problems with Aristotle's overall empiricism are best postponed until we reach a more detailed example of this kind of approach, such as that of J.S. Mill (which I discuss in chapter 3, sections 1–2). But Aristotle's treatment of infinity, in chapters 4–8 of book III of his *Physics*, is special to him, and deserves consideration. There is a convenient translation (with notes) in Hussey (1983). For further reading on this particular topic I recommend Bostock (1972/3) and Lear (1979/80). For a more general appraisal of Aristotle's position I suggest beginning with Mueller (1970) and Lear (1982). I have discussed Aristotle's philosophy of mathematics in greater detail in my (2009b).

Chapter 2

From Aristotle to Kant

1. Mediaeval times

In the nineteen centuries from Aristotle (died 322 BC) to Descartes (born AD 1596) philosophers paid little attention to the nature of mathematics. It was generally accepted that the truths of mathematics are eternal and unchanging truths, and deserve to be called necessary truths, but very few asked either about how we come to know them or about the ontological status of the objects that they apparently concern (i.e. the numbers, the geometrical figures, and so on). I therefore pass without comment over this gap in our subject. Instead I turn to a different issue, which did receive much attention then, and which has nowadays acquired an important role in philosophical discussions of mathematics, though in those days it was not regarded as in any way special to mathematics. This is the question of the ontological status of universals in general. No doubt numbers and geometrical shapes may well be regarded as special types of universals, but the mediaeval debate was conducted without any particular attention to these special cases.[1] The classic statement of the question is due to Porphyry (AD c. 232–c. 305). Do universals exist at all? If so, do they exist outside the mind, or simply as mental entities? If the former, are they corporeal or incorporeal? And do they exist *in* the things that are perceptible by the senses, or are they separate from such things?

In broad outline the *realist* answer to this question is that universals do exist outside human minds, and independently of all human thought. Plato and Aristotle were both realists in this sense, though Plato also held that universals existed separately from perceptible things, and independently of whether they had any perceptible instances, whereas Aristotle held that they existed only in their perceptible instances. In modern philosophy of

---

[1]   There is a convenient summary in Kenny (2005, pp. 121, 124–5, 137–8, 144–7, 150–2, 172–3, 208–11).

mathematics there are certainly realists of the Platonic type, e.g. Frege and Gödel, and one could cite J.S. Mill as a kind of Aristotelian realist, though there are differences as well as similarities.[2] The *conceptualist* answer is that universals exist only in the mind, and are indeed created by the mind, perhaps as a response to perceptions and perhaps independently of perceptions. Those who take a conceptualist line about *all* universals will usually accept that we may form universal ideas as a result of having various perceptions, but as we shall see this is often denied in the special case of mathematics. Modern conceptualists about mathematics – for example the twentieth-century school called 'intuitionists' – are apt to say that we ourselves create the numbers without reference to any perceptual stimulations. Finally, the *nominalist* answer is that universals do not exist in any way. This was not a popular answer in mediaeval times (though it did have a few adherents), but has become more popular recently, at least in the particular case of the numbers. This too has two varieties. One, which I shall call 'reductive nominalism', holds that words (*nomina*) which apparently name and refer to universals do not really do so, for they are merely used as shorthand for what could be said more long-windedly without such an apparent reference. So, in the case of simple arithmetic, the task is to show how sentences which apparently refer to numbers may be paraphrased in a way which eliminates this apparent reference. (Bertrand Russell took this line.) A different version of nominalism is what one may call an 'error' theory. On this view sentences which apparently contain a reference to a universal are just not true, because there are no such things as universals are supposed to be. When (like Hartry Field) one takes this approach to sentences which apparently refer to numbers, the task is then to explain why these sentences are so widely held to be true, and how they can be so useful in practical affairs even if they are not true. In later chapters I shall discuss these modern versions of realism, conceptualism, and nominalism, insofar as they concern the statements of arithmetic. Here I merely observe that the mediaeval debate was by no means restricted in this way; it concerned *all* universals, without discrimination.

In the last chapter I have introduced the two main varieties of realism in mathematics. The rest of this chapter traces the first major contributions to a conceptualist approach.

---

[2] More recent, and closer to Aristotle, is the position that Maddy espoused in her (1990), though she quite soon changed her mind. I add that a modern variation of Mill's approach is in Kitcher (1980) and (1984). I do not discuss either of these in this book, but I have said something about them in my (2009a).

2. Descartes

Descartes is often regarded as the founder of 'modern philosophy'. He certainly is the founder of a line of thought about the nature of mathematics that has proved very influential.

Descartes sought for certainty, and he began with the thought that in mathematics we do find certainty. His main philosophical endeavour was to try to extend this certainty to other areas too, but we need not pursue his effort in this direction. For it is the account of mathematics that is our concern. Here the central point is that Descartes asked *why* certainty can be attained in mathematics, though (at least at first sight) not elsewhere. For example, he thought that all our ordinary beliefs about the familiar physical world can rationally be doubted, and no certainty is to be found in them. This is because our beliefs about the physical world are all based upon the senses, but we all know that our senses sometimes deceive us, and we cannot prove (at least at first sight) that they do not always deceive us. Consequently one can even doubt whether there is a physical world at all. Descartes took it to follow that if mathematics enjoys a certainty that our physical beliefs lack, then the truths of mathematics cannot be truths about the physical world. They must instead be truths about *our ideas*, or perhaps about some feature of the mind of God that determines those ideas of ours.[3]

He wished to go further, for he also thought that the ideas with which mathematics is concerned are unlike most other ideas in being especially 'clear and distinct'. That is his explanation of why simple thoughts involving these ideas do strike us as ones that we cannot doubt, and why, from these simple starting points, we are able to carry out long deductions of more complex thoughts without loss of the initial certainty. It is the special lucidity of the basic ideas that makes this possible.[4] Consequently his guiding principle in the search for certainty in other areas of philosophy is to begin by seeking for further clear and distinct ideas in those areas too, in order to provide once more some similar simple starting points

---

[3] See further the text to n. 7, which takes up the second possibility.

[4] Descartes speaks of the 'clarity and distinctness' of ideas in all his published works, from the *Discourse on Method* (1637) onwards. Earlier in the unpublished *Regulae ad Directionem Ingenii* (1628) he had focused rather on the fact that mathematics 'deals with an object so pure and uncomplicated' (Rule II). But he had also claimed that these 'pure and simple objects' gave rise to 'clear and distinct propositions' (Rule XI). So this is more a change of terminology than a change of view.

that cannot be doubted, and from which deduction may safely proceed.[5] That is the general outline of his programme.

Descartes regards the clear and simple ideas on which mathematics is based as innate. In Meditation III he offers this general classification:

> Among my ideas, some appear to be innate, some to be adventitious [in the French version: foreign to me and coming from outside], and others to have been invented by me.

In elucidation he says that the ideas that one obtains in perception are probably to be classed as 'adventitious', for they do at least appear to come from something outside; the ideas of imaginary things such as sirens or hippogriffs are classed as 'invented by me'; and to explain 'innate ideas' he here says:

> My understanding of what a thing is, what truth is, and what thought is, seem to derive simply from my own nature.

These are no doubt intended as examples of ideas that Descartes finds clear and distinct, and it is natural to suppose that his phrase 'my understanding of what a thing is' is intended to include, *inter alia*, my understanding of what a number is, or of what a plane is, or a line, or a point, and so on. But both here and later in his *Replies to Objections* III he has glossed 'innate' simply as 'coming from my own nature', which is not a very adequate description of what is ordinarily meant by that word.

Elsewhere he is more forthcoming. In a passage that he had intended to be printed, though in fact it never was, he says of the soul of the infant in the womb:

> It has in itself the ideas of God, of itself, and of all such truths as are called self-evident, in the same way as adult humans have when not attending to them; it does not acquire these ideas later on, as it grows older. I have no doubt that if it were taken out of the prison of the body it would find them within itself.[6]

---

[5]   In the *Regulae* (Rule II) and the *Discourse* (Part II) it is clear that the aim is to extend what is special about *mathematics*. In the later *Meditations* (1641) the role of mathematics is downplayed, and the stated aim is to extend whatever it is about the *Cogito* that gives certainty to it. But again the desired feature is claimed to be the clarity and distinctness of the ideas involved (Meditation III).

[6]   Letter to 'Hyperaspides' of August 1941 (Kenny, 1981, p. 111). Descartes had intended his replies to 'Hyperaspides' to be published along with his other replies to objections to the *Meditations*, but in this case they missed the printer's deadline.

(The idea is that the infant soul is always distracted by the information that pours in on it from the senses, so it is not until maturity that it can set these aside and 'attend to' the clear ideas that have always been within it.) He also says early in Meditation V:

> I find within me countless ideas of things which, even though they may not exist anywhere outside me, still cannot be called nothing . . . They are not my invention but have their own true and immutable natures . . . For example there is a determinate nature, or essence, or form of the triangle which is immutable and eternal, and not invented by me or dependent on my mind.

This claim that 'there is a determinate essence' does not mean that there is something *outside* Descartes' mind which the idea represents, but only that the idea is not invented by him. Rather, it was implanted in him from outside, i.e. by God.[7] We may contrast this with Plato's early view that the source of such ideas is our experience in another world before this life. On Plato's account there must *be* such things as triangles, if not in this world then in another. But Descartes does not mean to imply this.

The comparison with Plato may usefully be elaborated. Their positive ontologies are quite different. In Plato's view mathematics is a theory of certain abstract objects, which exist independently of any kind of thought or minds. In Descartes' view mathematics is a theory of certain mental objects, called 'ideas', which can only exist in minds. But the main negative thrust is in each case similar: these objects do not owe their existence to anything perceptible, neither to perceptible objects (Plato), nor to perceptible ideas (Descartes). Consequently the epistemology is also similar, for here the negative claim is the main one: our knowledge of these objects does not in any way depend on our perceptual experience, but is wholly *a priori*. Moreover, both Plato and Descartes start with the thought that mathematical knowledge is certain and indubitable, though later they diverge. Both of them see that most mathematical knowledge is gained by proof, and both of them regard this as not subject to doubt. But both also saw that proof must have its unproved starting points, and this led Plato to doubt whether these starting points are known with certainty, whereas Descartes remained convinced that they were, and propounded a theory of 'clear and distinct ideas' that was supposed to explain how this was possible. But the

---

[7]　Recall the text to n. 3, which introduces this possibility.

theory leaves many questions unanswered, and this was left as a legacy to his successors.

3. Locke, Berkeley, Hume

By tradition, Descartes is classed as a rationalist, whereas Locke and Berkeley and Hume are on the opposite side, as empiricists. Nevertheless, on the most fundamental questions about the nature of mathematics, they are all agreed. On the ontological question they all hold that mathematics concerns our *ideas*, and on epistemology they all hold that mathematics is known *a priori*. Of course there are differences between them on less fundamental issues, but one should begin by noticing the underlying agreement.

I shall here pay little attention to Berkeley's views, because his overall ontology is special to him. He holds that the *only* things that exist are minds (or spirits) and ideas, so it is hardly surprising that he takes the objects of mathematics to be ideas. But since he also holds that chairs and tables, rocks and trees, lakes and mountains are equally ideas (or perhaps collections of ideas), this draws no interesting ontological distinction. By contrast, Locke (like Descartes) holds that there are both minds, or spiritual substances, and material substances, as well as ideas. So his claim that mathematics concerns ideas is to be understood in the more familiar way as carrying with it the claim that mathematics is not concerned with ordinary material things (save perhaps in a derivative way). Hume's position is similar, for he too distinguishes ideas from the material things that he calls 'bodies', and takes mathematics to be a theory of ideas rather than of bodies.[8] On ontology, then, there is widespread agreement.

As for the fundamental epistemology, I must begin with a disclaimer: none of the philosophers mentioned made explicit use of the phrase '*a priori* knowledge', but still the phrase as we use it does apply to their views. As we have seen, Descartes spoke of ideas so clear and distinct that the simple propositions formed from them could not be doubted. These

---

[8] For most of the time Hume apparently accepts the common belief that bodies exist independently of being perceived, whereas ideas do not. Indeed he says that no one can in practice avoid this belief, and so for present purposes we may regard this as his view. He does call it into question in his sceptical discussion of our belief in independent bodies (*Treatise* (1739) I, IV, 2, and *Enquiry* (1748), XII, part I), but this scepticism does not affect his view of mathematics.

provided the premises from which the mathematician deduces further consequences, by steps of reasoning which are again doubt-free. There is here no input from perceptual experience, though this is allowed a role elsewhere. As for Locke, he defines all knowledge as 'the perception of the agreement or disagreement of our ideas', and he certainly takes this to include our mathematical knowledge (*Essay* (1690), IV, i, 1–2). Moreover, he evidently considers the truths of mathematics to be *general* truths, and he thinks that general truths can be known only when the knowledge depends simply on the relationship between the ideas involved, without any reliance on what can be learnt from observation and experiment. For where observation and experiment do have to be invoked the most that we can say is that the general truth is probable, but not certain, and from this Locke infers that it does not count as known (*Essay* IV, iii, 28 and vi, 6–13). Hume also speaks of 'knowledge of relations of ideas', which he construes in much the same way as Locke does, but which he contrasts with 'knowledge of matters of fact and existence', thereby allowing for another kind of knowledge which is based on experience. But there is no disagreement over our knowledge of mathematics.

The main point of disagreement is on how we first acquire these ideas that are studied in mathematics. Descartes regards them as 'innate', as we have seen, whereas the empiricists suppose that – like all other ideas – they are obtained from experience. They offer different accounts of how this is done. Locke proposes a theory of 'abstraction', intended to apply to all general terms, which goes roughly like this. I see a number of particular men and receive from each a particular idea. I then compare these ideas, and 'leave out' what is not common to them all, thereby forming the abstract idea of a man. This abstract idea will then (if I have abstracted correctly) apply equally to all men, not only to those whom I have seen but also to those whom I have not. The same would apply to the abstract idea of three, which is presumably obtained by 'leaving out' what is not common to the various ideas of particular trios, and similarly the idea of a triangle, and so on (*Essay* III, iii, 6–10).[9]

Presumably Locke did not think of an abstract idea as a kind of mental image, but one that has become indeterminate where certain features have

---

[9] It is not clear how this notion of 'leaving out' some feature is to be understood. One recent defender of Locke, namely Mackie (1976, chapter 4), prefers a positive rephrasing: 'to leave out' what is not common is better understood as 'to attend selectively' to what is common.

been 'left out'. (Indeed, although Locke does sometimes talk as if ideas are mental images, at other times he evidently cannot be thinking in this way.) However, Berkeley did construe ideas as mental images, and so complained that Locke's 'abstract ideas' are simply impossible. For how could there be a mental picture of a triangle which pictured equally well all the different triangles that there are, of all different shapes and sizes? (*Principles*, Introduction, vi–xx.) Hume followed Berkeley on this issue. So, in his view, there is not one single abstract idea for all triangles, but instead a large number of different particular ideas, each an idea of some definite kind of triangle. This set of particular ideas is held together by each of the ideas being associated with the single *word* 'triangle'. In outline, his theory is that when this word is used just one idea of the set will come to mind, but the mind has a 'propensity' to conjure up other ideas of the set, if that is relevant to the thinking going on (*Treatise* I, i, 7). But we need not be concerned here with the details of just how experience is supposed to yield the ideas which are studied in mathematics. It is enough to say that the empiricists do claim that experience has an essential part to play, whereas the rationalists deny this.

In consequence we cannot understand the notion of *a priori* knowledge as meaning, for the empiricist, knowledge that is *totally* independent of experience. For all knowledge involves ideas, and in his view all ideas come – directly or indirectly – from experience. In Hume's terminology, simple ideas are just *copies* of the impressions that we get in experience, and complex ideas are put together by us by combining simple ideas in an appropriate way. So the criterion of *a priori* knowledge is now this: one knows a priori that *P* if and only if, *given* whatever experience is needed to form the ideas contained in '*P*', one needs no *further* experience to acquire the knowledge that '*P*' is true. (This is approximately the modern account.[10])

What are the supposed relationships between ideas that can give rise to *a priori* knowledge? None of the philosophers considered here provides any positive account, but Locke did introduce a distinction which Kant was to fix upon. Some propositions he described as 'trifling' or 'verbal', namely those which rely only on the fact that an idea is the same as itself (as in 'a man is a man'), or the fact that one idea is a part of the other (as in 'a man is an animal', or 'a horse is a quadruped'). These, he thought, are of

---

[10]   Further refinement is certainly needed, but the topic soon becomes controversial, so I cannot discuss it here. My (2009a) has more to say.

no serious interest, for they do not in any way advance our knowledge of the world, but at best have a use in teaching the meaning of words. But he contrasted these very simple relations of ideas with the examples that we find in mathematics, e.g. 'the external angle of all triangles is bigger than either of the opposite internal angles'. Here, he says, one thing is affirmed of another 'which is a necessary consequence of its precise complex idea, but not contained in it. . . . This is a real truth, and conveys with it instructive real knowledge' (*Essay* IV, VIII, *passim*). However, he says nothing positive about what these more complex relations are, or about how we discern them. Nor did either Berkeley or Hume improve upon his silence on this point, so this problem was bequeathed to Kant.

One might indeed say that Hume only makes the situation more obscure, by introducing further threads to his discussion. Here is his rather summary presentation of his position:

All objects of human reason or enquiry may naturally be divided into two kinds, to wit, *Relations of Ideas*, and *Matters of Fact*. Of the first kind are the sciences of Geometry, Algebra, and Arithmetic; and in short, every affirmation which is either intuitively or demonstratively certain. *That the square of the hypothenuse is equal to the square of the two sides*, is a proposition which expresses a relation between these figures.[11] *That three times five is equal to the half of thirty*, expresses a relation between these numbers.[11] Propositions of this kind are discoverable by the mere operation of thought, without dependence on what is anywhere existent in the universe. Though there never were a circle or triangle in nature, the truths demonstrated by Euclid would for ever retain their certainty and evidence.

Matters of fact, which are the second objects of human reason, are not ascertained in the same manner; nor is our evidence of their truth, however great, of a like nature with the foregoing. The contrary of every matter of fact is still possible; because it can never imply a contradiction, and is conceived by the mind with the same facility and distinctness, as if ever so conformable to reality. *That the sun will not rise tomorrow* is no less intelligible a proposition, and implies no more a contradiction, than the affirmation, *that it will rise*. (*Enquiry*, IV, 20–21)

---

[11] Note that a relation between figures, and a relation between numbers, is in each case taken to *be* a relation of ideas. This is because Hume takes figures and numbers each to *be* ideas, and the final sentence of the first paragraph hints at his reason, namely that the mathematical propositions in question would still be true even if there existed nothing other than ideas.

Note how, in the second paragraph, Hume simply runs together the four notions of being possible, of not implying a contradiction, of being distinctly conceivable, and of being intelligible. But they should certainly be distinguished. First, it is no doubt true that whatever implies a contradiction is impossible, but the converse may certainly be denied (and was denied by Kant). Second, it is no doubt plausible to claim that what can be clearly conceived – in the sense that Hume intends, namely what can be clearly imagined – will be possible, but, as we now realize, the converse is certainly false. Third, even explicit contradictions are intelligible, in the sense that if we attach a clear meaning to '*P*' then we also attach a clear meaning to '*P* and not *P*'. (For if it had no meaning how could it be false?) Kant does make some progress in sorting out these notions, though he is not completely successful.

Before I come to Kant, I add one further remark on the topic of 'rationalists *versus* empiricists'.[12] As already noted, both sides were agreed upon the *basic* ontology and epistemology of mathematics, but they did differ widely over its potential value.

Descartes had extended the traditional field of geometry by including temporal as well as spatial concepts, so that motion now fell within the sphere of mathematics. Moreover, he professed to offer an *a priori* deduction of the laws of motion, using only spatial and temporal concepts, so that the motion of bodies was now predictable by pure mathematics.[13] He thus hoped that mathematics would come to include all physics as a part of itself, and might eventually encompass all the other sciences too, so that all science would become a priori. His thought was that observation and experiment, though a useful heuristic in the early stages, should in the longer term be superseded by genuine a priori deductions.

Locke was altogether less hopeful. He was of course familiar with Newton's work, and was well aware that most of our scientific knowledge cannot be obtained by purely *a priori* methods. He was also unduly sceptical of the possibility of extending that knowledge, because he thought that this must depend upon discovering the nature of the minute corpuscles of which all bodies are composed, and that human beings would never be

[12] Of course, there were other rationalists besides Descartes (notably Leibniz and Spinoza), but I would say that their position on mathematics did not improve upon his.
[13] The deduction is given in his *Principles* (1644), II, xxxvi–liii. The supposed laws are of course mistaken, since Descartes is trying to get by without the Newtonian concept of mass, or anything equivalent to this.

able to go far enough beyond their ordinary perceptions to achieve this. But Locke did at least share Descartes' confidence that genuine mathematics was a real source of genuine and indubitable knowledge.

Berkeley and Hume had lost this confidence. The mathematics of their day included the so-called 'infinitesimal calculus' of Newton and Leibniz, which apparently invoked 'infinitesimal quantities' quite unlike anything that we could perceive. That led them to think that mathematics had here gone too far, and overreached itself, with the result that it was now a source of impossible contradictions. As we shall see more fully in chapter 4, there was reason on their side, and Berkeley's attack on the new calculus in his work *The Analyst* (1734) does contain a number of pertinent objections. Hume too supposed that mathematics could not treat of infinity without falling into paradox (*Treatise* I, II, 1; *Enquiry* XII, 124–5). If we may describe their position in Descartes' terminology, they thought (with reason) that the ideas here involved could not be described as 'clear and distinct', and that therefore there were no unambiguous 'relations between them' to form the basis of knowledge.

4. A remark on conceptualism

All of Descartes, Locke, Berkeley, and Hume supposed that mathematics is a theory of our *ideas*, but none of them offered any argument for this conceptualist claim, and apparently they took it to be uncontroversial. Yet on the face of it this is a strange identification. For one thing, the terminology that we use of numbers or geometrical figures we do not apply to ideas. For example, one does not naturally say that one idea is the cube root of another, or that when raised to the power of 3 it *is* (or perhaps: is equal to) that other. Or again, one does not naturally say that some ideas have 6 faces and 12 edges. This point is not by itself a very powerful objection, for it could be replied that this is merely a *façon de parler*. But I here mention two other lines of objection that look more serious.

(i) The suggestion that such a thing as a number simply *is* an idea was one that occurred to Plato. Putting it in his language, and in the more general form in which he raised it, it is the suggestion that (Platonic) forms are really just thoughts in men's minds, and have no other existence. But Plato quickly dismissed this proposal, on the ground that a thought is always a thought *of* something, and it is what the thought is a thought *of* that is the form, and not the thought itself (*Parmenides*, 132b–c). This fits well with our opening observation: for example, the idea of a cube is the idea *of* something that has 6 faces and 12 edges, but it does not itself have either faces or edges. But the response can also be bolstered in this way. Ideas are private to the person who has them, but what the idea is an idea of is not (usually) a private entity. Thus you perhaps have an idea which may be called an idea of three, and so do I, but your idea is in your mind, and mine is in mine, so they cannot be the same thing. (And if ideas either are or are associated with mental pictures, the pictures in question may be very different. Yours perhaps looks like this °₀°, while mine looks like this ₀ ₀ ₀.) But what each idea is an idea *of* is in each case the same thing, for it is the same thing that we are each speaking of when we each say 'Three is a prime number'.

(ii) The existence of such a thing as a number does not depend upon the existence of any mental entity, as may be argued in two ways. (a) There was a time before there were any thinking beings, and hence before there were any ideas. But we surely do not want to say that at that time there were no numbers, nor that an arithmetical truth such as '$3^3 = 27$' was at that time not true, and became true only when the relevant ideas were first formed. (b) Again, the number of ideas existing at any time, or that have existed up to any given time, is presumably finite. But, surely, there are infinitely many numbers? (Nor is one much comforted by the Aristotelian reply that the number-series is *potentially* infinite, for this accepts that at any particular time it is actually finite, which is not what is wanted.)

These objections are very straightforward, and really rather obvious,[14] but apparently no philosopher at the time thought them worth answering. I do not claim here that they cannot be answered, but I shall postpone further discussion. This is because the view that numbers should be regarded as being, or as depending on, some kind of 'mental construction' is with us still, and we shall have some more modern versions to consider later (in chapters 7 and 8). But it is worth bearing in mind that conceptualism tends not to square very well with our ordinary ways of talking and thinking.

5.   Kant: the problem

Kant believed that the truths of mathematics are both known *a priori* and synthetic, and his problem was to explain how such knowledge might be

---

[14]   They may all be found in Frege's *Grundlagen* (1884), in sections 26–7. (See also his Introduction, pp. v, vi, x.)

possible. In the Introduction to his *Critique of Pure Reason* he explains the two notions, and sets out his two claims that both apply to mathematical truths, but it must be confessed that he does not there, or elsewhere, offer very much by way of argument for these claims.[15]

Kant defines *a priori* knowledge in the expected way, as knowledge that is 'independent of experience' (B2), but what follows is unexpected. For he goes on to say that we now need 'a criterion by which to distinguish with certainty between pure [i.e. *a priori*] and empirical knowledge', and in fact he offers two such criteria. The first is that if a proposition is 'thought as necessary' then it is an *a priori* judgement, and the second is that if it is 'thought with strict universality' then the same will follow. In each case his main idea is that experience can teach us what is the case, but cannot show us what *must* be the case, and again that experience can tell us what has been the case so far, but not whether this universal connection will *always* hold. Kant claims that these two criteria must always coincide, thereby committing himself to the view that every truth known *a priori* is a universal truth (B3–4).

There is an obvious weakness to these two 'criteria', namely that a proposition may be 'thought as necessary' and 'thought with strict universality', even though it is not in fact either necessary or strictly universal, or even true. We may reasonably allow Kant to add as further conditions that the propositions in question must be true, and must be known to be true, but still a gap remains. For it will not follow that we are right in thinking that no exceptions are possible, and even if we were right about this it still would not follow that our knowledge of these truths would have to be *a priori* in the traditional sense, i.e. independent of experience.[16] But for the time being we may set these doubts aside, and suppose that in practice Kant understands the notion of *a priori* knowledge much as everyone else does. He goes on to claim that the propositions of mathematics are clear examples of propositions known *a priori*, but offers no argument, apparently assuming that this claim is not controversial. This is hardly surprising for – as we have seen – the view was widely accepted at this time. But one could wish that he had stopped to argue the point, and

---

[15] In this section and the next, references beginning with 'A' or 'B' are page references to Kant's *Critique of Pure Reason*, with 'A' for the first edition (1781) and 'B' for the second (1787).

[16] Kripke (1972) has given several examples of propositions where we know (*a priori*) that if the proposition is true then it is necessary, and we also know empirically that in fact it is true. So these are examples of propositions known to be necessary, but not known *a priori*.

given an argument more persuasive than just the observation that many people *think* of mathematical truths as 'necessary and strictly universal'.

However, what is distinctive of Kant's position is his claim that mathematical truths are synthetic (= not analytic), so let us move on to consider this. In a way, he had been anticipated by Locke, for what Kant means by 'analytic' truths is at least very closely related to what Locke had called 'verbal' or 'trifling' truths, and Locke had insisted that mathematical truths are not of that sort. But I here set Locke aside to concentrate on Kant's own account. This again is not very carefully done, and has frequently been criticized.

Where Descartes, Locke, and others, had spoken simply of 'ideas', Kant distinguishes between what he calls 'intuitions' on the one hand, which are singular mental representations of particular objects, and on the other hand 'concepts', which are general notions used to classify and describe objects. (They are certainly not to be regarded as 'mental pictures'.) Thus where Locke had spoken of one abstract idea being a part of another, Kant would talk not of ideas but of concepts, but he still finds apt the metaphor of one being *a part* of another. He uses this metaphor in his initial definition of an analytic judgement. Confining attention to (true) judgements of the form 'All A are B', he says:

> Either the predicate B belongs to the subject A, as something which is (covertly) contained in this concept A; or B lies outside the concept A, although it does indeed stand in connection with it. In the one case I entitle the judgement analytic, and in the other synthetic. (A6 = B10)

He proceeds to give 'all bodies are extended' as an example of an analytic judgement, for mere analysis of the concept of a body will reveal that the concept of being extended is contained within it, i.e. is a part of it, whereas 'all bodies are heavy' is synthetic (A7 = B11).

It is commonly objected that this definition is too narrow, since it applies only to propositions of the form 'All A are B', and that is a fair objection. In the next chapter I shall consider how it might be met. But for the moment I set the objection aside, since discussion of Kant himself may stay within the narrow framework envisaged. In any case, the main thought that accompanies this definition is quite clear, namely that it is easy to see how an analytic truth may be known *a priori*, for all that is needed is analysis of the concepts involved, and that does not depend upon experience. But synthetic truths cannot be known in this way, since the concepts in question are not there related as part and whole. However, Kant is also convinced that some synthetic truths can also be known *a priori*, and he claims that the truths of mathematics are clear examples.[17] So his problem is: how can *this* kind of *a priori* knowledge be explained?

Just as Kant seems to suppose that there will be general agreement that mathematical truths are known *a priori*, and so offers nothing useful by way of argument for this claim, so I believe that he thinks that it will readily be granted that mathematical truths are synthetic. He can hardly have supposed the point to be a familiar one, for the division of propositions into analytic and synthetic was not itself familiar at that time. But I imagine that he thought that, once the distinction was clearly drawn, it would seem obvious to all that (most) mathematical truths are not analytic.[18] However, we do get *a little* by way of argument in this case, for Kant discusses two examples, one from arithmetic and one from geometry.

(a) Arithmetic

The arithmetical example is '$7 + 5 = 12$'.[19] Kant says of this that, if we look closely, 'we find that the concept of the sum of 7 and 5 contains nothing save the union of the two numbers into one, and in this no thought is being taken as to what the single number may be which combines both' (B15). His idea is that to find this single number we must actually *do* the sum – e.g. by counting on our fingers – and this takes us beyond mere analysis of concepts. He adds that the point would be still more evident if we had taken larger numbers (B16). Since many have found this line of thought unconvincing, I add a further consideration in its support: the concept of 12 cannot actually be a *part* of the concept of $7 + 5$, or *vice versa*, since it is possible to possess either concept without possessing the other. It is obviously possible to have the concept of 12, but not of $7 + 5$, for this is the position of a child who knows how to count but has not yet learnt to add. But it is also possible to have the concept of $7 + 5$, but not that of 12, namely if one is familiar with small numbers but has no concept of larger numbers. Consider, for example, a primitive tribe who count as we do from 1 to 10, but count all pluralities of more than 10 members simply as

[17] He rashly says: 'all mathematical judgements, without exception, are synthetic' (B14). But of course the straightforward definitions of mathematical terms must be exceptions to this generalization, and in fact Kant himself notes some others at B16–17.

[18] As we have noted (pp. 40–1), it did seem obvious to Locke.

[19] I presume that Kant thinks of this proposition as having the form 'All *A* are *B*', i.e. as meaning something like: 'whenever a 5-group is added to a 7-group the result is a 12-group'.

'many'.[20] Such people may perfectly well be able to add, and so they will know that:

    7 + 1 = 8
    7 + 2 = 9
    7 + 3 = 10
    7 + 4 = many
    7 + 5 = many
    7 + 6 = many
    and so on.

Since this evidently is a possibility, the one concept cannot literally be a *part* of the other, and so the proposition is not analytic on Kant's definition.

The natural response to this is that the example only shows a defect in his definition. Indeed, it is curious that he thinks of applying the technique of 'analysis' only to the *subject*-term of a proposition, and not also to the *predicate*-term. This may make a difference, for someone who has both concepts, and can analyse them both, may thereby be able to see that they must coincide. To illustrate this, I switch (for brevity) to the example '2 + 2 = 4'. Leibniz had offered this demonstration.[21]

Rules of logic: (i) $a = a$
                  (ii) If $b = c$, then $a + b = a + c$ and $b + a = c + a$

Definitions: $2 = 1 + 1$; $3 = 2 + 1$; $4 = 3 + 1$.

Proof: $2 + 2 = 2 + 1 + 1$ (Def. 2)
                       $= 3 + 1$ (Def. 3)
                       $= 4$ (Def. 4)

The idea here is that *both* '2 + 2' *and* '4' may be analysed in terms of the same concepts, namely '1' and '+ 1', and when *both* are analysed in this way then the proposition will become a mere identity.

Now Kant may perhaps have known of this purported proof, and if so he may also have known of the standard criticism, which is that brackets should be inserted in the continued addition.[22] When we do this, we find that the proof assumes that:

---

[20] Interestingly, Locke already knows of such primitive peoples (*Essay* II.xvi.6).
[21] *Nouveaux Essais* (1765, IV, vii, 10).
[22] There is an interesting discussion of this possibility in Parsons (1969, pp. 21–3).

$$2 + 2 = 2 + (1 + 1) = (2 + 1) + 1 = 3 + 1$$

That is, in addition to the stated premises, it is also being assumed that addition is an associative operation. Could this in turn be proved, presumably from a definition of addition? I shall not pursue this question here, for it will recur in a more sophisticated form as we proceed. Meanwhile, I simply draw this moral: Kant's initial definition of analyticity is indeed over-simple, and not just because it confines attention to propositions of the form 'All *A* are *B*'. When this is realized, it becomes clear that the status of such arithmetical propositions as '7 + 5 = 12' cannot be settled quite so easily as Kant assumed, and more work is needed. Meanwhile, I turn to Kant's other example, taken from geometry.

(b) Geometry

Kant's geometrical example is: 'the straight line between two points is the shortest'. Nowadays one might protest that this just is the *definition* of a straight line, but Kant replies: 'my concept of straight contains nothing of quantity but only of quality' (B16). A first thought is that Kant's concept of a straight line probably begins from the thought that a straight line is one that looks like this:————————. No doubt, some refinements are needed. For example, (i) since the geometrical line is supposed to be one that has no thickness, we should think not of the whole two-edged line here drawn but just of one of its edges, say the upper edge, as our line; (ii) we should imagine this edge to be a perfectly definite edge, so that even under powerful microscopes it continues to be a clear and well-defined line, separating a white area above it from a black area below it; (iii) we should think of this line as continuing to 'look straight' no matter how microscopically it is viewed. That is, we begin from the idea of *looking straight*, and then try to idealize that. But then one can envisage an opponent who claims that *looking straight* and *being straight* are actually quite different ideas (as is witnessed by those well-known straight sticks that look bent in water), and so claims that no refinement of the first will yield the second. But there is also a different suggestion, which is that Kant is assuming a definition of straightness in terms of the concept of a direction, as was given in some mathematical textbooks of his time, namely this: a straight line is the line generated by a point that moves always in the same direction.[23] Here again one can easily imagine an objector who

[23] This definition is given in Barrow (1664, p. 612) and in Wolff (1739, p. 33).

argues that this definition gets things the wrong way round, since the notion of a direction is more complex than that of straightness.[24] Since Kant does not explain what his concept of straightness is, we do not know quite what to make of his example.

It is not worth pursuing this problem further, so I simply exchange Kant's example for another that is not controversial in the same way, namely: between any two points there is one and only one line that is the shortest. (On today's usual definition of straightness, this says that, for any two points, there is one and only one straight line that joins them, i.e. that two straight lines cannot meet twice, and so cannot enclose a space.[25]) I take it to be obvious that no *analysis* of the concept of one route being shorter than another could reveal to you that, for any two routes that are equally short, there must be another that is shorter than both of them. (This claim will be bolstered in the next chapter.) So Kant is right to say that here is a geometrical proposition which is not analytic. But now the doubt is whether it is one that can be known *a priori*. The next chapter will argue that it is not.

Let us sum up. Kant is surely right to say that many propositions of ordinary (i.e. Euclidean) geometry are not analytic: no analysis of the concepts involved will tell you that they must be true of the space that we perceive. But nowadays it is generally held that such propositions are not known *a priori*. In contrast, the usual (though not universal) view of arithmetic is that its truths do seem to be known *a priori*, and the doubt is with Kant's other claim, that they are not analytic. For the time being we must leave these questions open. But to understand the next section we must there assume that Kant is right to claim that the truths of both arithmetic and geometry are both synthetic and known *a priori*. For this evidently provokes the question 'how could that be?', and the next section considers Kant's answers.[26]

---

[24] It will be recalled that Frege considered defining the direction of a line in terms of all lines parallel to it (*Grundlagen*, 1884, sections 64–5), and of course the usual definitions of parallelism presuppose straightness.

[25] Kant gives this as an example of a geometrical proposition that is clearly synthetic at A47 = B65. It is a well-known geometrical axiom. (But I add, incidentally, that Euclid fails to list it explicitly as an axiom, though his proofs do quite soon invoke it, e.g. the proof of I.4.)

[26] Kant does not suppose that the propositions of mathematics are the *only* ones that are both synthetic and known a priori. He thinks that the same applies to the basic laws of physics (e.g. to Newton's laws of motion), and his main concern is to determine whether some propositions of metaphysics would also qualify. But we need not pursue these further issues.

6. Kant: the solution[27]

Kant's answer to the problem is given in two main parts. The first is the topic of the opening section of the *Critique*, which is entitled 'Transcendental Aesthetic' (A19–49, B33–73). The main idea here is Kant's claim that space is 'the form of outer sense' (and similarly time is 'the form of inner sense'). The second part comes very much later, towards the end of the book, in the section entitled 'The Discipline of Pure Reason in its Dogmatic Employment' (A712–38, B740–66). Here Kant attempts to say what is special about reasoning in mathematics, and the main idea is that it is based on what Kant calls 'intuition', which distinguishes it from reasoning in other areas. We may consider these two parts separately, especially since the relation between them is not at all clear.[28] In both it is the explanation of geometry that is the main focus of discussion. Indeed, in the 'Transcendental Aesthetic' arithmetic is never even mentioned.

(a) Space as the form of outer sense

There is a kind of experience which Kant calls 'outer sense', and is what normally occurs when we take ourselves to be perceiving (e.g. by seeing, hearing, touching, and so on). It is characteristic of this kind of experience that we take ourselves to be perceiving things that are 'outside' ourselves, and Kant claims that in this case we always perceive them as having spatial properties – e.g. size and shape, position relative to one another, and so on. His proposal is that this spatial organization of the things that we perceive is due to *us*, and not to the things themselves. This, he thinks, is a feature of *human* nature, and he concedes that other beings may perhaps perceive differently (though he gives no examples).[29] So he is not claiming that all 'outer' experience has to be spatial, or even that it has to have an organization somewhat similar to the spatial one, in order to count

[27] This section has benefited from discussions with Ralph Walker, who would not approve of my conclusions. Kant scholars often dispute the interpretation of the details of Kant's theory, but I shall try to avoid controversy by giving only the main outline. A helpful overview is Parsons (1992). The controversies may be pursued in Posy (1992).

[28] In the *Prolegomena* (1783), which was written between editions A and B of the *Critique*, the order of these two parts is reversed (§§6–14). We are still at a loss to know whether either is supposed to depend on the other, and if so how.

[29] For the limitation to human perceivers see e.g. A26–7 = B42–3, A42 = B59, B72. I guess that Kant thought that God's perception must be different, but had no views about other non-human creatures.

as being 'outer'.[30] He claims only that we human beings do in fact perceive in this way, and that is presumably uncontroversial. What is controversial is the further claim that it is *our* nature that imposes this spatial form on our experiences. Kant does not dissent from the usual view that in ordinary experience our consciousness is stimulated by data that come from elsewhere, but he also insists that the resulting appearances are only partly due to the nature of these data, and partly also due to the way our own minds react to this stimulation. His idea is that the spatial properties of appearances are contributed by us, and do not result from any spatial properties in the data. That is the central feature of his claim that 'space is the form of outer sense'. In his own words:

> Space does not represent any property of things in themselves, nor does it represent them in their relation to one another. That is to say, space does not represent any determination that attaches to the objects themselves, and which remains even when abstraction has been made of all the subjective conditions of intuition.[31] (A26 = B42)

Kant's main argument for this claim is based upon the opening premise that our knowledge of geometry is *a priori*. For he naturally takes geometry to be the study of space, and so this premise tells us that we have an *a priori* knowledge of the nature of space. But, he argues, if our experiences are spatial just because the data that we receive are themselves spatial, then our knowledge of the nature of space would have to be empirical. Consequently it would not be 'thought as necessary', or as 'strictly universal'. But, since this *is* how we think of it, the empirical theory must be mistaken (B41). He accordingly proposes his own rival theory as the only one that can explain our way of thinking. But now let us ask whether his own theory is any more successful at providing an explanation. One may look at the position in this way.

Locke (among others) had contrasted the so-called 'primary' and 'secondary' qualities of objects. One way in which he draws this distinction is by saying that the way that the secondary qualities appear to us

---

[30]   In a well-known discussion Strawson argued that, in a world of nothing but sounds (and a perceiver) the sounds could be thought of as 'outer' if they could be regarded as positioned in a one-dimensional space (Strawson, 1959, chapter 2; the example has been further elaborated by others, e.g. Bennett, 1966, chapter 3). But I do not see Kant as committed to any claim of this sort.

[31]   In the present context, for 'intuition' one can substitute 'experience'.

(e.g. colour, sound, smell) does not *resemble* anything in the objects that cause these appearances, whereas the primary qualities (which include size, shape, and other spatial properties) do appear to us as they really are. As one might say, the idea is that round objects usually appear round to us simply because they *are* round, and there is no further explanation, whereas red objects appear red because of the way that their surface atoms absorb and reflect light, a cause that has no intelligible connection with the appearance that is its effect. Kant's position may be understood as claiming that the apparent spatial properties of objects are equally 'secondary': there is no doubt some feature of an object that makes it appear to us to be round or square or triangular, but we do not know what it is, and we have no reason to suppose that it is anything like the appearance that it gives rise to. In note II to *Prolegomena* §13 Kant apparently accepts this parallel between his own view of spatial properties and the Lockean view of colours, sounds, smells, and so on.[32] It is worth pursuing this comparison.

Our way of perceiving sounds allows us to distinguish between simple and compound sounds. Each simple sound has a definite pitch, and the pitches form a one-dimensional scale from high to low. As we now believe, this scale reflects a one-dimensional scale of the wavelengths of the sound waves external to us, except that it covers only a part of that scale, excluding sounds that are too high or too low in pitch to be audible to us. By contrast, our way of perceiving colours does not provide any way of distinguishing simple and complex colours. Colours differ from one another in hue, but in this case the various hues do not fall into any one-dimensional scale. Rather, their relations are conveniently diagrammed in a two-dimensional map with the topology of the surface of a sphere. As we now believe, this way of ordering the hues mirrors no corresponding fact about the wavelengths of the light waves that give rise to them. There is no simple relationship between the different hues and the different combinations of light-waves impinging on the eye; indeed, many different combinations will produce exactly the same hue.

Let that suffice as a simplified description of our human way of perceiving colours and sounds. Now Kant supposes that something similar holds of our human way of perceiving space. For example, we perceive space as three-dimensional, but what gives rise to this perception need not be itself three-dimensional in any way. For the sake of argument, let us grant this

---

[32] I shall come later to the passage in the *Critique* which apparently denies this parallel.

theory.[33] Kant goes on to say that, because it is partly *our* contribution to experience that explains why points, lines, and shapes have the appearance that they do, the experience in question (which he calls an 'intuition') will count as an '*a priori* intuition'. This is an unusual use of the word '*a priori*', for normally it is *judgements* (propositions) that are said to be known *a priori*, and an experience ('intuition') is not itself a judgement. But we need not pay too much attention to this worry, because the notion of an '*a priori* intuition' figures only as an intermediate step towards the conclusion that he is aiming for. For he goes on to claim that, because the intuition is *a priori*, so too are the judgements to which it gives rise, and that is what had to be explained. The general idea is therefore this: if some feature of our experience is (partly) due to us, then that will provide *a priori* knowledge of the feature in question.

Now it appears that Kant *ought* to say just the same about sounds and colours as he has said about space. For again the way that we experience sounds and colours is partly due to our own human nature, so by parallel reasoning the experiences should count as '*a priori* intuitions', giving rise in the same way to *a priori* judgements about what is experienced. I genuinely think that this *is* what he ought to have said, and that it is *in a way* defensible. But it is not what he does say in the *Critique*, for at A28–30 and B44–5 he argues that our perception of space should not be regarded as relevantly similar to our perception of what Locke calls secondary qualities. The main point urged is that the two have to be different, just because we have *a priori* knowledge of the one but not of the other. This point simply begs the present question, for it does not explain how there could be this difference. In the A version two further points are made: (i) that colours and tastes are each perceived by just one sense, and (ii) that our perceiving something as coloured or tasty is not a necessary condition of our perceiving it as 'outer'. The first point on its own appears to be simply irrelevant, but when combined with the second it does suggest a contrast: the idea might be that in order to perceive things as 'outer' we must perceive them spatially, but do not have to perceive them as having any (other?) secondary qualities. However, even if this were true, it still would not explain how we can have *a priori* knowledge of the spatial properties

---

[33] I note in parenthesis that Kant's theory does have more truth in it than one is at first inclined to think. According to the theory of relativity, objects in the world do not really have what we think of as spatial properties, but only spatio-temporal properties which are relative to a frame of reference that needs to be specified.

that perceived objects always do exhibit to us, and besides one may very reasonably doubt whether it is true.[34] In the B version both of these points are dropped, perhaps because Kant saw them to be irrelevant to the problem in question. But apparently he has nothing else to put in their place, which might justify the supposed difference. One can only conclude that this first part of Kant's response to his own problem does not succeed.

To clarify matters further, let us ask whether we do have any kind of *a priori* knowledge of the nature of sounds, colours, or space. On the face of it, the answer is in each case 'no'. We know that heard pitches fall into a one-dimensional ordering, whereas seen hues do not, only from our experience of pitches and hues. The same is surely true of our experience of shapes and other spatial relations: we cannot know that they will not fit into a two-dimensional space, but will require three dimensions (at least), in advance of actually having such experiences. We apparently get the same answer if we apply Kant's own elucidation of *a priori* knowledge, i.e. as what is 'thought as necessary and as strictly universal'. We have no good reason to believe that all the sounds that we will ever hear *must* have a pitch within the familiar scale, or that all the colours that we will ever see *must* have one of the hues that we already know. (This is partly because we allow that human nature may evolve, e.g. so as to perceive colours caused by infrared light. But it is *also* because we allow that the way that we perceive things is only partly due to our own nature, and is also due to the nature of the things perceived. The second may change, even though the first does not.) But the same apparently applies to our experience of spatial properties and relations. Either our nature might change, or the received stimuli might change, perhaps because space becomes much more noticeably 'curved' in the small regions that we can perceive.[35] There appears to be no *a priori* knowledge here.

However, these considerations only show that the perceived features of sounds, colours, and space are not *in fact* necessary. But we may emphasize that Kant's question is whether they are *thought* as necessary, and then there is room to defend a different answer. If you ask the man in the street whether he can think of a different kind of experience of space, or of colours,

---

[34] Strawson's auditory world, mentioned in n. 30 above, would seem to provide a counterexample, for it is a world in which only sounds are perceived, but those sounds (or some of them) are perceived as 'outer'.

[35] I say something of the nature of a 'curved' space in section 1 of the next chapter. Of course, Kant was not aware of the possibility of non-Euclidean spaces.

or of sounds, he is quite likely to take you to be asking whether he can *imagine* it, and then he will very reasonably answer 'no'. We cannot imagine what it would be like to see two straight lines meeting twice, or to perceive a new colour, or even to hear a sound pitched much lower than the usual range. This does not distinguish, then, between our perceptions of spatial qualities and those of other qualities that Locke called secondary. But it is, I think, what really lies behind Kant's claim that our knowledge of space is *a priori*.

To strengthen this suggestion I turn now to the second passage where he discusses the nature of mathematical knowledge.

(b) The role of intuition in mathematics

In the section entitled 'The Discipline of Pure Reason in its Dogmatic Employment', Kant's overall aim is to *contrast* the reasoning employed in mathematics with reasoning in other fields, in particular metaphysics. He evidently thinks that mathematical reasoning has an advantage, and that that is due to the way in which it makes use of 'intuition'. This needs some elucidation, which I now attempt to supply. (I shall ignore what he says of metaphysical reasoning.) Once again, the emphasis is on geometry, so let us begin with this.

Earlier in the Transcendental Aesthetic Kant had claimed that space is what he calls 'an *a priori* intuition' (A23–5, B38–40). (Similar arguments are applied to time at A30–2, B46–8.) His claim here is that spatial (and temporal) concepts can be 'constructed in intuition'. So far as one can see, all that this means is that one can produce an example of the concept in question, either in imagination or perhaps drawn on paper, and that one can then pay attention only to those features of the example that are given by the concept in question. For example, to prove something about all triangles one begins by imagining or drawing a particular triangle, but one then makes use only of those of its properties that qualify it as a triangle (i.e. that it has three straight sides), so that the reasoning will generalize and apply to all triangles. But Kant's one detailed example of how 'construction in intuition' is supposed to help the mathematician *also* employs the notion of 'construction' in a further way, and I suspect that this is not accidental.

The example is of how to prove that the interior angles of a triangle always add to 180° (A716–7, B744–5). Our mathematician does not just meditate on the definition of a triangle, which would be the route to analytic truth, but begins by constructing a sample triangle. Then his next step is to construct some *further* lines, extending the original figure. In particular, he extends one side of the triangle, in order to create an exterior angle, and then divides this exterior angle by a line that is parallel to the opposite side of the triangle, like this:

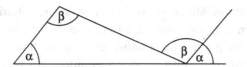

Using the known properties of parallel lines, he can then argue that the opposite interior angles are each equal to one part of the exterior angle, and this gives him his result. (This proof is given in Euclid I, 32.) Kant's comment on the reasoning is significant. He refers to it as 'a chain of inferences guided *throughout* by intuition' (*ibid*, my emphasis). So it is not just the initial step of imagining a triangle that is credited to intuition.

The example leaves us wondering whether Kant thinks that a geometrical proof *always* involves the construction of extra lines, over and above those of the figure originally given. Of course it is true that *most* geometrical proofs do employ this technique, but some do not. For example, Euclid's proposition I, 4 is that a triangle is completely determined by two sides and the angle between them. His proof asks one to imagine two triangles ABC and DEF, where it is given that the side AB is equal to the side DE, the side AC is equal to the side DF, and the angle at A is equal to the angle at D:

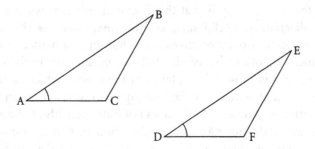

Euclid then 'proves' that the triangles are congruent by asking us to *move* the triangle ABC onto the triangle DEF, so that AB coincides with DE. He then argues that AC must also coincide with DF, and hence that BC must coincide with EF. This 'proof' requires us to imagine what will result from moving a geometrical figure from one place to another. It is often said to be a defect in Euclid's proofs that just occasionally he does have to employ this method of 'superposition', for it is an operation which his axioms do not provide for.[36] But I imagine that to Kant it would seem to be a perfectly good example of 'construction in intuition'. There is much to be said for the view that all interesting geometrical proofs do require some kind of 'construction'. Moreover, we do rely on 'spatial intuition' to see that the construction is always possible, and does have the result desired.

I illustrate this point with another example of how Euclid's proofs do frequently require a 'construction in intuition'. His very first proposition (I, 1) shows how to construct an equilateral triangle on a given base AB. The instruction is: draw a circle with centre A and radius AB, and then another with centre B and the same radius. These circles will intersect, say at the point C. Join AC and BC, and then ABC is the triangle required:

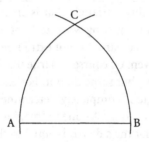

But how do we know that *there is* a point C where the two circles intersect? All that one can say is that this is entirely obvious when one constructs a diagram as Euclid says, i.e. we *cannot imagine* this diagram without imagining a point of intersection. But this is a matter of what *we* can imagine, and not a matter of what follows logically from Euclid's axioms. There is nothing in those axioms, taken by themselves, which requires the existence of such a point. Nowadays we say that this is a *defect* in Euclid's axiomatization of geometry, for he does not state explicitly all the assumptions that he needs. But Kant presumably thought that it would not be possible to state all the required axioms explicitly, so that nothing was left to our spatial imagination. Moreover, given the state of geometrical

---

[36]   See e.g. Heath (1925, pp. 249–50). I add that Kant approves of this method of superposition in *Prolegomena*, §12.

knowledge in his time, that thought is entirely reasonable. It was not until much later that spatial imagination was superseded by explicit axioms.[37]

Kant's position is, then, that geometrical reasoning requires the construction of figures in our imagination, and that the proofs rely upon what we can and cannot imagine. He never quite says this in so many words, and it should be noted that my last two examples are supplied by me and not by Kant. But this position seems to me to be comprehensible, and the best interpretation of what he does say. Also, I said of the earlier discussion in the Transcendental Aesthetic that the same idea seems to be the best way of explaining what he there says about geometrical knowledge being *a priori*. What he is thinking of is that we cannot imagine things otherwise. It is true that (in the way that is relevant) the same applies to colours, sounds, and other secondary qualities, whereas Kant will not accept this. So one cannot say that his position is entirely coherent. But it is fair to say that in geometry we cannot imagine things otherwise, and then the suggestion of the Transcendental Aesthetic is that this is to be explained by the hypothesis that the spatial nature of our (outer) experience is due to our human way of perceiving things, and not to the nature of the things experienced. Whether this is a good explanation is a question that will be taken up at the start of the next chapter.

Meanwhile I end this chapter with just a brief note of some other points in Kant's discussion.

In the *Critique* he seldom mentions any area of mathematics other than geometry, but he clearly does assume that what holds for geometry will hold for other areas too. As we have noted, in the 'Introduction' he says that we come to know that $7 + 5 = 12$ only by forming an image of 5 things, and adding them one by one to a given image of 7 things, and noting the result. This is presumably a 'construction in intuition'. But one wonders how he thinks that we know that $700 + 500 = 1200$, for it is surely not in this way. We must here be applying general rules for addition, though Kant seems not to acknowledge that there are such rules. At any rate he says that there are no *axioms* in arithmetic, apparently on the ground that each arithmetical equation is established by its own 'construction in intuition',

---

[37] I do not know whether Kant had read Euclid himself, but in any case the mathematical textbooks that were available in Kant's time were strongly influenced by Euclid, and their axioms were fairly similar to his. It was not until Hilbert (1899) that spatial intuitions were adequately replaced by explicit axioms.

and independently of every other (A164–5, B204–6). (He does not count what he regards as analytic propositions, such as 'equals added to equals yield equals', as qualifying for axiomatic status.) This is not a well-worked-out account of arithmetic.

I add that there is an occasional hint that, whereas geometry is concerned with space, arithmetic is somehow specially connected with time (e.g. A139 = B178). But this idea is not developed, and I think it can be ignored. There are also a few remarks on algebra as in some derivative way a 'construction in intuition' (A717 = B745, A734 = B762). I guess that this means that the operations on symbols which algebra permits are legitimate only because they represent operations on real magnitudes (of space, or number) which a genuine intuition can certify as correct. But in each case one has to say that we are only given a few hints, and proposals for a more fully elaborated theory would have to be very conjectural. The focus of Kant's thought about mathematics is geometry, and there he claims (with reason) that 'construction in intuition' is essential. He believes that the same will apply in other areas of mathematics too, but this belief is left without any detailed support.

Suggestions for further reading

Those who wish to pursue the mediaeval debate on realism *versus* conceptualism *versus* nominalism may begin by consulting Kenny (2005), first for the various passages mentioned in my note 1 above, and then for the many further references which Kenny gives. But I suggest that for present purposes it is best to skip this piece of history and instead to look at some more modern treatments of this debate. I recommend Russell (1912, chapter 9), and Quine (1948).

I expect that most readers of this book will already have some knowledge of Descartes, and of Locke, Berkeley and Hume. In that case, I would not suggest any further reading of these authors. For while Descartes is in fact strongly influenced by mathematics, still his writings do not emphasize this point; whereas with Locke and Berkeley and Hume mathematics is more of an awkward distraction from their real interest, which is in empirical knowledge.

By contrast, Kant's view of mathematics has been highly influential, and must be taken seriously. I suggest reading the passages of the *Critique of Pure Reason* that are mentioned in my discussion, i.e. B1–30, A19–49 =

B31–73, A713–38 = B741–66. A convenient overview is provided by Parsons (1992). Some 'classical' articles on Kant's philosophy of mathematics are Hintikka (1967), Parsons (1969), and Kitcher (1975). These are all reprinted in Posy (1992). The further articles in that volume explore particular problems that are of no central importance, except that the one by Posy himself (i.e. Posy 1984) makes some interesting connections between Kant and the modern school of intuitionism (which I discuss in chapter 7).

Chapter 3

Reactions to Kant

For almost two centuries after Kant's *Critique* (1781), i.e. until Quine's 'Two Dogmas' (1951a), a central question in the philosophy of mathematics was whether Kant was right to claim that mathematical truths are both synthetic and known *a priori*. Philosophers who thought of themselves as empiricists felt that this had to be denied, and so were forced into a dilemma: either show that these truths are analytic, or show that they are known only as a result of our experience. The first option was the more popular, but one who stands out as having defended the second is John Stuart Mill. He, like many philosophers before him, paid attention only to geometry on the one hand and the elementary arithmetic of natural numbers on the other. (At the time that he was writing this was rather clearly an inadequate view of the nature of mathematics – as my next chapter will show – but let us continue to accept it for the time being.) He defended the empiricist approach to both of these, but with much more success in the one case than in the other. In essence, he was absolutely right about geometry, but very clearly wrong about arithmetic, so I take these separately.

1. Mill on geometry[1]

Mill's main claim, stated at the outset of his discussion, is that 'the character of necessity ascribed to the truths of mathematics, and even (with some reservations . . .) the peculiar certainty attributed to them, is an illusion'. Like almost all philosophers before Kripke's *Naming &*

---

[1] J.S. Mill, *System of Logic* (1843, book II, chapter v). The section references that follow are to this chapter. (There is a further discussion of both geometry and arithmetic in book III, chapter XXIV. But I set this aside, as it introduces no new ideas.)

Necessity (1972), Mill runs together the ideas of necessary truth and *a priori* knowledge, so that his denial of necessity is at the same time a denial that knowledge of these truths is *a priori*. Indeed, his ensuing arguments are much more directly concerned with the nature of our knowledge than with the necessity or otherwise of what is known. And in fact they are mostly defensive arguments, claiming that the reasons given on the other side are not cogent.

The first is this. Some, he says, have supposed that the alleged necessity of geometrical truths comes from the fact that geometry is full of idealizations, which leads them to think that it concerns, not objects in the physical world, but ideas in our minds. Mill replies that this is no argument, because the idealizations in question cannot be pictured in our minds either; for example, one may admit that there is no line in the physical world that has no thickness whatever, but the same applies too to lines in our imagination (section 1). In fact this response is mistaken, since it is perfectly easy to imagine lines with no thickness, e.g. the boundary between an area which is uniformly black and a surrounding area which is uniformly white. A perfectly sharp boundary has no thickness. But that is of no real importance. It is clear that geometry *can* be construed as a study of the geometrical properties of ordinary physical objects, even if it does to some extent idealize, and I would say that it is better construed in this way than as a study of some different and purely mental objects. So here Mill parts company with the general tenor of the tradition from Descartes to Kant, and his revised (Aristotelian) ontology opens the path to his epistemology.

Mill very reasonably takes it for granted that geometrical knowledge is acquired by deduction, and that this deduction begins from axioms and definitions. In the present chapter he does not claim that there is any problem about our grasp of the deductions; that is, he accepts that if the premises were necessary truths, known *a priori*, then the same would apply to the conclusions. He also concedes (at least for the sake of argument) that there is no problem about the definitions, since they may be regarded as mere stipulations of ours, and hence as necessary truths and known *a priori*, just because we can know what we ourselves have stipulated. He is also prepared to grant that *some* of the propositions traditionally regarded as axioms might perhaps be rephrased as definitions, or replaced by definitions from which they would follow. (On this point he is somewhat over-generous to his opponents.) But he insists that the deductions *also* rely on genuine *axioms*, which are substantive assertions, not to be explained as concealed definitions. (The example that he most often refers to is: 'two straight lines cannot meet twice, i.e. cannot enclose a space'.)[2] So we can focus on the question of how axioms (such as this) are known (sections 2–3).

His answer is that they are known only because they have consistently been verified in our experience. He concedes that it is not just that we never have experienced two distinct straight lines that meet twice, but also that we cannot even in imagination form a picture of such a situation. But he gives two reasons for supposing that this latter fact is not an *extra* piece of evidence. The first is just the counter-claim that what we can in this sense imagine – i.e. what we can imagine ourselves perceiving – is limited by what we have in fact perceived, and there is clearly at least some truth in this, as we have seen (pp. 55–6). The second is that it is only because of our past experience, which has confirmed that spatial arrangements which we can imagine are possible, while those that we cannot imagine do not occur in fact, that we have any right to trust our imagination at all on a subject such as this. That is, the supposed connection between spatial possibility and spatial imaginability, which is here being relied upon, could not itself be established *a priori*. (I supply an example which Mill does not: we can certainly imagine an Escher drawing, because we have seen them. But can we imagine the situation that such a drawing depicts? If so, the supposed connection between possibility and imaginability cannot be without exceptions.) For both these reasons Mill sets aside as irrelevant the claim that we cannot even picture to ourselves two straight lines meeting twice. The important point is just that we have never seen it (sections 4–5).

But perhaps the most convincing part of Mill's discussion is his closing section 6 on the subject of conceivability. By this he means, not our ability to picture something, but our ability to see that it *might* be true. He concedes that we cannot (in this sense) even conceive of the falsehood of the usual geometrical axioms, but he claims that we cannot legitimately infer from this either that our knowledge of them is not based upon experience or that their falsehood is impossible. For what a person can conceive is again limited by what he has experienced, by what he has been brought up to believe, and by the weakness of his own creative thought.

---

[2] Strangely, Euclid's own text does not state this explicitly as a postulate, though he very soon begins to rely upon it. The gap was noted by his successors, and the needed extra postulate was added. For a brief history see Heath's commentary on Euclid's postulate I (Heath, 1925, pp. 195–6).

To substantiate these claims Mill cites several examples, from the history of science, of cases where what was once regarded as inconceivable was later accepted as true.

One of these is what we may call 'Aristotle's law of motion', which states that in order to keep a thing moving one has to keep applying force to it. It is clear that this seemed to Aristotle to be quite obviously true, and it is also clear why: it is a *universal* experience that moving objects will slow down and eventually stop if no further force is applied. Mill very plausibly claimed that for centuries no one could even conceive of the falsehood of this principle, and yet nowadays we do not find it difficult to bring up our children to believe in the principle of inertia. Another of Mill's examples is 'action at a distance', which *seems* to be required by the Newtonian theory of gravitational attraction. For example, it is claimed that the earth does not fly off from its orbit at a tangent because there is a massive object, the sun, which prevents this. But the sun is at a huge distance from the earth, and in the space between there is nothing going on which could explain how the sun's influence is transmitted. (To take a simple case, there is no piece of string that ties the two together.) The Cartesians could not believe this, and so felt forced into a wholly unrealistic theory of 'vortices'; Leibniz could not believe it, and said so very explicitly; interestingly, Newton himself could not – or anyway did not – believe it, and devoted much time and effort to searching for a comprehensible explanation of the apparent 'attraction across empty space' that his theory seemed to require.[3] But again we nowadays find it quite straightforward to explain the Newtonian theory to our children in a way which simply treats action at a distance as creating no problem at all. Of the several further examples that Mill gives I mention just one more, because it has turned out to be very apt, and in a way which Mill himself would surely find immensely surprising. He suggests that the principle of the conservation of matter (which goes back to the very ancient dictum '*Ex nihilo nihil fit*') has now become so very firmly established in scientific thought that no serious scientist can any longer conceive of its falsehood. Moreover, he gives examples of philosophers of his day who did make just this claim of inconceivability. But of course we from our perspective can now say that this principle too turns out to be mistaken, for Einstein's $e = mc^2$ clearly denies it. Indeed, we from our perspective could add many more examples of how what was

---

[3] But he never found an explanation that satisfied him, and so he remained true to the well-known position of his *Principia Mathematica*: on this question '*hypotheses non fingo*'.

once taken to be inconceivable is now taken to be true; quantum theory would be a fertile source of such examples.

I am sure that when Mill was writing he did not know of the development that has conclusively proved his view of the axioms of geometry to be correct, namely the discovery of non-Euclidean geometries.[4] These deny one or more of Euclid's axioms, but it can be shown that if (as we all believe) the Euclidean geometry is consistent then so too are these non-Euclidean geometries. We say nowadays that the Euclidean geometry describes a 'flat' space, whereas the non-Euclidean alternatives describe a 'curved' space ('negatively curved' in the case of what is called 'hyperbolical' geometry, and 'positively curved' in the case of what is called 'elliptical' geometry). Moreover – and this is the crucial point that vindicates Mill's position completely – it is now universally recognized that it must count as an *empirical* question to determine which of these geometries fits actual space, i.e. the space of the universe that surrounds us. I add that as a matter of fact the current orthodoxy amongst physicists is that that space is not 'flat' but is 'positively curved', and so Euclid's axioms are not after all true of it. On the contrary, to revert to Mill's much-used example, in that space two straight lines *can* meet twice, even though our attempts to picture this situation to ourselves still run into what seem to be insuperable difficulties.

For the curious, I add a *brief* indication of what a (positively) curved space is like as an appendix to this section. But for philosophical purposes this is merely an aside. What is important is that subsequent developments have shown that Mill was absolutely right about the status of the Euclidean axioms. There are alternative sets of axioms for geometry, and if we ask which of them is true then the *pure* mathematician can only shrug his shoulders and say that this is not a question for him to decide. He may

[4]   Mill tells us, in a final footnote to the chapter, that almost all of it was written by 1841. The first expositions of non-Euclidean geometry were due to Lobachevsky (1830) and Bolyai (1832), so in theory Mill could have known of them. But their geometry was the 'hyperbolical' one, in which it is still true that two straight lines cannot enclose a space, but another of Euclid's axioms is false, namely that there cannot be two straight lines, which intersect at just one point, and which are both parallel to the same line. Mill knows of this axiom, but does not take it as his main example, which he surely would have done if he had known that there is a consistent geometry which denies it. What he does take as a main example, namely that two straight lines cannot enclose a space, is false in 'elliptical' geometry, but that was not known at the time that Mill was writing. It is mainly due to Riemann (published 1867; proposed in lectures from 1854).

say that it is not a genuine question at all, since the various axiom-systems that mathematicians like to investigate are not required to be 'true', and we cannot meaningfully think of them in that way. Or he may say (as the empiricist would prefer) that the question is a perfectly good question, but it can only be decided by an empirical investigation of the space around us, and – as a pure mathematician – that is not *his* task. In either case Mill is vindicated. The interesting questions about geometry are questions for the physicist, and not for the (pure) mathematician. Consequently they no longer figure on the agenda for the philosopher of mathematics, and so I shall say no more about them. I close this section with a quick summary of what now counts as orthodoxy about geometry.

It is now generally agreed (i) that we all do have a fair grasp of what a Euclidean space is, and we can work out from our conception of it what has to be true in such a space. For example, we can tell what axioms are correct for it.[5] It is also agreed (ii) that our everyday experience of the space that we live in confirms the view that it is at least *approximately* Euclidean, though obviously our ordinary observations cannot tell us that it is exactly so. This no doubt explains why we all find the Euclidean geometry so very intuitive, and automatically interpret what we experience in conformity with it. For, of all the geometries that we now know about, it is the simplest, and is very effective for everyday purposes. But also (iii) it is not particularly surprising that the physicists should come to a different opinion about the overall geometry of space as a whole, for after all our everyday experience is restricted to only a tiny portion of it. If, at present, most of us cannot envisage a non-Euclidean geometry, that is no doubt for the reason that Mill gives. For example, concerning the axiom that two straight lines cannot enclose a space, it seems to be true that none of us can in fact form a mental picture which would show, in one glance, how an exception is possible. But nowadays we have no difficulty in saying how a *series* of experiences, when taken together, would be best interpreted on this hypothesis. The following appendix makes this clear.

Appendix: non-Euclidean geometry

Let us begin with the simple case of *two*-dimensional geometry, i.e. of the geometrical relations to be found simply on a surface. In this case it is easy to see what is going on. If the surface is a flat piece of paper, then we expect Euclid's axioms to hold for it, but if the surface is curved – for example,

---

[5] Compare my remarks on 'structuralism' in chapter 6, section 4.

if it is the surface of a sphere – then they evidently do not. For in each case we understand a 'straight line' to be a line on the surface in question which is the shortest distance, as measured *over that surface*, between any two points on it. On this account, and if we think of our spherical surface as the surface of the earth, it is easy to see that the equator counts as a straight line, and so do the meridians of longitude, and so does any other 'great circle'. (You may say that such lines are not *really* straight, for between any two points on the equator there is a shorter distance than the route which goes round the equator, namely a route through the middle of the sphere. But while we are considering just the geometry of a surface, we ignore any routes that are not on that surface, and on this understanding the equator does count as a straight line.)

Given this account of straightness, it is easy to see that many theses of Euclidean geometry will fail to hold on such a surface. For example, there will be no parallel straight lines on the surface (for, apart from the equator, the lines that we call the 'parallels' of latitude are not straight). Again, the sum of the angles of a triangle will always be greater than two right angles, and in fact the bigger the triangle the greater is the sum of its angles. (Think, for example, of the triangle which has as one side the Greenwich meridian of longitude, from the North pole to the equator, as another side a part of the equator itself, from longitude 0° to longitude 90°, and as its third side the meridian of longitude 90°, from the equator back to the North pole. This is an equilateral triangle, with three equal angles, but *each* of those angles is a right angle.) It is easy to think of other Euclidean theorems which will fail on such a surface. I mention just two. One is our old friend 'two straight lines cannot meet twice'; it is obvious that on this surface every two straight lines will meet twice, on opposite sides of the sphere. Another is that, unlike a flat surface, our curved surface is finite in area without having any boundary. Here is a simple consequence. Suppose that I intend to paint the whole surface black, and I begin at the North Pole, painting in ever-increasing circles round that pole. Well, after a bit the circles start to decrease, and I end by painting myself into an ever-diminishing space at the South Pole.

These points are entirely straightforward and easily visualized, but now we come to the difficult bit. We change from the two-dimensional geometry of a curved surface to the three-dimensional geometry of a genuine space, but also suppose that this space retains the same properties of curvature as we have just been exploring. A straight line is now the shortest distance in this *three*-dimensional space between any two points in it; i.e.

it is genuinely a straight line, and does not ignore some alternative route which is shorter but not in the space: there is no such alternative route. But also the straight lines in this curved three-dimensional space retain the same properties as I have been saying apply to straight lines in a two-dimensional curved space. In particular, two straight lines can meet twice. So if you and I both start from here, and we set off (in our spaceships) in different directions, and we travel in what genuinely are straight lines, still (if we go on long enough) we shall meet once more, at the 'other side' of the universe. Again, the space is finite in volume, but also unbounded. So suppose that I have the magical property that, whenever I click my fingers, a brick appears. And suppose that I conceive the ambition of 'bricking in' the whole universe. Well, if I continue long enough, I will succeed. I begin by building a pile of bricks in my back garden. I continue to extend it in all directions, so that it grows to encompass the whole earth, the solar system, our galaxy, and so on. As I continue, each layer of bricks that I add will require more bricks than the last, until I get to the midpoint. After that the bricks needed for each layer will decrease, until finally I am bricking myself into an ever-diminishing space at the 'other end' of the universe. That is the three-dimensional analogue of what happens when you paint the surface of a sphere.

Well, imagination boggles. We say: that *could* not be what would actually happen. The situation described is just *inconceivable*. And I agree; I too find 'conception' extremely difficult, if not impossible. But there is no doubt that the mathematical theory of this space is a perfectly consistent theory, and today's physicists hold that something very like it is actually *true*.

Inconceivability is not a safe guide to impossibility.

2. Mill versus Frege on arithmetic[6]

Mill's discussion of geometry was very much aided by the fact that geometry had been organized as a deductive science ever since Greek times. This allowed him to focus his attention almost entirely upon the status of its axioms. By contrast, there was no axiomatization of arithmetic at the time when he was writing, and so he had no clear view of what propositions constituted the 'foundations' of the subject. He appears to have

---

[6] Mill, *System of Logic* (1843, book II, chapter VI, sections 2–3). Frege, *Grundlagen* (1884, sections 5–9, 16–17). References are to these sections.

thought that elementary arithmetic depends just upon (a) the definitions of individual numbers, and (b) the two general principles 'the sums of equals are equal' and 'the differences of equals are equal' (section 3). Certainly these are two basic assumptions which are made in the manipulation of simple arithmetical equations, though as we now see very well there are several others too. Mill claims that the two general principles he cites are generalizations from experience, which indeed they would be if interpreted as he proposed, i.e. as making assertions about the results of *physical* operations of addition and subtraction. To one's surprise he *also* says that the *definitions* of individual numbers are again generalizations from experience, and this is a peculiar position which (so far as I know) no one else has followed. But we may briefly explore it.

First we should notice his ontology. He opens his discussion (in section 2) by rejecting what *he* calls 'nominalism', which he describes as 'representing the propositions of the science of numbers as merely verbal, and its processes as simple transformations of language, substitution of one expression for another'. The kind of substitution he has in mind is substituting '3' for '2 + 1', which the theory he is describing regards as 'merely a change in terminology'. He pours scorn upon such a view: 'The doctrine that we can discover facts, detect the hidden processes of nature, by an artful manipulation of language, is so contrary to common sense, that a person must have made some advances in philosophy to believe it'. At first one supposes that he must be intending to attack what we would call a 'formalist' doctrine, which claims that the symbols of arithmetic (such as '1', '2', '3', and '+') have *no* meaning. But in fact this is not his objection, and what he really means to deny is just the claim that '3' and '2 + 1' have *the same meaning*.[7] We shall see shortly how, in his view, they differ in meaning.

He then goes on to proclaim himself as what *I* would call a 'nominalist': 'All numbers must be numbers of something; there are no such things as numbers in the abstract. *Ten* must mean ten bodies, or ten sounds, or ten beatings of the pulse. But though numbers must be numbers of something, they may be numbers of anything.' From this he fairly infers that even propositions about particular numbers are really generalizations; for example '2 + 1 = 3' would say (in Mill's own shorthand) 'Any two and any one make a three'. But it does not yet follow that these

---

[7] Kant claimed that '7 + 5 = 12' was not analytic; you might say that Mill here makes the same claim of '2 + 1 = 3'.

generalizations are known empirically, and the way that Mill tries to secure this further claim is, in effect, by offering an interpretation of the sign '+'.[8]

He says: 'We may call "three is two and one" a definition of three; but the calculations which depend on that proposition do not follow from the definition itself, but from an arithmetical theorem presupposed in it, namely that collections of objects exist, which while they impress the senses thus, $\overset{\circ}{\circ}{}_{\circ}$, may be separated into two parts thus, $\circ$ $\circ$ $\circ$' (section 2). I need only quote Frege's devastating response: 'What a mercy, then, that not everything in the world is nailed down; for if it were we should not be able to bring off this separation, and 2 + 1 would not be 3!' And he goes on to add that, on Mill's account 'it is really incorrect to speak of three strokes when the clock strikes three, or to call sweet, sour and bitter three sensations of taste, and equally unwarrantable is the expression "three methods of solving an equation". For none of these is a [collection] which ever impresses the senses thus, $\overset{\circ}{\circ}{}^{\circ}$.' It is quite clear that Mill's interpretation of '+' cannot be defended.[9]

One might try other ways of interpreting '+', so that it stood for an operation to be performed on countable things of any kind, which did not involve how they appear, or what happens when you move them around, or anything similar, but would still leave it open to us to say that arithmetical additions are established by experience. For example, one might suppose that '7 + 5 = 12' means something like: 'If you count a collection and make the total 7, and count another (disjoint) collection and make the total 5, then if you count the two collections together you will make the total 12'. That certainly makes it an empirical proposition, but of course one which is false, for one must add the condition that the counting is correctly done. But this then raises the question of whether the notion of *correct* counting needs to be explained in empirical terms, and on the face of it that is not true.

For illustration I again change the example to '2 + 2 = 4', simply for brevity. In the orthodox notation of modern predicate logic we can define numerically definite quantifiers, e.g. thus:

---

[8]   The position outlined in this paragraph is very similar to the position I attribute to Aristotle, except that Aristotle would have begun 'there *are* such things as numbers in the abstract, but they exist only in what they are numbers of'.

[9]   Frege, *Grundlagen* (1884, pp. 9–10).

$(\exists_0 x)(Fx)$ for $\neg(\exists x)(Fx)$

$(\exists_1 x)(Fx)$ for $(\exists x)(Fx \wedge (\exists_0 y)(Fy \wedge y \neq x))$

$(\exists_2 x)(Fx)$ for $(\exists x)(Fx \wedge (\exists_1 y)(Fy \wedge y \neq x))$

and so on[10]

Then the suggestion about '2 + 2 = 4' could be put in this way: it means that, whatever 'F' and 'G' may be,

$$(\exists_2 x)(Fx) \wedge (\exists_2 x)(Gx) \wedge \neg(\exists x)(Fx \wedge Gx) \;\rightarrow\; (\exists_4 x)(Fx \vee Gx)$$

But we do not need the experience of counting to assure us of this truth, for it can easily be shown to be a theorem of the now familiar logic. So too is any other addition sum when treated in the same way.[11] This begins to approach Frege's own view of the nature of arithmetic. But before I come to that let us return to finish off his criticism of Mill, which now introduces the question of arithmetical equations involving large numbers.

For example, consider the equation 7,000 + 5,000 = 12,000. Surely we do not believe that this is true because we have actually done the counting many times and found that it always leads to this result. So how could the empiricist explain this knowledge? Well, it is obvious that the answers to sums involving large numbers are obtained not by the experiment of counting but by calculating. If one thinks how the calculation is done in the present (very simple) case, one might say that it goes like this:

$$
\begin{aligned}
7{,}000 + 5{,}000 &= (7 \times 1{,}000) + (5 \times 1{,}000) \\
&= (7 + 5) \times 1{,}000 \\
&= 12 \times 1{,}000 \\
&= 12{,}000
\end{aligned}
$$

---

[10]   Where 'n'' is the next numeral after 'n', the general scheme is:

$(\exists_n x)(Fx)$ for $(\exists x)(Fx \wedge (\exists_n y)(Fy \wedge y \neq x))$

[11]   To simplify the proposed definition, at the same time both making the second-level quantifiers explicit and representing the point that an equation works both ways, we may phrase it thus:

$$n + m = p$$

is to be taken as short for

$$(\forall F)((\exists G)(\exists_n x)(Fx \wedge Gx) \wedge (\exists_m x)(Fx \wedge \neg Gx)) \;\leftrightarrow\; (\exists_p x)(Fx))$$

Here the first step and the last may reasonably be taken as simply a matter of definition (i.e. the definition of Arabic numerals); the third step depends upon the proposition $7 + 5 = 12$, which we take as already established, together with the principle that equals multiplied by equals yield equals; the second step is perhaps the most interesting, for it depends on the principle of distribution, i.e.

$$(x \times z) + (y \times z) = (x + y) \times z.$$

But how do we come to know that *that* general principle is true? Of course the same question applies to hosts of other general principles too, and not only to the two that Mill himself mentions (i.e. 'the sums of equals are equal' and 'the differences of equals are equal'). If the knowledge is to be empirical in the kind of way that Mill supposes, it seems that we can only say that we can run experimental checks on such principles where the numbers concerned are small, and then there is an inductive leap from small numbers to *all* numbers, no matter how large. But if that really is our procedure, then would you not expect us to be rather more tentative then we actually are on whether these principles really do apply to very large numbers?

Let us sum up. Frege's criticisms of Mill may be grouped under two main headings. (i) Arithmetical operations (such as addition) cannot simply be identified with physical operations performed on physical objects, even though they may share the same name (e.g. 'addition'). One reason is that arithmetical propositions are not falsified by the discovery that the associated physical operations do not always yield the predicted result (e.g. if 'adding' 7 pints of liquid to 5 pints of liquid yields, not 12 pints of liquid, but (say) an explosion). Another reason is that the arithmetical propositions may equally well be applied to other kinds of objects altogether, where there is no question of a physical addition. As Frege saw it, the mistake involved here is that of confusing the arithmetical proposition itself with what should be regarded as its practical applications. It will (usually) be an empirical question whether a proposed application of arithmetic does work or not, but arithmetic itself does not depend upon this. (ii) We are quite confident that the general laws of arithmetic apply just as much to large numbers as to small ones, but it is not easy to see how the empiricist can explain this. For on his account we believe them only because they have very frequently been verified in our experience, and yet the verification he has in mind would seem to be available only when the numbers concerned are manageably small. To these objections made by Frege,

I add a third which (curiously) he does not make, but which we have seen bothered Aristotle: (iii) How can an empiricist account for our belief that there are infinitely many numbers? For, on the kind of account offered by both Aristotle and Mill, this belief seems in fact to be *false* (as is acknowledged by Aristotle, but overlooked by Mill).

I add here a few brief remarks on Frege's own position. He failed to learn from Mill's discussion of geometry, and he thought that on this subject Kant had been right, i.e. that the axioms of (Euclidean) geometry are both synthetic and known *a priori*.[12] But his serious work was on arithmetic, where he claimed that Kant was mistaken, since the truths of arithmetic are in fact analytic truths. In support of this claim he first generalized Kant's own definition of an analytic truth, explaining it as one that is provable just from pure logic, together with definitions of the concepts involved. He very reasonably claimed that this redefinition captured in a clearer way the essence of what Kant himself had really intended, while generalizing it to apply to propositions of all kinds. (As we have seen, Kant initially confined attention to propositions of the form 'All *A* are *B*', though he subsequently made use of the more general idea that an analytic truth is one whose negation is a self-contradiction, e.g. *Critique* B14. Frege develops this latter idea, identifying a self-contradiction as whatever can be proved false by logic alone. But just as Kant failed to define 'self-contradiction', so also Frege fails to define 'logic', and this is a gap which we must return to.)

What one may fairly call Frege's life work was the task of demonstrating that all arithmetical truths are in this sense analytic. I postpone a proper discussion of this to chapter 5, merely noting here (a) that Frege's criticisms of Mill's empiricism are decisive, and (b) that, in the light of what we have just said about '2 + 2 = 4', his own position is initially attractive. (But I add that Frege did not himself define addition in quite this way, for reasons which will emerge in chapter 5.) This naturally leads to my next topic: how *should* analyticity be defined, if Kant's position is to be proved wrong?

3. Analytic truths

Kant's initial thought is this: it is easy to see how 'analytic' truths may be known *a priori*, and known to be necessary, for all that is required is an

---

[12]   Later on he also failed to learn from Hilbert's (1899) axiomatization of geometry, and in the years that followed the two engaged in a somewhat futile correspondence on this topic. This correspondence is conveniently collected in Frege (1980).

'analysis' of concepts that we already possess. We may fairly say that Kant himself seems to have had an over-simple view of what counts as an analysis, namely as a process of dissecting a concept into its 'parts'. Frege spoke more neutrally of proofs that rely only on logic and on definitions, where it is quite clear that Frege's 'definition' is intended to capture what Kant had in mind as 'analysis', but is not so restrictive. Frege surely did not think that every correct definition specified the 'parts' from which a concept was constructed.[13] But others have claimed that even the notion of a definition is too narrow. On the traditional view, to 'define' an expression is to give a further (usually more complex) expression, which it simply abbreviates. Consequently a defined expression can always be eliminated from any context in which it occurs, and replaced by its defining expression. But in present-day mathematics the notion of a 'definition' is more relaxed than this. For example, we recognize that properties or relations or functions on the numbers can properly be introduced by what is called a 'recursive definition', even though this does not provide a rule of elimination.[14] However, others have argued, more radically, that the notion of an 'analysis' should be stretched even further, and beyond what anyone might wish to call a 'definition'. For example, many have wished to claim that 'everything red is coloured' is analytic, or again 'nothing is both red and green (all over, at the same time, etc.)'. But they claim this at the same time as admitting that 'red' (and 'green', and 'coloured') are not definable. So the suggestion here is that not only meaning-equations but also meaning-inclusions and meaning-exclusions should be permitted. But this proposed expansion of Frege's definition has no natural stopping-point. There is no end to the possible relations of meanings that would seem to be relevant (for example, 'a parent of an ancestor is an ancestor'), and so we would come to Carnap's suggestion that, for each language, a long list of its 'meaning-rules' should be provided, and there is no specified set of forms that these must take (Carnap, 1947, pp. 7–8).

But I wish to focus here on a different point about Frege's definition, namely that it includes as analytic not only the results of analysis but *also*

---

[13] His own definition of the 'ancestral' of a relation surely cannot be described in this way. Let R be any relation. Then its (improper) ancestral R^* is defined by:

$$R^*(a,b) \quad \leftrightarrow \quad \forall F(Fa \wedge \forall xy(Fx \wedge R(x,y) \rightarrow Fy) \rightarrow Fb)$$

As we shall see, this notion is crucially important for the definition of 'Natural Number'.

[14] For 'recursive definitions', and their justification, see p. 103 below.

what he calls logic. His account draws our attention to the fact that definitions, as ordinarily understood, do not *by themselves* ensure the truth of anything. To illustrate with a standard example, the definition of 'bachelor' may well tell us that this word means the same as 'man who is not married'.[15] Applying the definition, we see that

No bachelor is married

means the same as

No man who is not married is married

But what ensures that this *latter* is true? Here it is very natural to say that it is not any further definition that is wanted, nor even any more devious analysis of meanings, but something else altogether, namely *logic*.

For Frege, the truths of logic are automatically counted as analytic, and yet this opens up a gap in the simple Kantian thought that we began with. For the fact (if it is a fact) that it is not surprising that we can analyse our own concepts will not now by itself furnish an explanation of our a priori knowledge; it *also* needs to be explained how we can know a priori the truths of logic. On this, Frege himself has nothing to say. I turn, therefore, to a more ambitious account of analyticity, which aims not only to overcome the limitations recently mentioned (e.g. restricting 'analysis' to explicit definition) but also to encompass our knowledge of logic itself. It is this: an analytic truth is one whose truth is determined simply by the meaning of the words used to express it, and this applies to *all* analytic truths including the truths of logic. I take as an example A.J. Ayer's exposition, in his *Language, Truth & Logic*, chapter 4; for I guess that this will already be familiar to most English readers.[16]

Ayer claims that the truths of logic and mathematics are analytic, and he offers a familiar-sounding definition: 'a proposition is analytic when its

---

[15] Of course this traditional example is over-simple, but it will do to illustrate my point here.

[16] Ayer (1936). I shall quote from the second edition (1946). The basic idea was endorsed by all members of the so-called 'Vienna Circle' – a group whose leading members were Schlick, Carnap, and Hempel – and Ayer learnt it from them. (Some papers by Carnap and Hempel are reprinted in Benacerraf and Putnam, 1983. I add a remark on Carnap's version below, pp. 275–6).

validity depends solely on the definitions of the symbols it contains' (p. 78).[17] But it is at once obvious that his subsequent discussion scarcely anywhere appeals to what we ordinarily think of as a definition. In practice he speaks not of definitions but of what he variously calls 'the rules which govern the use of language' (p. 77), 'the function of words' (p. 79), 'the convention which governs our usage of words' (p. 79), 'our determination to use words in a certain fashion' (p. 84), and so on. This is the vocabulary that he uses in almost all cases, and in particular when discussing the laws of logic. For example, considering an instance of the law of excluded middle, he says:

> If one knows what is the function of the words 'either', 'or', and 'not', then one can see that any proposition of the form 'either p is true or p is not true' is valid, independently of experience. Accordingly all such propositions are analytic. (p. 79)

Similarly, discussing an instance of the syllogism in Barbara (i.e. 'if all A are B, and all B are C, then all A are C'), his comment is that he is 'calling attention to the implications of a certain linguistic usage', and in particular:

> I am thereby indicating the convention which governs our usage of the words 'if' and 'all'. (p. 79)

But there is not a word of *definitions* of such expressions as 'either', 'or', 'not', 'if', and 'all', either here or elsewhere.

The essential claim is, then, that the 'linguistic conventions' do, all by themselves, ensure that this and that proposition must be true. How so? Well, the only obvious suggestion is that the 'linguistic conventions' just *are* conventions that such-and-such is to be true. We might perhaps wish to add that *some* of these conventions, e.g. those corresponding to logically correct forms of inference, may be conditional in form: if such-and-such propositions are true (by convention, or otherwise) then such-and-such others are also to be true. But apparently at least some conventions must stipulate truth outright, if anything is to be true *just* by

---

[17] Since Ayer's official position is that propositions are expressed by sentences, and that sentences contain symbols whereas propositions do not, this is in any case a careless formulation. But the point is not worth dwelling on.

convention.[18] In any case, I remark *ad hominem* that Ayer himself pays no explicit attention to rules of inference, for his focus is on the *propositions* that are analytic, and which he claims to be true simply in virtue of our linguistic conventions. In fact he explicitly says, with logic in mind, that deduction is irrelevant.

> Every logical proposition is valid in its own right. Its validity does not depend on its being incorporated into a system, and deduced from certain propositions which are taken as self-evident. The construction of systems of logic is useful as a means of discovering and certifying analytic propositions, but it is not in principle essential even for this purpose. For it is possible to conceive of a symbolism in which every analytic proposition could be seen to be analytic in virtue of its form alone. (p. 81)[19]

The thought then is that the linguistic conventions do simply stipulate truth outright, and it is the propositions so stipulated that are the analytic ones. So they, and only they, are the propositions that can be known *a priori*.

But if the linguistic conventions just stipulate truth (or even conditional truth), what has this to do with meaning? The answer must be that by stipulating the truth of certain sentences we thereby confer meaning on the words in them, for the words in question are stipulated to have such a meaning as ensures that the sentences in question are true. Thus, linguistic rules do specify meanings, as we expect them to, but in a mildly roundabout way. For in the *first* place they specify what is to be true, and from this we have to work out what the meanings are. So the analytic truths are the ones that are stipulated, but they can still be said to be 'analytic' in something like the traditional sense, i.e. 'true in virtue of meaning', for what is stipulated is that they are to have such a meaning as makes them true. In fact I see no other explanation of how meanings can, all by themselves, ensure truth.

---

[18] One familiar with natural deduction (as, of course, Ayer was not; for when he was writing this method was scarcely known) might here object: it will be sufficient if nothing is stipulated to be true outright, but we allow for the principle of conditionalization in some such way as this. It is a convention that, if it is a convention that if '*P*' is true then so is '*Q*', then it is also a convention that 'if *P* then *Q*' is true. This involves in quite a complex way the idea of applying one convention to another, but I shall generally ignore this complication in what follows.

[19] I guess that he is thinking of a 'perspicuous notation' for truth functions, such as Wittgenstein develops in his *Tractatus* (1921), e.g. 4.27–4.442. (Ayer was not a logician; it would not have occurred to him that even the logic of first-level quantifiers might not be decidable, in this or any other way).

However, this doctrine will not do. One may fairly dispute whether it has an adequate view of the analysis of concepts, but here I shall concentrate upon what it implies about our knowledge of logic. For it supposes that the rules of logic are simply stipulated by us, and that what is so stipulated has to be true. But this is not so.

The simplest example to make this point is Prior's 'tonk' (Prior, 1960). Prior was writing at a time when natural deduction was in vogue, and so pointed his example in that direction. Imagine, then, someone wishing to introduce a new propositional connective 'tonk', which is to have the same grammar as the logician's 'and' and 'or', but a different meaning. Its meaning is stipulated to be such that it satisfies these two rules of deduction, the first an introduction-rule and the second an elimination-rule: for any propositions P and Q the following arguments are to be correct:

$$\frac{P}{P \text{ tonk } Q}, \qquad \frac{P \text{ tonk } Q}{Q}$$

We may, of course, transpose the example to Ayer's preferred setting by stipulating that these conditionals are always to be true:

If P, then P tonk Q
If P tonk Q, then Q

Clearly the stipulation cannot succeed. For, assuming that we retain our existing 'stipulations' about the correctness of arguments, or the meaning of 'if', it would have to follow that, whatever propositions P and Q may be, we always have as a correct argument:

$$\frac{P}{Q}$$

or as a true conditional:

If P then Q

So, on the same assumption, it is easily shown that with this addition of 'tonk' our language has become inconsistent (e.g. inconsistent in the sense of Post: every sentence whatever is now provable). The moral to draw is that it is no use *stipulating* that 'tonk' is to have such a meaning as makes these theses correct, for there is no such meaning for any sentential connective to have.

I pass straight on to another example, equally familiar, namely Grelling's paradox. Suppose that I attempt to introduce you to the new word 'heterological' by stipulating (i) that this word is to be an adjective, with the usual grammar of adjectives, and (ii) it is to be so interpreted that for *all* adjectives x

'Heterological' is true of $x \leftrightarrow \neg (x$ is true of $x)$

Again, if we assume the usual meanings for $\leftrightarrow$ and $\neg$, my attempted stipulation will lead to a contradiction, for it will follow that 'heterological' is true of itself if and only if it is not true of itself, and again we must infer that the stipulation fails. A contradiction cannot be true, and so no word could have such a meaning as would make one true. There are limits on what can be stipulated, for logic itself imposes such limits.

It is perhaps not very clear how logic could have such 'authority' over stipulations if logic is itself simply something that is stipulated, but that is not an issue that needs to be faced here. It is enough merely to observe that there is no guarantee that just *any* set of stipulations on meanings will be simultaneously satisfiable, even if we are sceptical about counting one part of the set as somehow 'authoritative' over the other. This is already a serious blow to the kind of conventionalism that Ayer espouses. It is serious because we are now deprived of the nice simple explanation of *a priori* knowledge that he was aiming for. The thought was that *a priori* knowledge must arise from knowledge of meanings, that meanings are simply stipulated by us, and it is not surprising that we can know what we ourselves have stipulated. But now we must add that even if we can know what we have stipulated, still there is something else we need to know too, namely whether those stipulations are successful, whether (in terms of this theory) they do succeed in introducing genuine meanings. And how we might know that is simply not explained at all. All that I have said so far is that logic surely comes into it somehow, and this brings me to my second objection: the conventionalist anyway cannot give a satisfactory account of our knowledge of logic.

The basic point is extremely simple. Let us grant for the sake of argument that what is stipulated to be true can be known *a priori* to be true. But then, on the usual view, we can *also* know *a priori* the logical consequences of what is stipulated. However, a consequence of what is stipulated is not *itself* something that is stipulated, so we have not yet been given an explanation of how *it* is known.

I illustrate the point with an extremely simple example from the ortho-dox propositional logic, choosing this because here it is not unreasonable to say that we do know what *is* stipulated; for example the meaning of the truth-functors is given by their truth-tables. Consider, then, a truth-table calculation, for example a simple calculation showing that the formula '$P \vee \neg P$' is always assigned the value 'true'. The calculation is represented graphically thus:

P	P	$\vee$	$\neg$	P
T	T	T	F	T
F	F	T	T	F

It represents the following reasoning (slimmed down to what is essential):

(i)   'P' is either true or false;
(ii)   If 'P' is true, so is '$P \vee \neg P$';
(iii) If 'P' is false, then '$\neg P$' is true;
(iv)   if '$\neg P$' is true, so is '$P \vee \neg P$'.

Hence:

(v)   In all cases, '$P \vee \neg P$' is true.

Each of the premises (i)–(iv) may be regarded, for the sake of argument, as something that is directly stipulated. ((i) is given by the convention for the use of sentence-letters, i.e. that they are to be understood as representing only what has a definite truth-value, true or false; (ii) and (iv) are given by the truth-table for disjunction, which defines the sign '$\vee$'; (iii) is given by the truth-table for negation, which defines the sign '$\neg$'). But the conclusion (v) is not itself something that is directly stipulated; rather, it is here *argued* to be a *consequence* of the initial stipulations. Granting that we know the premises to be true *a priori*, everyone who can follow this argument will suppose that they also know *a priori* that the conclusion is true. But while they are directly stipulated, it is not. The point would, of course, be even more obvious if I had chosen a more complex formula as an example. We can evidently understand the initial conventions without yet knowing their effect upon this or that particular formula, and when we do come to work it out we thereby obtain new knowledge which we did not have before. Yet apparently it is a priori knowledge, if the starting point was.

Someone may say, speaking in an 'ontological' mode, that the truth of the premises is by itself enough to 'determine', or to 'ensure', the truth of the conclusion. But my point is epistemological, rather than in this sense ontological, namely: what is it that gives us our *knowledge* of the conclusion? And here one must surely say that it includes not only our knowledge of the premises but also our ability to 'see' that the conclusion follows from them. I am granting for the sake of argument that knowledge of the premises may be explained as knowledge of what is directly stipulated (and that is perhaps why it counts as *a priori*), but the ability to 'see' what follows from them is not itself knowledge of a stipulation; it simply *does* follow, and we can 'see' this, but it is not *stipulated* to follow.

If argument is needed here, it may be given in this very general way. The idea is that the initial stipulations give meaning. So, if meaning is to be grasped (which surely it can be, in this case), these initial stipulations must be finite in number, and indeed relatively few. But they have infinitely many consequences, even in this very simple case. So it cannot be that all the consequences are themselves stipulations. The objection was presented in very much this way in Quine's well-known article 'Truth by Convention' (1936), but his argument there focused particularly on the problem of getting from a generalization to its instances. For, if the initial stipulations are to have infinitely many consequences, then they must be given as generalizations, e.g. (for my example) 'For *any* propositions "P" and "Q", if "P" is true then so is "$P \vee Q$" '. But our reasoning then goes on to infer from this that for any proposition 'P' there is another proposition '$\neg P$' that is its negation, and it will still hold in this case too that if 'P' is true then so is '$P \vee \neg P$'. This is an *inference*, from the general rule that we began with, applying it to a special case.[20] Quine's original article does not emphasize, as I do, that the inferences involved are not only those that take us from general stipulations to their particular cases, but also those that take us from a number of particular cases to a further conclusion deduced

---

[20] That there is inference involved here might be more obvious if we had begun from a metalogical formulation of the initial rule, i.e. 'For any formulae φ and ψ, if φ is interpreted as true then so is $\varphi \vee \psi$'. We then need the further premises that 'P' is a formula, and so is '$\neg P$', and we put all these premises *together* to infer that if 'P' is interpreted as true then so is '$P \vee \neg P$'.

from them. *Both* of these are very relevant considerations.[21] In any case, the overall situation is this. There can be only a finite number of meaning-giving conventions, if we are ever to be able to learn the meanings of words from them. But their consequences are potentially infinite, and apparently we can often 'see' that something is actually a consequence. This 'seeing' is traditionally held to be *a priori*, but then it must follow that *a priori* knowledge is not restricted just to knowledge of the meaning-giving conventions (whatever they are).

I think that this conclusion is nowadays very generally accepted, even by those who look with some favour on the idea that analyticity is an important source of *a priori* knowledge.[22] But there is one rather desperate attempt to avoid it, which is known as 'radical' conventionalism, and which perhaps deserves a brief discussion.

Ayer is commonly counted as a 'moderate' conventionalist, because he does not (explicitly) claim that what seem to be the consequences of conventions are in fact just further conventions. This is because he has not seen the problem. But the 'radical' conventionalist is one who has seen this problem, and attempts to meet it in just this way. His position is that on each occasion on which we would naturally say that we have discovered a new consequence, the truth of the matter is simply that we have decided to adopt a new convention. And since these conventions are supposed to be meaning-giving, he adds that we have thereby made a further determination of the original meanings of the expressions involved. (It is both asserted and denied that Wittgenstein may be counted as a 'radical conventionalist' in roughly the sense here outlined. I shall not enter into this dispute on Wittgensteinian exegesis.)[23]

It is at once obvious that such a radical conventionalism is extremely implausible. First, it is *very* implausible with regard to meanings, as a

---

[21] A different point, that is also made in Quine's 'Truth by Convention', is that one will naturally use the word 'if' when stating the rules for 'if' that are supposed to provide a meaning for the word. (For example: if 'P' is true and 'Q' is false, then 'If P then Q' is false.) So apparently we have to understand the word 'if' already, before we can understand the stipulation that is supposed to tell us what it means. This is also a strong objection to the conventionalist theory, but I shall not here ask whether with sufficient ingenuity we could get around it. (An attempt to get around it is offered by Boghossian in his 1996 and elsewhere, e.g. his 2000.)

[22] E.g. Peacocke (1993).

[23] For the attribution to Wittgenstein see e.g. Dummett (1959), and the long discussion in Wright (1980). For the denial see, e.g., Stroud (1965).

simple mathematical example will show. Suppose that you are asked 'is 167 a prime number?' Your first reaction is, I imagine, 'I don't know'. This is not because you do not understand the question. On the contrary, you know perfectly well what is meant by '167' and by 'prime', but you do not yet know whether 167 *is* prime. So you calculate, and you come to the correct conclusion that indeed it is. We would naturally say that you have here discovered a consequence of what you knew before you started. But the radical conventionalist must say that if this is an example of new *a priori* knowledge then it must be regarded as the adoption on your part of a new linguistic convention. And since the role of these conventions is to give meaning, this must imply that you now assign a new meaning to '167 is a prime number', over and above whatever meaning you initially assigned to it, before you knew that it was true. But what on earth could this supposed new meaning be? Is it the meaning of '167' that has now been made more precise? Or the meaning of 'prime'? Or perhaps the meaning of 'is'? It is quite clear that none of these suggestions is at all attractive. I add as an aside that we see on reflection that this theory has the strange consequence that you never can succeed in proving anything that you set out to prove. For if you do (as *we* would say) succeed in finding a proof, then on this theory you have thereby changed the meaning of the thesis that you set out to prove, so what you have proved is not the original thesis after all. As I say, the consequences of this position for the notion of meaning are *very* implausible.

Even if we forget about how meaning is supposed to be involved, still the radical conventionalist's position is highly counter-intuitive in another way: it denies 'the hardness of the logical *must*'.[24] For the initial thought is that (meanings, and hence) linguistic conventions are simply stipulated by us. It is up to us to decide what conventions to stipulate, for nothing compels us to develop our language in one way rather than another. So it is always 'up to us' to decide whether a proof should or should not be counted as establishing its conclusion. In principle, it is always open to me to say: 'I accept the premises, and I accept the rules of inference, but I do not accept the conclusion; I do not choose to adopt *that* convention, though I do adopt all the others'. Plainly this is not what anyone ever does say, for in practice we all do find that proof *compels* acceptance. Now it may perhaps be that our feeling that we are compelled is some kind of illusion, but the explanation of this illusion (if it is one) is surely not that of the radical conventionalist. For on his account there never was any compulsion in the

---

[24]   The phrase is from Wittgenstein (1964, p. 37).

first place, since it is always up to us to adopt whatever conventions we happen to fancy. We may adopt or reject them just as we please, for no choice or choices ever constrains any other. But, we protest, we certainly do *feel* constrained, and moreover we *feel* (perhaps wrongly) that this constraint is imposed from without, and is not of our own making. Perhaps there is a sense in which we 'make' the premises true, and possibly there is a sense too in which we 'make' the rules of inference, but beyond that we are sure that it is not 'up to us'; all that we can do is to recognize, or fail to recognize, a necessity that is already there.[25]

4. Concluding remarks

Kant introduced the suggestion that *a priori* knowledge of analytic truths is entirely comprehensible, and that what needs explanation is our *a priori* knowledge of synthetic truths. He did not persuade everyone that we do have such knowledge. He himself gave most of his attention to geometry, and we now see that he was quite wrong about this. When geometry is construed in the way that it always had been construed, i.e. as a theory of the space that we inhabit, its claims are certainly not analytic. But also they are not known a priori. On this topic Mill was quite right, though it was not until much later (i.e. with Einstein's revolutionary physics) that this was universally acknowledged. But on arithmetic Kant's claim is still outstanding, though everyone will now agree that his own arguments for it are quite unconvincing.

Frege set out to disprove this claim, by demonstrating in full detail that arithmetical truths must be admitted to be analytic. We have not yet explored Frege's attempted demonstration, and I postpone that topic until chapter 5. For it cannot fairly be tackled until we have first looked at some relevant advances in mathematics itself. But what I have tried to do in the last section is to draw attention to what is surely a lacuna in Frege's argument. He took it to be obvious that if arithmetical truths are analytic then they can be known *a priori*, but he never attempted to argue this point, and when one considers his own redefinition of analyticity two difficulties

---

[25] If Wittgenstein is to be counted as a radical conventionalist, then perhaps what he has to say about 'forms of life' should be viewed as an attempt to explain why we are (wrongly) convinced that the necessity is not of our own making. Cf. Stroud (1965). But I shall not explore that suggestion any further.

obviously stand out. First, what exactly is to count as an *analysis*, and how does one tell whether a proposed analysis is correct? So far I have only gestured at this problem, but as we proceed it will become obvious that it is a real problem. Second, how is our knowledge of logic to be explained? Mathematical knowledge may possibly be reduced to knowledge of analyses and of logic, but this is at best a *reduction* of the problem, and not yet a *solution* of it.

Suggestions for further reading

The chapters cited from J.S. Mill's *System of Logic* (1843) are entirely comprehensible without a commentary. But if a commentary is required I suggest Skorupski (1989). His chapter 7 contains a detailed discussion, which sometimes conflicts with mine. Another defence of Mill's ideas may be found in Gillies (1982, chapters 3–4). The passages of criticism cited from Frege's *Grundlagen* (1884) are also quite straightforward, but sooner or later you will have to read *all* of that book, and may feel that a commentary is desirable. In that case, use Dummett (1991). The chapter from Ayer's *Language, Truth & Logic* (1936) is again quite easy to read, and by way of riposte you might like to try Horwich (2000). But on the general topic of analyticity everyone should read Quine (1951a), an article which is highly suggestive though rather short of compelling argument.

Chapter 4

Mathematics and its Foundations

In this chapter I aim to sketch the history of some developments in mathematics which affect what is known as the 'foundations' of the subject, i.e. the basic principles on which all else rests. It may be disputed whether mathematics as a whole either has or needs such 'foundations', but my discussion is intended to suggest that the answer to both questions is 'yes'. In fact this answer will be assumed both in the next chapter and quite often elsewhere. I begin with geometry, because that is where foundations were first sought.

1. Geometry

We think of mathematics as beginning with the Greeks, because they were the first (so far as we know) to have thought of the subject in terms of *proof*. There was, of course, some knowledge of mathematics in earlier civilizations, e.g. in Babylon and in ancient Egypt, and we have a few records of it. But for the most part these records are exercises in how to solve numerical problems, and they do not contain anything that is recognizable as a proof. That is what is new with Greek mathematics.

The demand for proof can be looked at in two ways, either as a request for certainty or as a request for explanation. Nowadays it is probably the first that is uppermost in the minds of most mathematicians, but with the ancient Greeks I guess that it was mainly the second, at least to begin with. For otherwise it would seem strange that they thought it relevant to prove things which everyone had known for years, e.g. that a 3-4-5 triangle is right-angled. But also, it is the demand for explanation that fits better with other things that we credit to the Greeks, such as the beginnings of natural science and philosophy. For these too are strongly connected with the desire for explanation. (E.g.: Why do all animals breathe? Why do eclipses happen? Or again: Why ought one to keep promises? Why cannot contradictions be true?) Proofs do explain, as well as providing a guarantee of truth.

In either case, the demand for proof leads very naturally to the quest for axioms, i.e. for what one may regard as the basic principles of the subject. For it soon becomes obvious, when you start to look for proofs, that every proof has premises. So this naturally raises the question whether those premises can be proved in their turn, and so on. One is led on, therefore, to a quest for the *ultimate* premises which all proofs rely upon. (Nowadays these are called 'foundations', but Plato and Aristotle called them simply 'the starting points', though the word is often translated as 'first principles'.) I remark that the quest for proof does not always have this result, for there may be a sufficiently wide agreement to begin with on what one can assume without proof. An example of this is the elementary theory of the natural numbers, as pursued by the Greeks. Books VII–IX of Euclid's *Elements* contain many proofs in this theory, but he provides no special axioms for them, taking it that the needed assumptions are already implied in his more general axioms for geometry.[1] In fact there was no decent attempt to provide axioms for the theory of natural numbers until Dedekind (1888). But this did not prevent the discovery of many interesting proofs well before then.

With geometry it was different, and we can be sure that the search for suitable 'foundations' had been carried on for years before Euclid compiled the version that set the standard for many centuries. As I have said, we can only guess at what those earlier attempts came up with, for it is only Euclid's version that has survived. He begins with a number of definitions, e.g. of a circle, a triangle, and so on;[2] and continues with some theses which he calls 'postulates' and some that he calls 'common notions'. The latter are mainly what we might think of as laws of identity (e.g. 'things which are equal to the same thing are also equal to one another'). The former begin with the three existential claims cited earlier on p. 26, and then continue with two which together add up to Euclid's well-known 'postulate of

---

[1] Euclid thinks of the different numbers as represented by lines of different lengths (so for example the axiom that every number has a successor is implied by his postulate 2, stating that every straight line may be extended).

[2] A fair number of these definitions are *useless* definitions, in the technical sense that they are never invoked in subsequent proofs, for example 'A point is that which has no part', 'a line is breadthless length', 'a straight line is a line which lies evenly with the points on itself'.

parallels'. As I have observed, from our contemporary point of view these 'postulates' and 'common notions' leave unsaid a great deal that needs to be said, but they are clearly intended to fulfil the role that we think axioms should fulfil, i.e. to provide all the ultimate premises that are needed in this subject. All the rest of the thirteen books of Euclid's *Elements* purports to be a series of deductions just from them, and that is why the work is such a landmark in the history of mathematics. It also explains why Plato and Aristotle, writing before Euclid, as well as thinkers such as Galileo and Newton, writing a long time after, thought of all *proper* mathematics as being geometry; it was because in geometry everything has to be *proved*.

So far as we know, it was Euclid himself who discovered that his postulate of parallels, or some equivalent postulate, was needed, and seemed not to be deducible from anything simpler. But everyone admits that this postulate does not seem so obviously to be true as do the other postulates which he did state, or which he ought to have stated. So in later ages there were serious attempts to show that this postulate could after all be deduced from something simpler, and the upshot was the discovery that it could not. On the contrary, it could consistently be denied. That is to say: non-Euclidean geometries were discovered.

The natural effect was to encourage a broadly 'formalist' attitude to geometry. (I discuss formalism in more detail in chapter 6.) It had always been assumed that Euclidean geometry was the theory of the space of our universe, but when alternative geometries were discovered this assumption became doubtful. The pure mathematician now had to admit that he was not in a position to settle this question, and that it had to be resigned to the physicist. To put this in another way, it had become unclear whether Euclidean geometry did have its *intended* model. But at the same time it had been known ever since Descartes that it did have an *unintended* model, for the introduction of Cartesian coordinates provides one. That is, for plane geometry one may interpret a point simply as an ordered pair of real numbers, a straight line simply as a set of all such pairs as satisfy a linear equation (i.e. an equation of the form '$ax + by = c$'), and so on. This was not, of course, how the notions of a point and a line had originally been understood, but that did not worry the mathematician in the slightest, for in fact the new interpretation proved to be extremely helpful in the study of geometry. (Similar interpretations, in terms of pairs or trios of real numbers, are also available for non-Euclidean geometries.) This naturally encourages the idea that there is no serious question of whether the axioms of geometry are *true*, for all that one can say is that they may be interpreted in various ways, and are no doubt true under some interpretations and false under others. So the suggestion is that in mathematics we are concerned simply with what does or does not *follow* from the axioms, and their truth (under this or that interpretation) is neither here nor there.

As I say, this is quite a natural reaction to the discovery of non-Euclidean geometries, i.e. the thought that – so far as *mathematics* is concerned – there is no serious question over the truth of the axioms. No one so far has discovered what you might call a 'non-Euclidean arithmetic'. But might it happen? And, if so, would it have the same effect? Well, I will simply leave that question hanging, in order to return to the topic of this chapter, which is to sketch some history of mathematics.

2.   Different kinds of number

In classical Greek mathematics there is officially only one kind of number, namely what we call the natural numbers (or, better, the counting numbers, for the series begins with 1 and not with 0). We might naturally say that in order to do geometry you need also the rational and the irrational numbers, to measure such geometrical quantities as length, area, angle, and so on. But the Greek procedure does not assign numbers to particular lengths. It speaks instead of the ratios between lengths, and compares these to ratios between natural numbers. Thus one length might be $^3/_4$ as long as another. We would naturally assign the number 1 to one of these lengths and the number $^3/_4$ to the other. But they would put the same point by saying that the ratio between the lengths was the same as the ratio of the number 4 to the number 3. Similarly for ratios in area, volume, and so on.

It seems very probable that in early days it was taken for granted that any ratio between (e.g.) lengths could be equated in this way to a ratio between natural numbers, but then it was discovered that this was not so. For example, the ratio in length between the side of a square and its diagonal cannot be equated with any ratio between natural numbers (because, in our language, $\sqrt{2}$ is irrational). So it became clear that there was a need for a general theory of ratios, free from this assumption, and this was eventually provided (by Eudoxus; the theory is preserved for us in Euclid, book V).

The central and crucial idea is this. Let '$a{:}b$' stand for the ratio of a to b (e.g. in length, or in whatever other respect we are considering);

similarly, let '*c:d*' stand for the ratio of *c* to *d* (in the same or in a different respect). Then the central notion is written

a:b :: *c:d*,

which is pronounced

a is to *b* as *c* is to *d*.

The definition of this notion assumes that any length (or area, etc.) can be multiplied by a natural number, so that '*n·a*' will be a length that is *n* times as long as the length *a*. Then, where '*n*', '*m*' range over the natural numbers, the definition is

$$\forall nm(n{\cdot}a \gtreqless m{\cdot}b \;\leftrightarrow\; n{\cdot}c \gtreqless m{\cdot}d)$$

It is easier to see what this comes to if it is rephrased in terms of division rather than multiplication:

$$\forall nm(\frac{a}{b} \gtreqless \frac{n}{m} \;\leftrightarrow\; \frac{c}{d} \gtreqless \frac{n}{m})$$

Now think of '$\frac{a}{b}$', i.e. the ratio of *a* to *b*, as *being* a real number, and similarly '$\frac{c}{d}$'. Then this definition says that the real numbers $\frac{a}{b}$ and $\frac{c}{d}$ are the same number, if and only if they each bear all the same relations to the rational numbers $\frac{n}{m}$. That is: any rational number greater than the one is greater than the other; any rational number equal to the one is equal to the other; and any rational number less than the one is less than the other. To paraphrase once more: the real numbers in question are the same (or equal) if and only if they each make exactly the same 'cut' in the rational numbers. So what Dedekind is famed for saying in 1872 is almost the same as what Eudoxus had proposed in around 350 BC. What a pity that in the 22 centuries that separate them the basic idea was lost to view.

It must be admitted that the Greek theory of ratios is not as complete as the modern theory of real numbers, based on Dedekind's work. Both say that ratios, i.e. real numbers, are the same if and only if they 'cut' the rational numbers at the same place, but the Greek theory cannot show that, for every such 'cut' in the rationals, there always is a ratio, i.e. a real number, which makes it. Yet there are many ways in which the theories are similar. For example, the Greek theory explained how to add one ratio to another, how to multiply one ratio by another, and generally all the operations on real numbers that are familiar to us can be mirrored on ratios as understood in Greek geometry. It is mostly, if not completely, true that whatever you can do with the one theory you can also do with the other. But the point that I wish to make here is that the Greeks *did* provide a 'foundation' for their theory of ratios, whereas our theory of real numbers had to wait many years for its 'foundation'.

This is an example of a more general truth: as time went on more and more kinds of things were recognized as 'numbers', but there was no attempt to state precise axioms for these expanding theories. As I have said, the Greeks themselves failed to provide suitable axioms for their natural numbers, and apparently did not feel the need for them. The first expansions to the Greek theory were the addition of zero as a number, and of the negative integers. (Both of these seem to have been Hindu innovations, and they reached Europe, via the Arabs, in the late middle ages.) The recognition of zero was a boon for practical calculation, for it made possible the positional notation that is everywhere used today, and that we still call the system of 'Arabic numerals'. (With Greek numerals even the addition of largish numbers is a serious problem. With Roman numerals addition is more straightforward but multiplication is still a problem. With Arabic numerals the rules for both are easy to state and easy to apply.) The introduction of negative numbers met with some initial resistance, but they were soon proved to have many useful applications in practice. Naturally, rules of thumb for handling these new numbers were quickly adopted, e.g. that the result of multiplying any number by zero is zero, and that the result of multiplying two negative numbers together is positive. But still apparently there was no demand for a precise deduction of all such rules.

The next important conceptual innovation was the gradual replacement of the Greek use of ratios by what we are accustomed to think of not as ratios but as numbers, namely the rational numbers and the real numbers. This was helped by the fact that simple fractions, such as $^1/_2$ or $^3/_4$, had always been familiar, and a fraction both behaves in ordinary ways of thinking much as a natural number does – e.g. '$2^1/_2$ apples' – and at the same time is at once suggestive of a ratio. In classical Greek mathematics fractions are not counted as numbers, and ratios are used instead, but it is hardly surprising that this distinction was not rigorously preserved in

actual practice. The change was also much assisted by the growing importance of another branch of mathematics that we owe to the Arabs, namely
algebra. The variables of an algebraic equation were often taken to represent geometrical quantities, such as lengths or areas, but they could also
be interpreted as representing the natural numbers, or the fractions, or indeed
the negative numbers. So there is quite a natural tendency to say that in
all cases these variables should be understood as ranging over things called
'numbers', for – so long as real numbers are included – that does not stop
us applying algebraic reasoning to geometrical problems. On the contrary,
it encourages this application, and the technique turns out to be extremely
productive, as became quite clear both from the pioneering work of
Descartes and from later developments.

The concept of a number, then, comes to include not only the original
natural numbers but also zero, the negative integers, and the positive
rational and real numbers. There is clearly nothing to prevent the further
addition of the negative rational and real numbers, especially as the same
old ways of manipulating algebraic equations continue to work. And the
way is open for yet further kinds of number, e.g. the so-called 'imaginary'
numbers, if algebraic problems seem to require them.

Naturally, the developments that I here sketch in a very superficial way
did not occur all at once, and not without some controversy, but I will leave
that aside as not important for our purposes.[3] For the point that I am aiming to bring out is just this. When mathematics began in the Renaissance
to recover from the doldrums of the Dark Ages and the Middle Ages, and
as it flowered once more in the great developments of the seventeenth century, the concept of a number was very greatly expanded. But no one sought
a 'foundation' for this expanded theory. Indeed, the topic of 'foundations'
seems to have been completely ignored, until the subsequent development
of what was called 'the calculus' began to show that foundations really were
needed. That is a good many years on.

3. The calculus

An important precursor to the development of the calculus was the extension of geometrical techniques, by Descartes and Fermat, to cover time as

---

[3] There is a conveniently brief account in Mancosu (1996, chapter 3). He provides many
further references for those who wish to pursue the history in proper detail.

well as space. This meant that motion was now within the province of mathematics, and could be represented geometrically by a line on a graph with one axis for time and one for distance. The line itself represented a *function*, and the study of such functions (from real numbers to real numbers) became more and more important in the development of mathematics. But the initial idea is very simple. The points on the line give the positions of the moving object for each time, so that a uniform motion will be represented by a straight line, and an accelerating or decelerating motion will be represented by a curved line. More importantly, the instantaneous velocity of the object at any time will be represented by the slope of the curve at that time, i.e. by the slope of its tangent at that point. So for the first time mathematicians had an effective way of thinking of the *rate* of a change at any specified instant. In principle, the technique can be applied to any kind of continuous change, but the change of position over time which is motion is the simplest example, and I shall stick to this. For instance, if we know how far a falling object has travelled in any stretch of time, we can now say just how fast it was falling at any point on its route. For this has now become a question of finding the tangent, at any point, to a curve that is already specified.

Of course, different kinds of curves, and their properties, had always been of interest to mathematicians, but a new factor that entered with Descartes' work was the realization that they were highly relevant to problems in physics. I think that it is fair to say that the discovery of the calculus was largely (though not entirely) motivated by the fact that it was needed to provide answers to questions in physics. Certainly, this kind of application weighed heavily with the two men who are usually credited with introducing it, namely Newton and Leibniz.[4] They each saw that many useful results could be obtained in this way, though neither of them understood how the method actually worked. I give a simplified illustration of the kind of thing that was said at the time.

Suppose that we have a curve given by a function $y = f(x)$, where (for example) y represents the distance travelled and x the time taken. For definiteness, let us take a particular and very simple function:

---

[4] Newton and Leibniz were both building on the work of others. (There is a useful account of their predecessors in Mancosu, 1996.) Their followers made an issue of the question of priority, and the truth seems to be that Newton was the first to *have* the basic ideas but Leibniz was the first to *publish* them. In any case it is clear that each was working independently.

$$y = x^2 \tag{1}$$

We wish to find the slope of the tangent to the curve for each value of x, and the method recommended is this. Think of dx as a small addition to the value of x, increasing it to $x + dx$. Let dy be the corresponding addition to the value of y, so that

$$y + dy = f(x + dx)$$

Noting that we began with $y = f(x)$, this can be rewritten as

$$dy = f(x + dx) - f(x)$$

Now seek to express the right-hand side of this equation as a multiple of dx. With our particular example of f this is easy, for we have

$$\begin{aligned} dy &= (x + dx)^2 - x^2 \\ &= (x^2 + 2x(dx) + (dx)^2) - x^2 \\ &= dx(2x + dx) \end{aligned} \tag{2}$$

As the next step we divide both sides of the equation by dx, to obtain

$$\frac{dy}{dx} = 2x + dx \tag{3}$$

Finally we argue that dx may be taken as small as we please, so that the term dx on the right 'vanishes', and we are left with

$$\frac{dy}{dx} = 2x \tag{4}$$

This says that, for any value of x, the slope of the tangent at that point of the curve is $2x/1$, which gives us the answer required.[5]

In this simple case you may check the answer by familiar and old-fashioned methods, and you find that it is *correct*. Similarly in other cases where old-fashioned checks can be brought to bear. But isn't this amazing? For the argument as I have given it must obviously be condemned

[5] The notation used here is based on Leibniz, and is close to that used today. But Newton's version was not very different.

as quite incoherent. To obtain equation (3) we assume that dx is not zero, for division by zero is not a legitimate operation. But to obtain equation (4) we apparently assume that dx is zero, and that is why its occurrence on the right-hand side can be disregarded. A consequence of this is that dy is also zero, and so the ratio dy/dx is the ratio $0/0$, which makes no sense at all. Nevertheless we *did* reach the right answer. And, in all other cases where the answer is quite easily checked by old-fashioned methods, we find that the answer given by this method of arguing is again the right one. But why should that be? Newton did not know, and Leibniz did not know, and nor did anyone else for almost 200 years.

Leibniz simply recommended the method, without attempting anything much by way of an explanation of why it worked. Mathematicians on the continent mainly followed this recommendation, and used the method to make many useful discoveries, but without any proper understanding of what they were doing. Newton was worried by the lack of an explanation, and this worry was inherited by the English mathematicians, who continued the search for an explanation, but without success. (In consequence, the continental mathematicians were rather more successful in obtaining new and significant results.) The trouble was that the symbol dx, as just used, seemed to be *both* an expression for a positive quantity *and* an expression for zero, and so it came to be thought of as denoting something between the two, namely an 'infinitesimal quantity'. This was supposed to be something greater than zero, but also less than any finite quantity, and that seems to be incomprehensible.[6] It was a fair gibe on Berkeley's part (in *The Analyst*, 1734) that these infinitesimals were 'the ghosts of departed quantities'. At the time no one knew how to make sense of the role that they were apparently playing.

This situation continued for many years, in which the method was further developed and fruitfully employed to yield many new and significant results, though still it was not understood. I shall not attempt to summarize the history, save in these few brief remarks.[7] The development of the method led naturally to the use of infinite series, and to the problem of how to sum such a series, and this is where serious doubt began to set in. Eventually it was recognized that an important feature of such a series was that it should be *convergent*, for where this was not so the standard techniques broke down. It was Cauchy (1821) who first gave a clear

---

[6]   I mention on pp. 155–6 a *modern* proposal for making sense of this idea.

[7]   For a conveniently brief, but also informative, account, see Kitcher (1984, chapter 10).

definition of convergence, using the notion of a *limit*. At the same time he also stressed the importance of this notion, both for defining what it is for a function to be a continuous function, and for explaining what was really going on in the familiar, but still uncomprehended, methods of the calculus.[8] What is basically needed, and had until then been absent, is this notion of a limit. But unfortunately Cauchy was not completely successful in his use of this notion, for he thought of it as *legitimizing* talk of infinitesimals, and continued to think in those terms, in ways which sometimes led him into error. It was really Weierstrass who was the first to be completely clear about these matters, in writings from 1857 on, who gave us the important definitions that we use today, and who banished 'infinitesimals' for ever.

Let us sum up. The notion of the infinitesimal can be traced back to Archimedes (c. 250 BC), who explains it in an interesting treatise called 'The Method' as a useful heuristic device.[9] His attitude was that it might usefully *suggest* conclusions, but those conclusions then had to be proved by traditional methods of argument. When the notion became prominent once more, in the seventeenth century, Newton himself hoped for a similar explanation, i.e. for an argument to show that whatever could be proved by the new methods of the infinitesimal calculus could *also* have been proved by more traditional methods. It was a vain hope, but one which shows how unsatisfied he was with a technique which was not properly understood, though it did seem to yield many new and correct results. What then happened was that the technique was used, for almost two centuries, but still without any good understanding of why it worked. Descartes' desire for 'clear and distinct ideas' had been set aside, until eventually mathematicians found that there were real problems arising which could not be resolved simply by instinct. That sent them back to a search for proper 'foundations', and we can now resume that theme.

4. Return to foundations

After Weierstrass the theory of the real numbers – i.e. what is now called 'analysis' – was very much more coherent, but still incomplete. The new

---

[8] In the example above, do not interpret $^{dy}/_{dx}$ as a ratio between (infinitesimal?) quantities, but as an expression for the limit to which $2x + dx$ tends, as dx tends to 0.

[9] The treatise is translated in Heath (1912).

approach had stressed the importance of there being *limits*, and so one now needed to be able to prove the existence of such limits. But, since there was still no explicit account of what the real numbers are, there was no clear way of doing this. What had in practice happened, ever since the Greeks, was a reliance on 'geometrical intuition'. It had always been taken to be obvious that the real numbers corresponded to the points on a line, and so in practice mathematicians had relied upon their understanding of what a geometrical line was supposed to be. (Similarly, the notion of a 'continuous function' was understood via the geometrical notion of a continuous curve, i.e. – roughly speaking – a curve that you can draw without taking your pencil off the paper.[10]) The work of Cauchy and Weierstrass had gone a long way towards 'de-geometrizing' analysis, for they had provided explicit definitions, which proofs could appeal to. But the final step in this process was to tackle the question 'what *is* a real number?', and to banish geometrical intuition from that question too.

Dedekind gave one way of doing it in his (1872). In that work he is quite explicit that his object is to free the theory of real numbers from any reliance upon geometry. For example, he complains that at present no one can give a *proof* of such a simple thesis as

$$\sqrt{2} \cdot \sqrt{3} = \sqrt{6}$$

It is obvious that you can easily give a *geometrical* proof, i.e. one that relies upon the usual laws of Euclidean geometry.[11] But Dedekind wanted a proof which had nothing to do with geometry.

As I have said, his basic thought is the same as that of Eudoxus, over 2000 years before:[12] a real number makes a 'cut' in the rational numbers, i.e. it separates the rational numbers into two (non-empty) groups, all those in the one being less than all those in the other. It is the least upper bound of all those in the left group, and the greatest lower bound of all those in the right group. As Eudoxus had claimed, two distinct real numbers cannot both make the same cut in the rationals, for any two real numbers must be separated from one another by a rational number. As Eudoxus had not

---

[10] Until the work of Riemann (1854) and Weierstrass (1872), it was not even clear that continuity needed to be distinguished from differentiability.

[11] Reformulate the thesis as '$\sqrt{1} : \sqrt{2} \ :: \ \sqrt{3} : \sqrt{6}$'. Interpret $\sqrt{n}$ to mean the length of the side of a square with area n. Then the proof comes out at once.

[12] Dedekind does acknowledge a debt to Eudoxus (i.e. to Euclid V) in his (1888, pp. 39–40).

claimed, for every such cut there is a real number which makes it. This gives a complete theory of the real numbers, for it determines their structure uniquely. One can go on to define suitable operations on the real numbers – e.g. addition, multiplication, exponentiation, and so on – in terms of similar operations on the rational numbers. For example, if x and y are any two real numbers, then $x + y$ is to be the real number which has in the lower half of its cut just those rational numbers $r + s$ for $r < x$ and $s < y$. It is easy to prove that there is such a cut, and to show that addition, defined in this way, has all the expected properties. (Of course we are assuming here that the arithmetic of the rational numbers is already known.)

There are two ways of looking at Dedekind's approach. One, favoured by Dedekind himself, is to say that it *postulates* a real number for each cut in the rationals. It does not *identify* real numbers with cuts, or with anything else, but just claims that *there are* some things that are related to the rational numbers as stated.[13] An alternative, favoured later by the logicist approach, is simply to identify a real number with a cut. In practice it proves more convenient to define a cut in such a way that the lower half of a cut never has a highest member, and then to identify the real number with that lower half. That is, the real number is to be regarded as *being* the set of all rational numbers less than it. But that is a detail of no importance. For the present, let us just say that both approaches are possible, and each is equally adequate for mathematical purposes.

Dedekind's way of thinking of the real numbers is not the only one, and at the same time (1872) Cantor was introducing a different way. His basic idea was to define what is to count as a convergent sequence of rational numbers, and then to postulate that, for each such sequence, there is a real number which is the limit to which it converges.[14] To pursue this idea we shall of course need to give a criterion, *not* assuming the existence of limits, for when two such sequences converge to 'the same limit'. But in view of Cauchy's work that is now relatively straightforward. So again one can *postulate* that there is a thing called a real number, which functions as the limit of all converging series which converge to the same limit.

---

[13] As Dedekind said: 'Whenever, then, we have to do with a cut produced by no rational number, we *create* a new, an irrational number, which we regard as completely defined by this cut' (p. 15, my emphasis).

[14] This generalizes the idea behind our usual decimal notation. For example, the notation for the real number π, which begins 3.14159 . . . can be regarded as representing a convergent series of rational numbers, namely $\langle 3.1, 3.14, 3.141, 3.1415, 3.14159, \ldots \rangle$.

Or one can *identify* the real number with the set of all series which converge to it. As before, either approach works perfectly well for mathematical purposes.

As Cantor also showed, there is yet another way of doing things, which is simply to lay down a set of axioms for the real numbers, and this can be done without any explicit mention of the rational numbers. I do not give the axioms, for some of them are quite complex.[15] We need only to notice that this also is an approach which can be pursued.

In all of these ways a proper 'foundation' for the theory of real numbers can be provided, and was provided in the latter half of the nineteenth century.

A proper foundation for the theory of the real numbers was *needed*, because mathematicians had run up against a number of problems (e.g. about the existence of limits) which could not be resolved without it. There was no similar need to 'found' the theory of other kinds of number, for no such problems had been encountered. But the bug had bitten, and further research into 'foundations' very soon followed.

It is quite easy to see how the theory of rational numbers may be 'reduced' to the theory of pairs of integers. The starting point is to specify the appropriate criterion of identity, i.e. the criterion for when the rational number $^x/_y$ is to be counted as the same number as $^z/_w$. The answer is obvious at once: it is when $x \cdot w = y \cdot z$. So we can introduce an equivalence relation between pairs of integers, which mirrors the identity of the rational numbers formed from them, i.e.[16]

$$\langle x,y \rangle \approx \langle z,w \rangle \quad \text{iff} \quad x \cdot w = y \cdot z$$

Then the rational number $^x/_y$ can simply be identified with the set of all pairs of integers which are in this sense equivalent to $\langle x,y \rangle$. It is easy to go on to define operations on rational numbers – such as addition, multiplication, and so on – in terms of the corresponding operations on integers.

In the same way the theory of signed integers can be 'reduced' to the theory of pairs of natural numbers. In fact there are many definitions which will do the trick. The simplest is perhaps just to identify the integer $+n$

---

[15] For those who happen to know about such things: the axioms are those for a complete ordered field. Cantor showed that this set of axioms is categorical, i.e. that it uniquely describes the structure of the real numbers.

[16] I use 'iff', as usual, to abbreviate 'if and only if'.

with the pair $\langle 0,n \rangle$ and the integer $-n$ with the pair $\langle n,0 \rangle$. But if we prefer $+n$ can be identified with the set of all pairs of natural numbers $\langle x,y \rangle$ such that $x + n = y$ (i.e. $n = y - x$), and $-n$ with the pairs $\langle x,y \rangle$ such that $x - n = y$ (i.e. $n = x - y$). It is obvious that the proposed identifications are somewhat arbitrary, but for mathematical purposes they all work perfectly well.[17]

In this way we reach the position that Kronecker was describing when he said: 'God made the natural numbers; all else is the work of man'.[18] That is: we *start* with the natural numbers, which are in some way 'given', but from these we can 'construct' the signed integers (e.g. as pairs of natural numbers), and from these in turn the rational numbers (e.g. as sets of pairs of integers), and then from these the real numbers (e.g. as sets of rationals determined by some 'cut' in the rationals). It is clear that one can go on to complex numbers, or yet further kinds of numbers, in a similar spirit of 'construction'. But what of the natural numbers, from which we start? Can these also be provided with a 'foundation'?

The answer is 'yes', and it is due to Dedekind (1888). The structure of the series of natural numbers is that it has a *first* member, which we may call zero, and then for each member it has a *next*, which we may call its successor. It is easy to write down a number of axioms concerning these notions, e.g. that zero is not the successor of any number, that every number has one and only one successor, that no two numbers have the same successor, and so on. But the crucial thought is this: *every* natural number can be obtained from zero by iterating the successor function some finite number of times, and there are no *others*. How shall we say this in formal terms? Dedekind's idea is very straightforward: the set of natural numbers has zero as a member, and also has as a member the successor of each of its members, and it is the *smallest* set satisfying this condition. That is, it is a subset of every set that satisfies the condition, or in other words it is the intersection of all such sets. Using '0' for zero, 'x'' for the successor of x, 'N' for the set of all natural numbers, and 'α' as a variable for sets, we therefore have

(a) $0 \in \mathbb{N}, \quad \forall x(x \in \mathbb{N} \rightarrow x' \in \mathbb{N})$

(b) $\forall \alpha((0 \in \alpha \wedge \forall x(x \in \alpha \rightarrow x' \in \alpha)) \rightarrow \mathbb{N} \subseteq \alpha)$

---

[17] For details one may consult Russell (1919, chapter 7).

[18] This well-known saying by Kronecker is first cited in his obituary, by A. Weber, in *Jahresberichte der Deutschen Mathematiker-Vereinigung* 2 (1893): 5–31.

It is quite easy to see that these two conditions determine the set $\mathbb{N}$ uniquely.

They will not do the job all by themselves, for we do need some further information on zero and on the successor function. For example, the conditions (*a*) and (*b*) just given would be satisfied if $\mathbb{N}$ had just one member, namely 0, which was its own successor. So the set of axioms which has now become canonical is this:

1   $0 \in \mathbb{N}$
2   $\forall x(x \in \mathbb{N} \rightarrow x' \in \mathbb{N})$
3   $\forall xy(x' = y' \rightarrow x = y)$
4   $\forall x(0 \neq x')$
5   $(0 \in \alpha \wedge \forall x(x \in \alpha \rightarrow x' \in \alpha)) \rightarrow \mathbb{N} \subseteq \alpha$

The last is, of course, the postulate of mathematical induction, for it tells you that whatever is true of 0, and is true of x' whenever it is true of x, must be true of all numbers.[19] This axiom-set is standardly known as 'Peano's postulates' for natural numbers, but in fact they come directly from Dedekind's work, as Peano himself acknowledged. Dedekind himself regarded these axioms as characterizing any 'simply infinite system', i.e. what we now call a progression, with an arbitrary function f (for 'successor') and zero defined as the sole member which is not a successor. On his account the natural numbers are obtained by 'abstraction' from all such simply infinite systems.

The axioms as given here say nothing about addition, multiplication, or other operations on the natural numbers. If the background logic to Peano's postulates is taken to be a first-level logic, with no quantification over sets, then postulate (5) becomes a *schema*, generating infinitely many postulates, with the predicate-schema '... $\in \alpha$' replaced in turn by each

---

[19]   The postulate is first explicitly formulated in the late sixteenth century (see e.g. Kline, 1972, p. 212). The equivalent 'method of infinite descent' is stated and used by Fermat in 1640 (Kline, 1972, p. 275). At this stage I am not distinguishing between variables that range over sets and variables that take the place of predicates. So postulate 5 could equally well be written as

$$(F(0) \wedge \forall x(F(x) \rightarrow F(x'))) \rightarrow \forall x(\mathbb{N}(x) \rightarrow F(x))$$

Compare n. 13 of the previous chapter.

well-formed formula of the system. In that case one adds the usual recursive equations for addition and multiplication, namely

6 $x + 0 = x$
7 $x + y' = (x + y)'$
8 $x \cdot 0 = 0$
9 $x \cdot y' = x \cdot y + x$

But if, as Dedekind assumed, the background logic is taken to be a second-level logic, so that postulate (5) quantifies over all subsets of the domain, then these (and other) functions can be explicitly defined. They are in effect defined as the *minimum* functions which satisfy the above equations, i.e. the intersection of all functions that satisfy them. Dedekind generalized this idea, to yield what is called 'the recursion theorem' for the natural numbers, showing that any function introduced by a pair of such recursive equations is legitimately introduced, for there always is a unique function that satisfies them.[20] This is a crucial result.

5. Infinite numbers

At the same time as Dedekind was providing a foundation (i.e. an axiomatization) for the theory of natural numbers, and thereby – as it seemed – for *all* of traditional mathematics, Cantor was discovering a *new* area of mathematics, which certainly could not be obtained from that foundation. This was his theory of infinite numbers (1895). I give a brief explanation of the central ideas, starting with the theory of the infinite cardinals, since this extension of the ordinary notion of a number is the most natural.

If we consider any of the 'constructions' of new kinds of number described on pp. 98–101, we see that the first problem is to provide a *criterion of identity* for numbers of the new kind. For example, we need to say when two different notations for a rational number, say '$^1/_2$' and '$^3/_6$', should be understood as denoting the same number. Similarly with notations for the real numbers, e.g. '0.9' and '1.0'. So the theory of infinite cardinal numbers must begin in the same way: when shall we say that we are

---

[20] His proof relies upon a second-level logic as background logic. So does his proof that the 'Peano' axioms are categorical, i.e. that they describe a unique structure. I shall say more about the distinction between first and second level logics in the next chapter.

dealing with *the same* cardinal number? Cantor takes cardinal numbers to apply to sets, and answers that two sets have the same cardinal if and only if there is a relation which correlates their members one-to-one, i.e. what is currently called a 'bijection' between the two sets. This is now so much a part of orthodox thinking that we do not often stop to ponder it, but it deserves a little thought.

The idea obviously introduces this question: when shall we say that a relation *exists*? The first thought is that a relation will exist whenever we can specify it, i.e. say just how it relates the objects involved. Let us assume this, for the time being. Then it is easy enough to show that Cantor's criterion for 'the same number' gives the right results for the *finite* cardinals, for we can always (in principle) specify a relation that correlates two finite sets which each have the same number of members. (If necessary this can be done just by *listing* the pairs to be related, for in the finite case such a list is always possible in principle.) Conversely, it is easy to see that there cannot be a one-to-one relation between two finite sets that do not have the same number of members. So, as I say, it is clear that in the finite case Cantor's proposed criterion does give the right results. Cantor extends the criterion to all cases, whether the sets in question are finite or infinite.

This at once leads to results which are surprising to common sense. For example, it seems obvious to common sense that half of the natural numbers are even, and half are odd, so there should be more natural numbers altogether than there are even numbers. But according to the proposed criterion that is not so, and there are the same number of each, since one can easily specify a relation that correlates the two. (Just correlate each number with its double.) Similarly, common sense is likely to say that there must be more rational numbers than there are natural numbers, since between any two natural numbers there are infinitely many rationals, but again this is not so according to the criterion proposed. One may consider the rational numbers as set out in a two-dimensional array like this:

$$1/1, 1/2, 1/3, 1/4, \ldots$$
$$2/1, 2/2, 2/3, 2/4, \ldots$$
$$3/1, 3/2, 3/3, 3/4, \ldots$$
$$4/1, 4/2, 4/3, 4/4, \ldots$$

and so on.

But any such two-dimensional array can be correlated with the (one-dimensional) series of natural numbers, e.g. in this way:

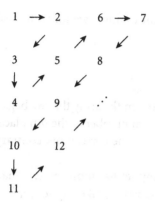

Faced with these, and similar, examples, one might quite naturally swing to the opposite opinion, and suppose that *all* infinite sets can be correlated with the natural numbers. But Cantor showed that this is not so; there cannot be a one-to-one correlation between the real numbers and the natural numbers. This is his famous 'diagonal' argument, which has since been used in many ways. In the present case it works like this. Consider just the real numbers between 0 and 1, expressed in the decimal notation,[21] and suppose that they could be correlated one-to-one with the natural numbers. Then we can consider the supposed correlation like this: in the left column write the natural numbers (in their natural order), and in the right column write their correlated real numbers, e.g. thus:

1   0.123412312 ....
2   0.234523423 ....
3   0.345634534 ....
4   0.456745645 ....
     and so on.

Now 'go down the diagonal' of this supposed array of all the real numbers between 0 and 1, and write down a new expression for a real number, by putting in each place a digit that is different from the one on the diagonal. For example, put '1' wherever the digit on the diagonal is even, and

[21] Choose a unique decimal representation. E.g. always write '0' in place of an unending string of '9's.

'2' wherever it is odd, so that with the present example we form the real
number

0.2222 . . .

This cannot be the same number as any of those in the array that we began
with, since, for each *n*, it differs from the *n*th real number at the *n*th place.
So there is a real number that is not included in the correlation, contrary
to our initial assumption.

This argument generalizes, to show that any set has more subsets than
it has members. To see how the generalization works, take any set {*a*, *b*, *c*,
d, . . .}, and suppose that there is a correlation between its members and
its subsets. We can think of it in this way. On the left list the members of
the set, and on the right the subsets with which they are correlated,
specified by first writing out all the members of the set and then striking
out the ones that are not in that particular subset:

a   {*a̸*, *b̸*, *c*, *d*, . . .}
b   {*a̸*, *b*, *c*, *d̸*, . . .}
c   {*a*, *b*, *c*, *d*, . . .}
d   {*a*, *b̸*, *c*, *d̸*, . . .}

Now form a new subset by 'going down the diagonal', and putting a member in the subset if and only if it is not in the subset with which it is correlated. So in the present example the new subset would be

{*a*, *b̸*, *c̸*, *d*, . . .}

As before, this is a subset which differs from all of those in the correlation, and has therefore been omitted from the correlation.[22] Now we
assume that for every set there is another, called its 'power set', which contains as members all and only the subsets of the first set. So, by what we
have just proved, for every set there is another which has more members
than it does, which is to say that there is no end to the series of infinite

---

[22] My illustration assumes that the members of the set in question, i.e. *a*, *b*, *c*, *d*, ..., can
all be listed, and hence that there cannot be more of them than there are natural numbers.
But this feature of the illustration is not at all essential to the argument, which can be framed
in general terms so as to apply to any set whatever.

cardinal numbers. Cantor introduces a notation for the infinite cardinals which is based on the Hebrew letter aleph, furnished with appropriate sub-scripts, so the series begins

$$\aleph_0, \aleph_1, \aleph_2, \aleph_3, \ldots$$

and it has no end.

Here there is an interesting question that arises. It is quite easy to show that the number of all the natural numbers is the *least* infinite number, i.e. is $\aleph_0$. (One needs only to argue that every subset of the set of natural num-bers either is finite or can be correlated with the series of all the natural numbers.) It is also not difficult to show that the number of the real num-bers is the same as the number of all the subsets of natural numbers, which can be written as $2^{\aleph_0}$. The question is: is this the *next* infinite number, i.e. is it the case that $2^{\aleph_0} = \aleph_1$, or are there other infinite numbers in between? The hypothesis that it is the next is Cantor's 'Continuum Hypothesis' (so-called because the number of the real numbers is the num-ber of the continuum). Cantor tried hard to prove this hypothesis, and failed. It is still neither proved nor disproved, and is probably the most well known of all the unsolved problems of his theory.[23] It will play some role in the chapters to follow.

Let us return to our starting point. We have learnt to live with these results of Cantor, and his criterion for the sameness of infinite numbers in terms of one-to-one correlations, but we have lost sight of the opening question: when does such a relation *exist*? We have said that if you can specify a cor-relating relation then there is such a thing, and if you can prove that no correlation is possible then of course there is no such thing, but perhaps we can do neither of these. The problem is nicely illustrated in an exam-ple given by Russell (1919, chapter 12). It concerns only the smallest infinite sets, i.e. those which have as many members as there are natural numbers, which we also say have 'denumerably many' members.[24]

Suppose that we have denumerably many *pairs* of boots, and denumer-ably many *pairs* of socks. We wish to show that it follows that we have

---

[23] The 'Generalized Continuum Hypothesis' is the hypothesis that, for *each* subscript α, $2^{(\aleph_\alpha)} = \aleph_{\alpha+1}$. It is still true that no case of this has yet been either proved or disproved. (I assume here a way of defining $\aleph_\alpha$, for any ordinal α, which will be given later, on p. 111.)

[24] A collection is said to be 'countable' if either it has finitely many members or it has denumerably many members.

denumerably many boots, and denumerably many socks. Russell's point is that there is no problem over the boots. For we are given a correlation between the pairs of boots and the natural numbers, and we can simply say: within each pair take the left boot first and the right boot second, and this obviously yields a correlation between the boots and the natural numbers. But we have a problem with the socks, for socks are not distinguished as left socks and right socks, so we cannot specify a correlation in this way, and there is no guarantee that we can specify any correlation at all. But, if Cantor's criterion for sameness of number is to be retained, we have to say that there must *be* a correlation, even if we cannot specify it. For it is obvious that in this example there must be just as many socks as there are boots. Cantor's criterion forces the existence of relations that we cannot specify.

The relevant axiom providing for their existence is what is nowadays called 'the axiom of choice'.[25] There are various ways of formulating it, and the version most relevant to Russell's example is called 'the trichotomy'. It says that, for any sets x and y, *either* there is a relation which correlates all members of x with some subset of the members of y, *or* there is a relation which correlates all members of y with some subset of the members of x, in symbols:[26]

$$x \leqslant y \ \lor \ y \leqslant x$$

One or other of these relations must exist, whether or not we can specify it. It is clear that we cannot accept Cantor's criterion unless we also accept this axiom, though Cantor himself seems not to have realized this.[27]

I add as a footnote that Cantor himself does not say what kind of a thing a cardinal number *is*. He assumes that each set has a cardinal number, and

---

[25] Any textbook on set theory will give a number of equivalent ways of formulating it. (Russell himself, thinking mainly of a different application, called it 'the multiplicative axiom'.)

[26] The name 'trichotomy' is explained thus. Let us define:

$$x < y \ \leftrightarrow \ x \leqslant y \ \land \ y \nleqslant x; \qquad x \approx y \ \leftrightarrow \ x \leqslant y \ \land \ y \leqslant x$$

Then the axiom states:

$$x < y \ \lor \ x \approx y \ \lor \ y < x$$

[27] Cantor never gives any version of this axiom. Some of his proofs do assume it, but he seems not to have noticed this.

that sets which can be correlated one-to-one have the same cardinal number, so there must be such things. But he does not tell us what they are. It is a later development which claims that cardinal numbers are themselves sets of a special kind.

For future reference, it will be useful to say something here of the infinite *ordinal* numbers. The familiar natural numbers can be thought of as cardinals, providing answers to 'how many?', or they can be thought of as ordinals, measuring the length of sequences, and allowing us to 'number off' the members of a sequence as '1st, 2nd, 3rd, 4th . . .' and so on. In the usual view, a sequence is given by an ordering relation,[28] and we now wish to give serious consideration to infinite sequences. We confine attention to those that are *well*-ordered, which means that every sub-sequence of the sequence has an earliest member.[29] An alternative definition is that, in a well-ordered sequence, if you start with any member and go back to an earlier member, and then back to another that is earlier still, and so on, then you will reach the bottom in finitely many steps. There are no infinite *descents* in a well-ordered sequence, though there may perfectly well be many infinite *ascents*. Ordinal numbers are used to measure the length of well-ordered sequences, and to number off their members.

For example, let us start with the sequence of the natural numbers in their natural order. This is the shortest infinite sequence, and is said to have length ω, for ω is the smallest infinite ordinal. Any other sequence which can be matched with this one in an order-preserving way, i.e. the nth member of the one always correlated with the nth member of the other, is also said to have length ω. (For example, the series of even numbers in their natural order, and equally the series of prime numbers, both have length ω.) Now let us extend this infinite sequence by adding something else,

---

[28]   A relation '<' counts as an ordering relation if and only if it satisfies these three theses:

1. $\forall x \neg(x{<}x)$ (Irreflexive)
2. $\forall xyz(x{<}y \wedge y{<}z \rightarrow x{<}z)$ (Transitive)
3. $\forall xy(x{<}y \vee x{=}y \vee y{<}x)$ (Connected)

Note that on this account an ordered sequence cannot contain repetitions of the same item.

[29]   To continue the previous footnote, the further condition is:

4. $\forall\alpha(\exists x(x{\in}\alpha) \rightarrow \exists x(x{\in}\alpha \wedge \forall y(y{\in}\alpha \wedge x{\neq}y \rightarrow x{<}y)))$ (Well-founded).

perhaps ω itself, to the end of it, i.e. to come after all the natural numbers, thus:

$$\langle 0, 1, 2, 3, \ldots ; \omega \rangle$$

This sequence has a different length, namely length $\omega + 1$. It has the same *cardinal* number of members (since $\aleph_0 + 1 = \aleph_0$), but a different *ordinal* to describe its length, since the whole sequence cannot be matched just with its first part in an order-preserving way. We can go on to add something further, say $\omega + 1$, at the end of what we have so far, as in

$$\langle 0, 1, 2, 3, \ldots ; \omega, \omega + 1 \rangle$$

This is a sequence with length $\omega + 2$. Clearly, we can continue the process to sequences with length $\omega + 3$, $\omega + 4$, and so on. Here is a brief sketch of how the series of all infinite ordinal numbers begins

$$\omega, \omega + 1, \ldots, \omega + n, \ldots, \omega + \omega \; (=\omega \cdot 2)$$
$$\omega \cdot 2, \omega \cdot 2 + 1, \ldots, \omega \cdot 3, \ldots, \omega \cdot n, \ldots, \omega \cdot \omega \; (= \omega^2)$$
$$\omega^2, \omega^2 + 1, \ldots, \omega^3, \ldots, \omega^n, \ldots, \omega^\omega$$
$$\omega^\omega, \ldots, \omega^{\omega^\omega}, \ldots, \omega^{\omega^{\omega^{\cdot^{\cdot^{\cdot}}}}} \; (= \epsilon_0), \ldots$$

These are all different ordinal numbers, each measuring the length of the sequence of all ordinals less than it, and clearly the series of all ordinal numbers continues without end. Every ordinal given here is the ordinal of a sequence with equally many members, namely $\aleph_0$ in each case. But *eventually* the sequence of ordinals has a cardinal greater than $\aleph_0$, namely $\aleph_1$. Unfortunately we cannot in any informative way say where this break comes, for if we knew that then we could probably provide a solution to the continuum hypothesis. But it must come somewhere.[30] And we can go on, for the sequence of ordinals must eventually exceed all the cardinals $\aleph_2$, $\aleph_3$,

[30]   The simplest argument is this. It is a consequence of the axiom of choice that every set can be well ordered, so this applies to the set of all sets of natural numbers, which has at least $\aleph_1$ members. (It has exactly $\aleph_1$ members if the continuum hypothesis is correct.) But every well-ordered set has an ordinal, and if the set has at least $\aleph_1$ members then its ordinal must have at least $\aleph_1$ predecessors.

and so on.[31] In fact the usual procedure these days is to *identify* a cardinal number with the earliest ordinal number that has that number of predecessors, and an ordinal number is itself *identified* with the set of all ordinal numbers less than it. But this is a post-Cantor development. As with cardinal numbers, so also with ordinal numbers, Cantor insists that such things exist, but he does not tell us what they are.

6. Foundations again

In my outline of Cantor's theory of infinite numbers I have used the notion of a *set* in a quite uncritical way, as Cantor himself does in his published writings.[32] But it was soon found that this leads to problems.

The first problem to emerge was the Burali-Forti paradox (1897). Since this involves some quite complex assumptions about ordinal numbers and well-ordered series, the general view at the time was that it could probably be resolved by revising those assumptions in a small way. The paradox goes like this. The series of all ordinal numbers is itself a well-ordered series. But the idea is that every well-ordered series has an ordinal, namely the ordinal that comes next after the ordinals which number off the members of the series. In this case these ordinals which number off the members are those very ordinals themselves. So the ordinal number which is the ordinal of the whole series should be an ordinal number which is greater than *all* the ordinal numbers. This is clearly a contradiction.

Other and simpler paradoxes soon came to light, and they were seen to affect, not just Cantor's ideas about infinite *numbers*, but also the very notion of a set. Here is a very simple one. Consider a 'universal' set, which is supposed to be the set of all things whatever. In that case all subsets of this set must also be members of it, and yet Cantor has proved that every set has *more* subsets than it has members. This is a contradiction. By meditating on this contradiction, Bertrand Russell was led to an even simpler one. The universal set (if there is such a thing) would have to be a member of *itself*, and this idea already generates a paradox. For consider

---

[31] We use ordinal numbers as subscripts to the letter aleph to generate an unending series of alephs.

[32] His private correspondence is more cautious, and shows that he did have some appreciation of the difficulties.

the set of all sets that are *not* members of themselves – all 'normal' sets, as one might say. This would have to be a set w such that

$$\forall x(x \in w \ \leftrightarrow \ \neg x \in x),$$

from which it follows that

$$w \in w \ \leftrightarrow \ \neg w \in w.$$

This is a contradiction, and one which has nothing to do with the question of whether or not there are such things as infinite numbers.

The moral must be this: we need a proper investigation of the notion of a set, or else we must try to do without it. Cantor's theory evidently presupposes an underlying theory of sets, but he never attempted to formulate such a theory himself, and at this date there was no one else who had tried to do so. But it was now apparent that such a theory is *needed*, if we are to avoid the contradictions that arise from a 'naïve' use of this notion. One remembers at this point that Dedekind's construction of the real numbers, i.e. as infinite sets of rational numbers, is also presupposing a theory of sets. So too are the usual ways of constructing rational numbers from pairs of integers, and of constructing integers from pairs of natural numbers. It seems that a theory of sets is everywhere being presumed, but has nowhere been explicitly formulated. This is a further call for 'foundations'.

This describes the situation in around 1900. Foundations for *traditional* mathematics had apparently been provided by Dedekind, but foundations for Cantor's theory of infinite number were lacking, and it had become clear that they were needed. All philosophy of mathematics from now on is well aware of this problem, and partly for this reason the theories become more and more dependent upon logical techniques. I begin the next chapter with Frege, whose main work was done before 1900, and who was not aware of the problems just mentioned. But still, Frege belongs in the twentieth century, and not in the nineteenth.

It is usually said that the twentieth century begins with three great 'isms' in the philosophy of mathematics, namely logicism, formalism, and intuitionism. I wish to add a fourth, namely predicativism. My next four chapters are devoted to these four 'isms'.

Suggestions for further reading

For those with a historical interest it is worth looking at the initial definitions, postulates, and axioms of Euclid's Elements (Heath, 1925, vol. 1, pp. 153–240), and at what we would call the beginnings of the theory of real numbers in his book V (Heath, 1925, vol. 2, pp. 112–87). Otherwise, I suggest only two much more modern but seminal works, both easily available, namely Dedekind (1872, 1888) and Cantor (1895, 1897). A useful and very comprehensive history of all mathematics, which may be consulted on particular topics, is Kline (1972). A much shorter history, which nevertheless does cover all the topics mentioned in this chapter, is Bell (1999). A point of particular interest is the development of what used to be called 'the infinitesimal calculus'. I suggest Mancosu (1996) for the period leading up to Newton and Leibniz, and Kitcher (1984, chapter 10), for later developments.

Chapter 5

Logicism

The previous chapter has been concerned with developments *within* mathematics, rather than with philosophical thoughts *about* mathematics, but that is not because the people mentioned had no such thoughts. On the contrary, Dedekind might be classed as a kind of 'logicist', since he himself says, in the preface to his (1888): 'In speaking of arithmetic (algebra, analysis) as a part of *logic* I mean to imply that I consider the number-concept entirely independent of the notions or intuitions of space and time' (p. 31, my emphasis). This clearly shows his opposition to Kant. On the other hand one might reasonably say that the main thrust of his work (in 1888) is to characterize the *structure* of the series of natural numbers, and on this account he may be dubbed a 'structuralist'. But at the same time he did claim that numbers are 'free creations of the human mind' (p. 31), created by abstracting from the relevant structure, and that is surely a 'conceptualist' thought. I shall say no more of Dedekind's views, which are perhaps not entirely coherent,[1] but will come directly to logicism as it is nowadays understood.

The logicist thesis is that mathematics is analytic in roughly Frege's sense, i.e. that it is simply a matter of logic plus definitions. Frege himself did not hold this to be true of *all* mathematics, for he still thought of geometry as a theory of space, and agreed with Kant that it relied upon some kind of spatial intuition. But he did attempt to argue, in full detail, for the logicist claim about elementary arithmetic.[2] Logicists after Frege have

---

[1] There is a nice discussion of Dedekind's views in Hellman (1990, pp. 309–13), and a full-length treatment in Potter (2000, chapter 3). I remark that Cantor also had philosophical views, which I touch upon in pp. 300–2 below.

[2] He intended to argue for the same claim about the real numbers, but I shall pay no attention to this, since his treatment is left incomplete.

usually claimed that all of mathematics is just logic, without regarding geometry as an exception, because they do not think of it as a problem for *mathematics* to determine which geometry (if any) might apply to actual space. But in practice they too have concentrated mainly on elementary arithmetic. They will hope that Cantor's theory of infinite numbers can be accommodated too, perhaps in some cut down form, but the theory of the natural numbers is where it all begins, so that is where most effort is directed.

The study of logicism must begin with Frege, and then go on to Russell. Which later developments should be taken as a contribution to logicism is a matter for debate.

1. Frege

Frege's first, and (in my view) greatest, contribution to our subject is that he invented modern logic. It is presented in his *Begriffsschrift* of 1879. It is true that he made use of a somewhat awkward two-dimensional notation, which no one else has ever followed. It is also true that, partly because of the notation, his logic did not at first receive the attention that it deserved. In consequence, some of the basic ideas had to be redis-covered, not long after, by Peano (1889) and then Russell (1901), both working independently of Frege. But after a little, i.e. in 1903, Russell himself did recognize the importance of Frege's work in logic, saw that it was superior to his own, and recommended it. So Frege's first and greatest achievement then became generally known. In this book I have to assume that the reader has already been introduced to modern logic, and is famil-iar with the basic symbols that are now used. But I do start by giving a *brief* description of what Frege counted as logic, though it is an updated description, and one that Frege himself would have found very surprising.

The logic falls naturally into five parts, namely: (i) the logic of the truth-functors $\neg$, $\wedge$, $\vee$, $\rightarrow$; (ii) the logic of the first-level quantifiers $\forall x$ and $\exists x$; (iii) the logic of the identity-sign $=$; (iv) the logic of the second-level quantifiers $\forall F$ and $\exists F$; (v) a disastrous assumption about sets. I assume that the reader will be familiar with (i) and (ii) and (iii), and I shall here say just a little about (iv).[3] (The topic will keep coming up as we continue.)

---

[3] Alternative titles: (i) is also called propositional, or sentential logic; (ii) is also called first-level, or first-order, predicate logic; similarly (iv) is also called second-level, or second-order, predicate logic. (In this context I shall use 'level', rather than 'order', for a reason which I will not elaborate until chapter 8.)

For the present I shall set aside (v), until it becomes needed for Frege's over-all position.

In all of (i)–(iv) the crucial notion is that of *validity*. A formula is valid when it comes out true in all (permitted) interpretations – or, as is also said, in all 'structures'. A sequent, with some (or no) formulae to the left, and one (or no) formulae to the right, is valid when it preserves truth in all (permitted) interpretations, i.e. when there is no (permitted) interpretation in which all the formulae on the left (if any) come out true, and the formula on the right (if any) comes out false. For this purpose we permit as an interpretation any way whatever of assigning truthvalues to the sentence-letters which are used in the logic of truthfunctors, but only the intended interpretations of the truthfunctors, as given by their truthtables. Where subject-letters, predicate-letters, and quantifiers are involved, we permit any choice of a domain of objects (save that the empty domain is usually excluded[4]), and any way of interpreting the relevant subject-letters and predicate-letters on that domain. That is, each subject-letter is inter-preted as denoting some member of the domain, and each predicate-letter is interpreted as true of some members of the domain, or of none, and as false of all the rest (if any).[5] (This is an *extensional* interpretation of predicate-letters, i.e. one simply specifies which objects they are true of. From a logical point of view there is no more to an interpretation than this.[6]) But we insist that only the intended way of interpreting the quantifiers is permitted, i.e. that $\forall x(—x—)$ is to be interpreted as true if and only if $(—x—)$ is already interpreted as true for *every* way of interpreting x as denoting some member of the domain. Similarly for $\exists x(—x—)$. For what is called a 'normal' interpretation we also insist that $=$ is given its intended interpretation as identity. Finally, we also specify an intended interpretation for the second-level quantifiers, namely that $\forall F(—F—)$ is to be interpreted as true if and only if $(—F—)$ is already interpreted as true for every way of interpreting F on the domain. Similarly for $\exists F(—F—)$. The definition of *validity* remains the same in all

---

[4]   This makes for a simpler and smoother logic, which is entirely adequate for the philo-sophy of mathematics. (I have argued in part III of my 1997 that *in other contexts* the simplification is without justification.)

[5]   Strictly speaking, this explanation is appropriate only for the one-place predicate-letters. Suitable adjustments have to be made for predicate-letters of two or more places, but I expect that the reader can supply them.

[6]   I am not reproducing Frege's view that predicates have to be understood as referring to 'gappy' entities called concepts, but I do remark that he construed concepts extensionally, i.e. as everywhere interchangeable if they are true of exactly the same objects.

cases: a formula (or sequent) is valid if and only if it is interpreted as true (or truth-preserving) in all permitted interpretations.

The next task is to search for *rules of proof*, which we desire to be both sound (i.e. everything provable is valid) and if possible complete (i.e. everything valid is provable). Frege gave rules of proof for each of the areas of logic that I have labelled (i)–(iv). (The rules for the second-level quantifiers are just the same as those for the first-level quantifiers, though it may reasonably be said that their effect is rather different.[7]) In all cases his rules of proof were sound, and in cases (i)–(iii) they were also complete, but that is not possible in case (iv). I explain why in the next chapter. I remark that Frege's own rules of proof would not look at all familiar to anyone who has learnt their logic fairly recently, but that is immaterial. They provide proofs for just the same formulae (or sequents) as today's rules do.

As I have said, Frege himself would have been astonished at the description of his logic that I have just given. He did not define validity. He never thought of validity as we do, i.e. as truth in all interpretations in all domains. This is partly because he thought of the domain as being always the same for all of logic, namely the domain of *all objects*. But more importantly it was because he always thought of logic in terms of *proof*. We now know that, in a second-level theory, proof and validity may not coincide, but that is not an idea that ever occurred to Frege.[8] (I further remark that Frege thought that a predicate *referred* to a special kind of thing that he called a *concept*, and hence that a predicate-quantifier needs the notion of 'all concepts'. You will notice that my explanation has avoided this idea altogether, and there is no need to adopt this aspect of his logical theory.)

[7] The rule of substitution applies both to name-letters and to predicate-letters, as should be familiar to those who know the theory of the first-level quantifiers. But in the second case they are much more powerful, as may be seen from the existential assumptions that they reflect, respectively:

$\exists x(x=a)$
$\exists F \forall x(Fx \leftrightarrow (\text{---}x\text{---}))$

The latter is to hold for any formula $(\text{---}x\text{---})$ containing x free (but not F free), for any such formula represents a (complex) predicate. But in elementary logic we do not need to permit complex names.

[8] For example, Frege defines an analytic truth as one that is *provable* just from logic plus definitions. But he would have done better to say that it is one which must be *valid*, given the logic and the definitions.

So much for Frege's logic. Let us now turn to the use that he made of it in his construction of arithmetic. I take this in two stages.

Frege never comments directly on Dedekind's important work on the foundations of the theory of natural numbers. He does notice the book, in the preface to volume 1 of his *Grundgesetze* (1893), where he calls it 'the most thoroughgoing work on the foundations of arithmetic that has lately come to my notice' (p. 4). But otherwise he merely observes ('not as a reproach') that Dedekind is able to go a long way in a short space mainly because he never formulates his underlying logic and never gives proofs in full detail. However, one of Frege's own aims was to do just this, i.e. to set out his proofs in full detail, so that one could be sure that they nowhere relied upon an unacknowledged intuition. He nowhere comments explicitly on Dedekind's characterization of a 'simply infinite system', which underlies what we call Peano's postulates, nor on his proof that these postulates are categorical, nor on his proof of the recursion theorem, which allows us to introduce functions and relations on the natural numbers by recursion. As a matter of fact Frege does have his own version of all these theorems, but they do not appear until near the end of the first volume of the *Grundgesetze*, and so are little known.[9] From our point of view it will do no harm to ignore them, and to say that the task is just to provide a basis for the deduction of Peano's postulates. Everything else will follow from there. For this purpose the main text is his *Grundlagen* (1884).

We must begin by asking what kind of a thing a number is. Frege notes that in arithmetic the numerals are quite naturally construed as names, with the function of denoting objects, whereas in ordinary everyday language they frequently occur in a different way, e.g. as adjectives or as part of a numerical quantifier such as 'there are 2', 'there are 3', and so on. He observes that the latter use can always be paraphrased in terms of the former. Thus instead of 'Jupiter has 4 moons', or 'there are 4 moons of Jupiter' one can always say 'the number of Jupiter's moons is 4'. The last is explicitly a statement of identity, and for Frege identity is always to be construed as a relation between *objects*. This is really all that he gives by way of argument for his claim that numbers are objects (in sections 55–7 of the

---

[9]   A convenient exposition is Heck (1995). It is not clear whether these theorems of Frege's owe something to his reading of Dedekind. At any rate, he does not acknowledge any such debt.

Grundlagen), and it is not very convincing. For although it is true that statements using numerical quantifiers can be rephrased as statements which apparently refer to numbers as objects, Frege fails to ask whether one could equally well paraphrase in the opposite direction, so that statements which apparently refer to numbers as objects are replaced by statements which use numerical quantifiers. (As we have noted on p. 70, J.S. Mill clearly took the second view, and we shall find that others do too.) It must therefore be observed that there is also a further reason for Frege's view that numbers are objects, though it is a reason which he never acknowledges, namely that the deductions which he proceeds to give do rely on it, and would not be possible without it.[10] We shall see why when we come in the next section to Russell's views. But for the present let us continue with Frege's idea.

We may abbreviate 'the number of the objects x such that Fx' to '$[Nx{:}Fx]$'. This is itself an expression for an object, and Frege assumes that there always is such an object, whatever predicate we put in place of 'F'. He offers the same criterion of identity for such objects as Cantor does, namely that the number of the Fs is the same as the number of the Gs if and only if there is a relation which correlates the Fs one-to-one with the Gs. Let us abbreviate this to '$F{\approx}G$'. Then Frege's basic assumptions about numbers are:[11]

HP(i) $\exists y(y = [Nx{:}Fx])$
HP(ii) $[Nx{:}Fx] = [Nx{:}Gx] \leftrightarrow F{\approx}G$

What is often called 'Frege's theorem' is the proof that, in the context of a second-level logic, Peano's Postulates can be deduced from these assumptions.

For the deduction we shall need definitions of '0', 'successor', and 'natural number'. Writing 'nSm' for 'n succeeds m', and '$\mathbb{N}x$' for 'x is a natural number', Frege's definitions are these:

0   for $[Nx{:}x{\neq}x]$
nSm for $\exists F\exists y(Fy \wedge n=[Nx{:}Fx] \wedge m=[Nx{:}Fx \wedge x{\neq}y])$
$\mathbb{N}x$ for $\forall F(F0 \wedge \forall nm(nSm{\rightarrow}(Fm{\rightarrow}Fn)) \rightarrow Fx)$

---

[10]   This point is elaborated in some detail by Dummett (1991, pp. 131–40).
[11]   'HP' is for 'Hume's Principle'. Frege somewhat generously credits it to Hume at *Grundlagen*, §63. He never states HP(i) explicitly, but his rules of inference do assume it. I think that an explicit statement is helpful.

The first two of these one might quite naturally expect. The third says that the natural numbers are those things for which mathematical induction holds, starting from 0 and proceeding via the successor-relation. The idea is the same as Dedekind's idea that the natural numbers form the *least* set that contains 0 and is closed under the successor-function. Frege's definition is easily obtained from Dedekind's principles (*a*) and (*b*) as given on p. 101, but with a quantification over predicates replacing the quantification over sets.

It is obvious that from this third definition we can obtain Peano's fifth postulate, the postulate of mathematical induction. In fact from these definitions together one can quite easily obtain all of Peano's postulates, except for the claim that every number has a successor. This requires a more complicated argument, for we must in effect prove a version of what Russell was to call the axiom of infinity, i.e. the claim that each natural number is the number of something:

$$\forall n(\mathbb{N}n \;\rightarrow\; \exists F(n=[Nx{:}Fx]))$$

Frege's tactic is to prove that each number n is the number of the numbers less than it:

$$\forall n(\mathbb{N}n \;\rightarrow\; n=[Nx{:}\; \mathbb{N}x \wedge x{<}n])$$

Given Frege's assumptions, the argument is a straightforward application of the principle of mathematical induction. For clearly 0 is the number of the numbers less than it, and if there are n numbers less than n then there must also be $n+1$ numbers less than $n+1$. This gives the result, and thereby the proof of the missing postulate. As one might say, the trick is done by using the numbers to count those very numbers themselves.[12]

Until this point everything seems to be going very nicely, but Frege's ambitions cannot be content with what we have had so far. For one thing, he himself notes that HP does not tell us *which* objects the numbers are. (This has become known as 'the Julius Caesar problem', i.e. the problem

---

[12]   The argument requires a definition of 'less than'; I leave that to the reader. I add, incidentally, that Frege defines each individual number on the same plan. Thus 1 is defined as the number of those numbers that are identical with 0, 2 is defined as the number of those numbers that are identical with 0 or with 1, and so on.

of whether on this account Julius Caesar might be a number.[13] I shall discuss it in chapter 9, under the heading 'Neo-Frege'.) But from his own perspective it is more important that he cannot simply assume HP without further justification. For what he wants is a deduction of arithmetic just from logic, plus auxiliary definitions, and the assumption HP is neither itself a definition nor provable in what Frege counts as logic. But it is the attempt to go further that leads to disaster.

Frege assumes that there are such things as sets. (He himself speaks of the extensions of concepts, but I shall use the terminology that is nowadays more familiar.) He assumes that a set counts as an object belonging to his original domain of objects, and that, for any condition whatever, there always is a set of just those objects that satisfy the condition. He also assumes the standard criterion of identity for sets, i.e. that sets are the same if and only if their members are the same. This gives him his notorious axiom V:[14]

V(i) $\exists y(y=\{x:Fx\})$

V(ii) $\{x:Fx\}=\{x:Gx\} \leftrightarrow \forall x(Fx\leftrightarrow Gx)$

He does this because he thinks that the notion of a set is a very general and 'topic-neutral' notion, which has a good claim to belong to pure logic, and the above axiom just represents what has always been assumed about sets. Moreover we can (somewhat artificially) define the number of the Fs to be the same thing as the set of all objects y which are the extensions of concepts equinumerous with the concept F, i.e.

$[Nx:Fx]$ for $\{y:\exists G(y=\{x:Gx\} \land G \approx F)\}$

This definition allows one to deduce the assumption HP that we began with, and so apparently completes Frege's project.[15]

---

[13] The label 'Julius Caesar' arises from putting together sections 56 and 66 of the *Grundlagen*.

[14] As with assumption HP, Frege does not state V(i) explicitly, but his rules of inference do assume it. (I have also simplified by giving a version of V that speaks only of sets. Frege's own version uses his more general notion of what he calls a 'value-range'.)

[15] Frege himself prefers to say that the number of the Fs is the set of all *concepts* equinumerous with the concept F, and he assumes that a set of concepts is an object. Given his background logic, this makes no difference.

But here there has been a crucial error, for the axiom that Frege introduces for sets is inconsistent, and leads straight to a contradiction (as Russell pointed out, in a letter to him of 1902). To see this most simply, let us define the relation of membership, as Frege himself does:

$$x \in y \quad \text{for} \quad \exists F(y=\{z:Fz\} \land Fx)$$

Then it is easy to deduce the unrestricted principle of set abstraction:

$$\forall F \exists y \forall x(x \in y \leftrightarrow Fx)$$

As a special case, taking Fx as $x \notin x$, we then have:

$$\exists y \forall x(x \in y \leftrightarrow x \notin x)$$

From which follows the contradiction:

$$\exists y(y \in y \leftrightarrow y \notin y)$$

There must be something here that is very wrong indeed. But can logicism somehow be rescued?

Frege himself attempted a rescue in an appendix hurriedly added to his *Grundgesetze* vol. 2, which proposed a modification to his axiom V(ii). The modification is in itself implausible, and Frege found soon afterwards that anyway it did not actually work, i.e. that the modified version would not allow him to prove what he wanted to prove. Years later it was shown that his modified version did not even restore consistency. (See e.g. Quine, 1955.) I shall not pursue it, for – with hindsight – it is clear that what needs modifying is V(i) and not V(ii). So the task passes to others.

2. Russell[16]

Frege's logic quite deliberately departs from our ordinary way of talking in its use of quantifiers. Ways of expressing quantification in English, and in other natural languages, seem to have grown up in a quite haphazard way, and have no rhyme or reason behind them, whereas Frege's formal

---

[16]   I have given a fuller account of Russell's position, and its rationale, in my (2009b).

language is crystal clear, and makes for much greater simplicity in the expression of complex thoughts. But Frege is retaining our ordinary habits in his over-generous assumption about the existence of sets. We ordinarily think nothing of turning a predicate into a noun-phrase, which apparently introduces a reference to an object of some kind, and it never occurs to us to doubt whether there really is such an object. For example, the predicate '. . . is a man' can be replaced by the noun-phrases 'the set of all men', 'the property of being a man', or simply 'humanity'.[17] Russell is more radical than Frege in this respect, for he eventually rejected *any* way of turning a predicate into a noun-phrase, i.e. of 'nominalizing' that predicate. But it took him several years of difficult thought to reach this position.

His first reaction to the contradiction which had defeated Frege is given in an appendix to his *Principles of Mathematics* (1903), and it is a somewhat hasty and ill-digested sketch of a (simple) theory of types, as applied to classes. (Russell always talks of classes, rather than sets, so in the present section I shall follow his usage. But no distinction between classes and sets is implied.) The theory sketched in the *Principles* differs in several ways from what we now call Russell's simple theory of types, but I here ignore the differences, and give what became of Russell's initial theory as his thought developed.[18] This is a version of what we now call his simple theory of types. (Throughout this chapter I shall be concerned only with the 'simple' theory of types, and its descendants. I postpone discussion of the full 'ramified' theory to chapter 8.) It can be described thus.

We begin with some things that are not classes. Russell calls them 'individuals'. These are each of level 0. We then form classes which have individuals as members, and these are classes of level 1. Then we can go on to form new classes of level 2, which have classes of level 1 as members. Then classes of level 3, which have classes of level 2 as members. And so on. There are classes of every finite level, but no classes of an infinite level. The hierarchy is a 'strict' hierarchy, in the sense that a class of any level can

---

[17] One may also note that the words with which this last sentence opens, namely 'the predicate ". . . is a man"' have themselves formed a noun-phrase from a predicate, and even this way of nominalizing is not without problems, as is shown by the predicate '. . . is not true of itself'.

[18] The principal differences are: (i) that in the early version the sum of any two types is itself allowed as a type, which in effect gives a cumulative rather than a strict hierarchy of types, though it is not clear that Russell realized that at the time; and (ii) that this version admits an infinite type, coming after all the finite types, but apparently it admits only one infinite type. Russell soon dropped both of these proposals.

have as members only items of the next lower level. There are distinct variables for the classes of each level, and the simplest way of marking this distinction is to attach to each variable a subscript to indicate the level of the items which form its domain. (When actually working with this theory one usually omits the subscripts, for they do become very tedious in practice, but officially they are always present.) There are no variables which range over classes of more than one level. At each level there is an unrestricted abstraction axiom, or rather an axiom-schema generating infinitely many individual axioms, of the pattern

$$\exists y_{n+1} \forall x_n (x_n \in y_{n+1} \leftrightarrow (-x_n-))$$

Here $(-x_n-)$ may be any well-formed formula containing x_n free (but not containing y_{n+1} free). There is also, of course, an axiom of extensionality for each level:

$$x_{n+1} = y_{n+1} \leftrightarrow \forall z_n (z_n \in x_{n+1} \leftrightarrow z_n \in y_{n+1})$$

That is almost all, but there is still one point that is not yet decided.[19]

The contradiction is blocked by the stratification into levels. As a matter of fact Russell *always* wanted to block it in another way too, by declaring that '$x \in x$' is *meaningless*, but in the present theory that is quite unnecessary. The theory is explained to us as requiring that, whatever the subscript on x, '$x \in x$' cannot be true. So the simplest thing to do is to say that it is false, and hence that '$x \notin x$' is always true, just like '$x = x$'. Russell's theory does provide, at each level $n+1$, a class containing all the items of the next lower level:

$$\exists y_{n+1} \forall x_n (x_n \in y_{n+1} \leftrightarrow x_n = x_n)$$

By what I have just said, he could equally have accepted

$$\exists y_{n+1} \forall x_n (x_n \in y_{n+1} \leftrightarrow x_n \notin x_n)$$

It would be the same class y_{n+1} in each case. But no contradiction results from this, for the contradiction comes from taking the special case in which

---

[19] The theory has to handle relations too, which introduces some complications of detail. But I here set them aside, since they introduce no new principle.

x_n is y_{n+1}, and that is now clearly illegitimate. For the variables x_n and y_{n+1} range over quite different domains. I would say, then, that this theory is best completed by adding an axiom-schema, to hold wherever $m \neq n$, stating that:

$$x_m \notin y_{n+1}$$

Russell, however, preferred to say that '$x_m \in y_{n+1}$' was not a formula at all unless $m = n$. In practice the decision on this point does not make any significant difference to what can be done with this simple theory of types.

It would appear that Russell was never satisfied with this theory, as I have stated it here, though he never says what his reasons are. In the light of what happens next the best conjecture is that he found the restrictions on what can be said in this theory to be both awkward and lacking in motivation. For example, *why* can you not quantify over all classes at once? For in our ordinary way of thinking we often do, and so do mathematicians. And *why* can you not have a class with members of different levels? No mathematician is likely to find this a natural restriction. Or again, on Russell's version of the theory, *why* must '$x \in x$' be condemned as meaningless? It is clear that there is nothing wrong with it from a purely grammatical point of view, and it is not hard to suggest examples where it seems to be true. What one can say in favour of this theory is that at least it does avoid what Russell always called 'the contradiction', but that is a rather indirect recommendation, and a weak one if there are also other theories that would equally avoid it. So for the next few years Russell did explore some other theories, as he tells us in his (1906a).

The theory that he there looks upon most favourably he calls a 'no-classes' theory, and this is the ancestor of his final theory. In its first version, i.e. in (1906a), it was not only a 'no-classes' theory but also a 'no-propositional-functions' theory, for it permitted quantification only over individuals and propositions and nothing else.[20] But he quite soon discovered that quantifying over all propositions led to contradictions, and his first thought was that propositions must therefore be distinguished into types.

---

[20] I say more on propositional functions in a moment. As Russell notes, the arguments for and against classes apply equally to propositional functions, when they are considered 'as separable entities distinct from all their values'. He infers that 'anything said about a propositional function is to be regarded as a mere abbreviation for a statement about some or all of its values' (1906a, p. 154n.). I think that he in effect retains this view even when quantification over propositional functions is re-introduced.

But he then realized that quantification over what he called 'propositional functions' would be needed too, and then that this was all that was needed. (In *Principia Mathematica*, 1910, there is no quantification over propositions.) So this gives us a new version of the (simple) theory of types that he began with, but now with classes replaced by propositional functions. As a result the theory now has a much more convincing rationale.

A propositional function is something which yields a proposition when you supply it with an argument of the appropriate type. It is simplest to think of a propositional function just as a predicate (but I shall come back later to reconsider this point). Nowadays we are all familiar with the notion of a first-level predicate, and how that is combined with quantifiers that govern variables which take the place of names to yield a first-level logic. We now need to be more explicit about the notion of a second-level predicate, and how that is combined with quantifiers governing variables which take the place of first-level predicates. This is essentially Frege's theory, which I outlined earlier (pp. 115–17), but we should now attend to some details omitted there, so that we can see how to go on to add third-level predicates, with quantifiers governing variables which take the place of second-level predicates, and then fourth-level predicates, and so on for as far as we wish to go. Again, for simplicity, I shall for most of the time confine my attention to monadic predicates only.

To a first approximation the simple hierarchy of monadic predicates can be thought of as obtained in this way. Begin with a sentence that mentions an individual, e.g.

Socrates is a man

Drop out the reference to that individual, and substitute a gap in its place, as in

. . . is a man

This is a first-level predicate. Now consider another sentence which contains that same predicate in a different context, e.g. where the reference to a particular individual has been replaced by a quantified variable ranging over individuals, as in

$\exists x \, (x$ is a man$)$
$\forall x \, (x$ is a man $\rightarrow \ x$ is mortal$)$

We can now drop out that first-level predicate and leave a gap in its place, as in

$\exists x \ (— \ x \ —)$
$\forall x \ (— \ x \ — \ \rightarrow \ x$ is mortal$)$

These are second-level predicates. Then again we can put a variable 'F' into the gap that they contain and introduce a quantifier to bind it, as in

$\forall F \ \exists x \ (Fx)$
$\exists F \ \forall x \ (Fx \rightarrow x$ is mortal$)$

Once more we can drop out the second-level predicates, leaving a gap in their place, to form

$\forall F \ (\text{---} \ F \ \text{---})$
$\exists F \ (\text{---} \ F \ \text{---})$

These are third-level predicates. Then again we can introduce a suitable variable to fill these gaps, and bind it with a quantifier, and so on and on indefinitely.

There are various ways of doing this. Let us in all cases use Frege's letter $\mathcal{M}$ to take the place of an arbitrary second-level predicate. Then we have as three different notational conventions

Frege: $\mathcal{M}_x(Fx)$
Russell: $\mathcal{M}(F\hat{x})$
Church: $\mathcal{M}(\lambda x{:}Fx)$

Church's lambda-notation is the simplest in practice, and so it is most often used today. Russell's cap-notation can be seriously ambiguous, and I do not think that it has ever been followed,[21] but it is *intended* to do the

---

[21] For a simple example, consider $\forall$ and $\exists$ explicitly as second-level predicates. Then the same formula is rendered in these three ways by our three notations:

Frege: $\forall_x \exists_y (Rxy)$
Church: $\forall(\lambda x{:} \exists(\lambda y{:} Rxy))$
Russell: $\forall(\exists \ (R\hat{x}\hat{y}))$

Clearly the Russellian version is ambiguous, for it cannot show whether the initial '$\forall$' governs the capped 'x' or the capped 'y'.

same thing as both Frege's notation and Church's notation: it allows one to show how a second-level predicate, such as '$\forall x(\text{—}x\text{—})$' will attach to its first-level subject '$F\ldots$'. In this respect, I am sure that Frege's notation is the least misleading, from a philosophical point of view. For Church's λ-notation strongly suggests that the first-level predicate '$F\ldots$' has to be turned into a noun '$\lambda x{:}Fx$' before a second-level predicate can be applied. Russell's cap-notation has the same suggestion (especially given Russell's own informal explanations of how to understand it). But there is no such suggestion in Frege's notation, which does all the same work as Church's does, and of course the suggestion is quite misleading. You do not have to turn the predicate '$F\ldots$' into a noun in order to say such things as '$\forall x(Fx)$'. The whole point of Russell's (revised) theory of types is that it is a theory of predicates of different types, and these predicates *remain* of different types all through. They are not replaced by nominalizations (i.e. noun-expressions, apparently referring to individuals) whenever you try to say that something is true of them. Russell should not be understood as intending his cap-notation in that way.

Some, such as Quine, will no doubt protest that we are treating predicates as noun-phrases when we use quantifiers that bind predicate-variables.[22] This is again because *in English* the quantifiers are expressed by such words as 'every', and English grammar requires that these words be followed by a noun or noun-phrase.[23] But the language of the theory of types is not English, and its quantifiers are not bound by the same restriction. We may nevertheless quite easily understand them. As I have said earlier: just as a quantifier over individuals, '$\forall x\ (\text{—}x\text{—})$', may be explained as saying that '$\text{—}x\text{—}$' comes out true for all ways of interpreting 'x', so equally '$\forall F\ (\text{—}F\text{—})$' may be explained as saying that '$\text{—}F\text{—}$' comes out true for all ways of interpreting 'F'. In the first case one interprets the letter (on a given domain) by saying what object (in that domain) it is to refer to, and in the second case by saying which objects (in that domain) it is to be true of. But this does not treat a predicate-letter as if it named an object, and in the theory of types there is no way of doing this: a predicate always occurs with the syntax of a predicate, and not that of a name.[24]

---

[22] See e.g. Quine (1950, section 38) and (1970, chapter 5).

[23] There are exceptions. In such words as 'everywhere', 'anyhow', 'sometimes' the quantifier is followed not by a noun but by an adverb.

[24] I should make it clear that this account of the meaning of '$\forall F$' is mine, and not Russell's (or indeed Frege's).

So far I have described only the hierarchy of monadic predicates. The full theory of types, of course, includes dyadic predicates, and more generally predicates of any polyadicity. This also introduces a complication into what I have described as the *level* of a predicate, for in the full theory there are also what one naturally calls 'mixed level' relations, taking one argument of one level and another of a different level. This certainly complicates things in practice, but no difference of principle is involved, so again I pass over all the details. It is still the case that a predicate of any type can significantly occur only with arguments of the appropriate lower type(s). The rationale for this is now completely straightforward, and can be illustrated just from the monadic hierarchy that we began with. At the lowest level are names of individuals. Next come the first-level predicates, which contain gaps where such a name may be slotted in to make a sentence, or a variable which takes the place of those names to make an open sentence. Next come the second-level predicates which contain gaps where a first-level predicate may be slotted in. And so on up. But a gap that may be filled by the name of an individual (say 'Socrates') cannot be filled instead by an expression of any other type, e.g. by '... is a man' or by '$\exists x$ (— x —)', for the result would simply be ungrammatical, e.g.

... is a man is a man

$\exists x$ (— x —) is a man

These are obviously not well-formed sentences, for they contain unfilled gaps, and the same continues to apply as we move further up the hierarchy.

Russell himself gave just this line of argument in his (1906b, pp. 177–8), and (1906c, pp. 201–2). I think that he means to repeat it in section IV of chapter II of the Introduction to *Principia Mathematica*, which is entitled 'Why a given function requires arguments of a certain type'.[25] My explanation spoke of the 'gaps' which expressions for propositional functions contain, whereas in this section Russell speaks of the 'ambiguity' that is essential to a function. It would be fair to say that what we each have in mind is that an expression for a function contains a free variable, and so the function cannot occur in a proposition unless either it contains some

---

[25] I regard this section as applying to the simple theory of types. Of course Russell elsewhere invokes the vicious-circle principle to justify his type-restrictions, and that is the appropriate procedure for the ramified theory. But in this section Russell means by 'type' just what I mean by 'level'.

higher-level function (such as a quantifier) which binds this variable, or the variable is supplanted by a constant of suitable type. In Russell's terms, the 'ambiguity' must be eliminated if a genuine proposition is to result. Moreover, he explicitly says that in

$\hat{x}$ is a man is a man

the ambiguity is *not* eliminated, evidently because there is here nothing to bind the variable 'x' (p. 48). There are times when he shows quite clearly that his cap-notation is not intended to turn a predicate into a name, grammatically suited to function as a subject-expression to a first-level predicate.

In ordinary English there are many ways of nominalizing what starts as a predicate, but Russell's focus is on mathematics, and in the vocabulary of mathematics there is only one nominalizing technique that is at all widespread, namely that which prefixes 'the class of . . .'. However, Russell's theory is still a 'no-class' theory, and contains no such nominalizing device. It does introduce a notation which *looks* as if it refers to classes. Using modern symbols, the definition is that formulae containing class-descriptions

— $\{x{:}Fx\}$ —

are to be taken as short for the associated formulae[26]

$$\exists G \, (\forall x(Fx \leftrightarrow Gx) \wedge \text{— } G \text{ —})$$

I add that Russell's reason for introducing this definition was that he understood his predicate-letters intensionally, but wished to ensure that the class-expressions functioned extensionally. The definition does achieve this result, i.e. it allows us to deduce the usual axiom of extensionality:[27]

$$\{x{:}Fx\} = \{x{:}Gx\} \ \leftrightarrow \ \forall x(Fx \leftrightarrow Gx)$$

[26] In the *Principia* version the variable 'G' is here restricted to range only over 'predicative' functions of x. This is a complication required only by the ramified theory of types, to be discussed in chapter 8, so I here ignore it.

[27] The deduction assumes that identity is defined for predicate-letters, e.g. by

$$F = G \quad \text{for} \quad \forall \mathcal{M}(\mathcal{M}_x Fx \leftrightarrow \mathcal{M}_x Gx).$$

But this does not create the problems that Frege's notorious axiom V does, even though it looks exactly the same as his axiom. Problems are avoided just because the expressions '$\{x:Fx\}$', which look as if they refer to classes, are construed by Russell as having the syntax of predicates and not of names.

As a matter of personal preference, I would rather construe the predicate-letters extensionally in the first place. In the orthodox first-level logic we assume the correctness of Leibniz's law:

$$a = b \;\rightarrow\; (Fa \leftrightarrow Fb).$$

Its correctness is required by what is now the standard way of explaining validity in this logic. A valid formula is one that is true in all domains under all permitted interpretations, and the permitted interpretations simply assign denotations. (That is, a name-letter is interpreted by assigning to it some object of the domain that it refers to, and a predicate-letter by assigning to it some objects which it is true of.) On this account, Leibniz's law must be taken to be valid. As we all know, there are examples in a natural language when it appears not to be, because in these cases the *senses* of the expressions involved are making a difference, and not only their denotations. But in order to keep our logic simple we just shrug and say that in these cases the logic fails to apply. I would advocate a similar treatment of logic at the next level, i.e. that we only allow as interpretations of the letter '$\mathcal{M}$' those that verify the formula[28]

$$\forall x(Fx \leftrightarrow Gx) \;\rightarrow\; (\mathcal{M}_xFx \leftrightarrow \mathcal{M}_xGx).$$

On this understanding Russell's definition of the usual notation for classes would serve no purpose, and where ordinary mathematics does apparently refer to classes we simply paraphrase this directly as a use of predicates – or, in Russell's preferred terminology, of propositional functions. Of course this was not Russell's own view in the first edition of *Principia Mathematica*, but he did adopt it in the second (1925), possibly because of its recommendation by Ramsey (1925).

Where I have spoken of predicates (of various levels) Russell speaks always of propositional functions, so let us finally come to the question: what *is* a propositional function? There are basically two alternative answers. The

---

[28] Hence an interpretation of the letter '$\mathcal{M}$' will simply specify which interpretations of 'F' it is to be true of.

first is that they are just the same as predicates, as we normally understand them, i.e. they are linguistic expressions of a certain sort. In simple cases one can think of them as gappy sentences, but in more complex polyadic cases it is clear that the pattern in which the gaps are to be filled must somehow be indicated. This is normally done by putting free variables into the gaps, and so forming what is called an open sentence. In any case, the first alternative is that a propositional function is a linguistic item. The second is that it is not itself linguistic but is what the linguistic item means, or refers to, or expresses, or something of that sort. This second approach would allow us to suppose that there are also further propositional functions which no actual predicate expresses, and at first sight it looks as if Russell must be committed to this. For there cannot be more than denumerably many predicates in any (learnable) language, as Russell knew (e.g. 1906b, pp. 184–5), but there are more than denumerably many real numbers, as he also knew. So if he hoped to analyse talk of the real numbers in terms simply of logic, and if logic is a theory of propositional functions, it looks as if there have to be more propositional functions than there are predicates.

In most of his writings up to and including *Principia Mathematica*, Russell is vague on just what propositional functions are. My guess is that he started, e.g. in *Principles* (1903), by thinking of them as non-linguistic entities, for the general attitude of that work is that although language is 'our guide' still it is never what logic is about. But it seems that he gradually shifted towards the view that they are merely linguistic entities, and that is why what he says in print so often leaves this question open. But he tells us in his autobiography of (1959), referring to the time when he was working with Whitehead on *Principia*, that 'Whitehead and I thought of a propositional function as an expression' (p. 124). In later writings this identification is explicit, e.g. (1918, pp. 185 and 196), and (1919, p. 195).[29] It certainly seems to be the simplest view, and it does not in fact give rise to the problem just mentioned, as we can see in this way.

The standard use of the schematic letters '*a*', '*b*', '*c*', . . . in ordinary first-level logic is to stand in place of names (or other referring expressions), but when such a name is replaced by a quantified variable, as in '∀*x*(—*x*—)', we do not think that it follows that the quantification only concerns names. That is, we do not take it to follow that the truth-condition

---

[29] Such statements also occur earlier, e.g. 'A propositional function of *x* is any *expression* φ!*x* whose value, for every value of *x*, is a proposition' (1906a, p. 30, my emphasis). But you can also quote passages on the other side.

for '$\forall x$(—x—)' is that '(—x—)' should come out true for every name in place of 'x'. On the contrary, the usual truth-condition is that '(—x—)' should be interpreted as true for every way of interpreting 'x' as denoting some member of the domain. And it may well be that there are more objects in the domain, and hence more ways of interpreting 'x', than there are names. I take it that this point is uncontroversial. But now all that we have to do is to apply the same idea to the letter 'F'. This is used as a schematic letter to stand in place of a predicate, but when it is treated as a variable bound by a quantifier we do not need to suppose that the quantification is concerned only with the predicates in whatever language is in question. We can say, as before, that '$\forall F$(—F—)' is to be true if and only if '(—F—)' is interpreted as true for every way of interpreting 'F' on the domain, and there may well be more ways of interpreting 'F' than there are predicates in the language. If there is no problem in the first case, then equally there is none in the second, and this resolves the supposed difficulty. Although the language will contain no more than denumerably many expressions, by means of the quantifiers we can still use it to speak of non-denumerable totalities.[30]

Let this suffice as a description of what Russell's simple theory of types is. As with my earlier description of Frege's logic, the account does not pretend to be just the account that Russell gave himself (for that contained many gaps), or the account that he would have given if questioned. But it seems to me to be the *best* way of understanding his theory. I now proceed to some comments.

The most obvious problem with this theory of types is that, by comparison with ordinary English, it has a limited vocabulary and a highly restrictive grammar. This means that you cannot say those things that would lead to the known contradictions. But there are also many *other* things which cannot be said in this language, and which we do wish to be able to say. This applies in particular to the notion of *number*, which is our central topic. How can numbers be represented within type theory?

Russell had begun by taking numbers to be classes. In fact he had adopted the same identification as Frege did, taking the number n to be the class of all n-membered classes. This identification evidently cannot survive his adoption of a 'no-classes' theory, so let us see what happens to

---

[30] The same line of thought evidently applies to predicate-variables of higher types, from '$\mathcal{M}$' upwards for as far as we wish.

it in the light of the definition of class-notation that is adopted in *Principia Mathematica*. The idea there is that what is at first glance said about a class should be paraphrased as saying the same thing about some propositional function that defines the class. So what is said about the number n is now paraphrased as saying something about a propositional function true of just the n-membered classes, i.e. true of the propositional functions that define them, i.e. of just those propositional functions which are true of just n individuals. That means that what is said about the number n is to be paraphrased as saying the same thing about some second-level propositional function equivalent to the numerical quantifier 'there are n ...', for this is a second-level propositional function true of just those first-level propositional functions that are true of just n individuals. So, in effect, the theory of the natural numbers becomes the theory of the numerical quantifiers 'there are n ...'. I add that the same happens to the theory of the infinite cardinal numbers, for that becomes the theory of the infinite numerical quantifiers 'there are $\aleph_0$ ...', 'there are $\aleph_1$ ...', and so on. But it is enough for the present to confine attention to the finite case, for even here there are two serious difficulties.[31]

The first is familiar: how can one deduce that there are infinitely many distinct numbers? On Russell's approach, this is the problem of proving that there are infinitely many distinct numerical quantifiers, and in the logical theory that he presents the only way of *proving* that two propositional functions are distinct is by proving that they are not equivalent, i.e. that one is true of something that the other is not.[32] This evidently means that each numerical quantifier has to be true of something, i.e. that, for each (finite) number n

$$\exists F \exists_n x (Fx)$$

---

[31]　One can work either with what I call the *weak* numerical quantifiers 'there are at least n ...', or the *strong* numerical quantifiers 'there are exactly n ...'. The first is technically simpler, but the second is perhaps more intuitive, and from a philosophical point of view there is no significant difference. So I shall here use the strong quantifiers throughout. The definitions follow the pattern

$$\exists_0 x(Fx) \quad \text{for} \quad \neg \exists x(Fx)$$
$$\exists_n x(Fx) \quad \text{for} \quad \exists x(Fx \wedge \exists_n y(Fy \wedge y \neq x))$$

[32]　As I have noted, in the first edition of *Principia* Russell does not assume axioms of extensionality (though he does in the second). But such axioms can consistently be added, and hence it is only a difference in extension that will force a non-identity.

Clearly this requires there to be infinitely many individuals.[33] Now Frege was able to prove that there are infinitely many individuals by taking the numbers themselves to be individuals, but this course is not open to Russell. For (in his eyes) it would require us to turn an expression for a numerical quantifier, which is a second-level predicate, into an expression for an individual. But a central feature of the theory of types is that it does not allow any such nominalization.

There are other ways of trying to prove that there are infinitely many individuals, and I shall mention some of them later on, in section 4 of the next chapter. But none of them could be regarded as proofs in pure logic. So Russell has to adopt an axiom for the purpose, and since even he admits that this axiom is not itself a truth of logic, this is to admit that he has failed in his overall project. For his project, like Frege's, was to show that the theory of numbers is really just logic (plus definitions). But Frege's attempt failed because he took *too much* to be logic, for his axioms turned out to be inconsistent, and now Russell's attempt fails because he takes *too little* to be logic, for he cannot deduce the result that he is aiming for without adding a non-logical axiom of infinity.

A second problem is this. The last two paragraphs have assumed that the relevant numerical quantifiers are propositional functions of *second* level, but that seems to be an arbitrary choice. For the numerical quantifiers can be applied wherever the ordinary quantifier $\exists$ and the notion of identity apply, and on Russell's account that is at *every* level above the first. For example, we also have at the next level:

$$\exists_0 F(\mathcal{M}_x Fx) \leftrightarrow \neg\exists F(\mathcal{M}_x Fx)$$
$$\exists_{n'} F(\mathcal{M}_x Fx) \leftrightarrow \exists F(\mathcal{M}_x Fx \wedge \exists_n G(\mathcal{M}_x Gx \wedge G{\neq}F))$$

where '$G{\neq}F$' is defined so that

$$G{\neq}F \leftrightarrow \neg\forall \mathcal{M}(\mathcal{M}_x Gx \leftrightarrow \mathcal{M}_x Fx)$$

Clearly the same idea can be repeated at all higher levels.[34]

---

[33] This point holds even if we consider numerical quantifiers of higher levels, which I shall do shortly. For if the lowest level is finite then (assuming extensionality) every other level will be finite too.

[34] Frege held that identity applied only to objects. That is because he is relying on the ordinary English (or German) grammar of the predicate '. . . is the same as . . .', which requires a noun-phrase in each of its gaps. But Russell has now fought free of the grammar of ordinary language, and has seen that the Leibnizian definition of identity applies at every level.

What this shows is that the simple theory of types, even when afforced by the axiom of infinity, cannot provide an acceptable theory of the numbers, construed as numerical quantifiers. For, as Frege claimed (*Grundlagen*, §24), numbers can be used to count things of *any* kind. It is true that on his approach an item of any kind either is an object or can by nominalization be replaced by an object, so all that is needed is the ability to count objects. But Russell rejects such reductions by nominalization, and therefore has to admit that numbers can be used to count not only individuals but also propositional functions of all levels. Moreover, we are convinced that it is the *same* numbers that are being used, whatever kind of item is being counted. Yet the theory of types cannot accommodate this, for it cannot admit any kind of item that applies unchanged at all levels. It follows that the theory cannot provide a satisfactory account of the numbers, either as they are used in daily life or as they are studied by the mathematician.

It is time to look elsewhere.

3. Borkowski/Bostock

The theory of types insists that a predicate of any level can take as its subject (or argument) only an expression of the next lower level, for (to put it in a simplified way) a predicate is a sentence with a gap in it, and predicates of different levels are sentences with different kinds of gap, in each case a gap suited to expressions of the next lower level and to them only. As we have seen, there is a perfectly good rationale for this restrictive formation rule, though it does prevent us from saying many things that we want to say. The basic idea of this section is to overcome the difficulty by *adding* to the theory of types some further vocabulary which is 'type-neutral' in the sense that its expressions do apply in all types without change of sense. The idea is used by Borkowski in his (1958), and I attempted to use it myself in my (1974), but have put forward what I hope is an improved version in my (1980).[35]

To explain this approach we may begin by observing that no one finds anything strange in the idea that exactly the same truthfunctors apply at

---

[35] For the record: I was trying out this idea in papers which I wrote in the 1960s, and it was only later that I learnt (from Arthur Prior) that it had already been put to use by Borkowski. I add that there is a favourable word for this programme in Heck (1997a, p. 307, n. 54), but I think that he has not seen my (1980).

each level of the theory. It is true that in *Principia Mathematica* Russell did find himself forced to deny this, and to claim that their sense is changed from one type to another in what he called a 'systematically ambiguous' fashion (Introduction, chapter II, section 3). But what forces his hand is the fact that the official theory of *Principia* is not the simple but the ramified theory of types, and I am postponing the ramified theory for consideration in chapter 8. While we remain with the simple theory there is absolutely no ground for supposing that the truthfunctors somehow change their sense from one level to another, and I shall regard this as uncontroversial. What is more controversial is the next step. The familiar quantifiers '∀' and '∃' are also applied at all levels of the theory, and we do not usually think of them as changing their sense from one level to another. Indeed, the explanation that I have given, namely that '∀α(—α—)' is true if and only if '(—α—)' is true for every permitted interpretation of 'α', is one that holds whatever the type of the variable 'α'. What counts as a permitted interpretation of the variable will of course be different for different types of variable, but the meaning of '∀' can surely be regarded as constant. One way of thinking of this can be made vivid by changing from Frege's notation for the theory of types to Church's, in which '∀' is always followed by a λ-abstract, in the form '∀(λα:—α—)'. We may view this formula as breaking into two either at its first bracket or at the colon, and either way is just as good, for it means the same in either case. It is when the formula is viewed in the first way that the symbol '∀' stands out as a symbol that is repeated from one type to another without change. It is, therefore, our first example of what I call a 'type-neutral predicate'.

If the truthfunctors and the quantifiers can each be seen as type-neutral, then we may also extend this recognition to whatever can be defined in terms of them. So, since identity can be thus defined (i.e. in Leibniz's manner), identity will count as type-neutral. Further, since the (finite) numerical quantifiers can be defined in terms of the ordinary quantifiers and identity, they too will count as type-neutral. (We may think of them as appearing in a context such as '∃₂(λα:—α—)'.) But of course there are many other predicates which can be defined in these purely logical terms. In *Principia* Russell himself notes that 'practically all the ideas with which mathematics and mathematical logic are concerned' are subject to what he calls 'systematic ambiguity', which is when what appears to be the same idea crops up over and over again in one type after another (p. 65). He has to claim that this is an ambiguity, for his theory forbids him to say that the ideas are really the same; they can only be 'analogous'. But our reformed theory will say that many predicates can also be seen as genuinely type-neutral predicates, which are indeed the same whatever type they are applied to.

I hope that this gives a convincing introduction of the general idea. The details are not so easy. First, one needs to give a general criterion for when a predicate can be recognized as a type-neutral predicate, and in working out this idea one finds that there are several cases where intuition gives no very clear guidance. I have explored this question in my (1980, pp. 233–44, 386–93), and I shall not try to summarize the results here. But some general criterion must be given if we are to be able to use quantified variables in the place of our type-neutral predicates. Moreover, one also needs to consider the predicates of type-neutral predicates and the use of quantified variables in place of these, if one is to produce an adequate theory of the (finite) numerical quantifiers. It is easy enough to define the first such quantifier $\exists_0$, and to define for any quantifier, say Q, another which is its successor Q′. But to define what a finite numerical quantifier is we have to be able to say (i) that $\exists_0$ is one, (ii) that for any Q, if Q is one then so is Q′, and (iii) that there are no other finite numerical quantifiers than those given by (i) and (ii). The difficulty is over the last clause. In effect both Dedekind and Frege resolved it by invoking the principle of mathematical induction, which in our case would be: whatever is true of $\exists_0$, and is always true of Q′ if it is true of Q, is true of all the finite numerical quantifiers. In order to be able to say this we need to be able to quantify over whatever may be true of a type-neutral quantifier. But provided that we can say it then – as we have seen – the principles (i)–(iii) can easily be transformed into an explicit definition of the finite numerical quantifiers, as logicism demands. Then what may be called the 'arithmetic' of these quantifiers can be pursued.

As before, both in Frege's logic and in Russell's simple theory, our logic will allow us to *prove* a non-identity only when we can prove a non-equivalence. So to show that the series of numerical quantifiers is infinite, i.e. that each is non-identical with every one that precedes it, we have to show that each is true of something. Russell had to take this as an axiom, but in our revision of his theory it can be proved. The simplest way of doing so is by observing that there must be infinitely many items within the simple theory of types, which is our background theory, for if there are n items at one level then there must be 2^n at the next level, and we have infinitely many levels. So, for any numerical quantifier, there must be some level at which it is true of something, and it is not difficult to construct a proof of this. A more daring, but also more informative, way of obtaining the result is to follow Frege's plan. He used numbers to count numbers, proving that any number n is the number of the numbers less than it. To adapt this to the present approach, one proves that any (finite) numerical quantifier $\exists_n$ is true of the numerical quantifiers preceding it. Both Borkowski and I have taken this route, but of course it is even more of a departure from the simple theory of types than we have already had. For our type-neutral predicates are now being applied, not just to what occurs in the orthodox theory of types, but also to themselves (and to their own predicates, and so on). The idea is that the type-neutral predicates themselves fall into (non-orthodox) types, and one should expect a predicate that is genuinely type-neutral to be applicable to *all* types, both orthodox and unorthodox.

The above account is obviously very sketchy, and I shall leave it as such. The details of a deduction of Peano's postulates for the type-neutral quantifiers are given in one way in Borkowski, and in a different and simpler way on pp. 21–5 of my (1974), and in yet another different way on pp. 2–9 of my (1979). But while the central moves in the deduction seem (to me) to be perfectly clear, still one has to admit that the background logical theory is in each case not so clear. In Borkowski's case the logic is not specified at all precisely, but if we add what seems to be quite a natural way of making it more precise then that logic gives rise to paradox, as I showed on pp. 373–85 of my (1980). I also found that the logic assumed in my (1974) gives rise to paradox, as pp. 275–82 of the same work reveals. So far as I am aware, the same malady does not afflict the later TN-systems proposed in my (1980), and I there provide an explicitly formulated background logic for the deduction of Peano's postulates on pp. 241–57. But I do not have very much confidence in this logic. The task of harmonizing what certainly *seem* to be quite evident principles concerning type-neutral predicates, with the overriding need to steer clear of paradox, has turned out to be not at all simple. While I expect that others could make a better job of it than I have done, still the logic that they end up with will certainly be much more complex than is the simple theory of types that has provoked it. This is especially so if (as in my 1980) what is aimed for is a general theory of type-neutral predicates of all kinds, rather than a concentration on the one particular case of the finite numerical quantifiers. I add that such a general theory will naturally cover the more elementary parts of the theory of infinite cardinal numbers (construed as infinite numerical quantifiers), and of infinite ordinal numbers (construed similarly). This is outlined on pp. 397–412 of my (1980). But just how much of Cantor's classical theory of infinite numbers could be reconstructed in a theory of type-neutrality is currently an open question.

To sum up: the familiar elementary arithmetic *can* in my view be reconstructed as a theory of type-neutral numerical quantifiers, within a theory which may be argued to be just 'pure logic'. (I shall return to this question shortly.) But such a 'logic' is certainly much more complex than the simple logic of the truthfunctors and the first-level quantifiers, which is now familiar to many.

As a footnote to this discussion, I add that there is another and much more familiar way of extending the basic logic that we begin with, namely by adding the modal operators 'it is necessary that' and 'it is possible that', symbolized by '□' and '◊'. This is not irrelevant to our general topic, as we shall see more fully in the next chapter. But here I simply note the effect that this has on one of the problems in Russell's account, namely his axiom of infinity. If we redefine identity for the numerical quantifiers as *necessary* equivalence, then in order to prove non-identity it is enough to show a *possible* case where the one quantifier applies truly and the other does not. The axiom required is then that for each numerical quantifier it is possible that there is something of which it is true, i.e. for each n

$$\Diamond \exists F \exists_n x (Fx)$$

This has the advantage over Russell's simpler version that – as Russell himself admitted – it is not at all obvious that there actually *are* infinitely many individuals, but it certainly does seem that this is a *possibility*. However, we still have trouble with the logicist thesis, for logicism would demand that the formula just displayed is a truth *of logic*, which does not seem to be at all likely.

So much, then, on attempts to uphold the logicist thesis by concentrating upon the numerical quantifiers, as Russell does. Let us now turn to a different approach.

4. Set theory

Frege took numbers to be objects of the lowest type, i.e. what Russell called individuals, and he identified them with classes. But his assumptions about classes proved to be contradictory, and that set a problem for those who came after him. In an attempt to avoid the contradiction Russell was led into a type theory which was also a no-classes theory, and the effect was that arithmetic had to be construed as a theory of certain 'propositional functions'. But, at the same time as Russell was working out his theory, Zermelo was working out the opposite theory, i.e. a theory which retains the view that numbers are classes, and that classes are objects (of the same logical type as any other objects). Zermelo's theory aims to avoid the contradiction by being much more cautious about which purported classes do actually exist. Both published their thoughts in the same year, i.e. Russell (1908) and Zermelo (1908), and they are each quite different ways of reacting to the same problem. (Since it is usual to translate the German word 'Menge' as 'set' rather than 'class', I shall return to speaking of sets rather than classes in this section. But no distinction in meaning is intended.[36])

Zermelo's aim was to formulate a theory of sets which was strong enough to accommodate 'all' mathematics, including Cantor's theory of infinite numbers, but at the same time not so strong as to fall into contradiction. In his own words, the task is 'to seek out the principles required for establishing the foundations of this mathematical discipline [i.e. set theory]. In solving the problem we must, on the one hand, restrict these principles sufficiently to exclude all contradictions and, on the other, take them sufficiently wide to retain all that is valuable in this theory' (Zermelo, 1908, p. 200). What he then proceeds to do is to propose a set of axioms, and initially he gives no reasons for supposing that these axioms, rather than some others, are the appropriate ones. But he then does go on to show how several important results can be deduced from them.

His axioms begin with the usual principle of extensionality for sets, namely:[37]

1. $\forall xy(\forall z(z \in x \leftrightarrow z \in y) \rightarrow (x=y))$

The remaining axioms then provide for the existence of various sets. The first of them assumes that there is such a thing as the null set (i.e. the one

---

[36] I shall introduce at the end of this section the distinction that is now usual.

[37] Axiom 1 holds only when x and y are sets (and not when they are individuals). In this theory identity may be defined, both for sets and for individuals, by

$\forall xy((x=y) \leftrightarrow \forall z(x \in z \leftrightarrow y \in z))$

and only set which has no members), that for any object there is such a thing as its unit set (i.e. the set which has just that object as a member, and no others), and for any two objects their pair set (i.e. the set which has just those two objects as members). This is the starting point. But Zermelo goes on to add axioms providing for the subsets of a given set, the power set of a given set (i.e. the set of all its subsets), and for the union of the sets that are members of a given set. This yields the axioms (for any sets a, b)[38]

2. $\exists y \forall x(x \in y \leftrightarrow (x=a \lor x=b))$ (Pair set)

3. $\exists y \forall x(x \in y \leftrightarrow (x \in a \land Fx))$ (Subsets)

4. $\exists y \forall x(x \in y \leftrightarrow x \subseteq a)$ (Power set)

5. $\exists y \forall x(x \in y \leftrightarrow \exists z(x \in z \land z \in a))$ (Union set)

These may be counted as the basic axioms.[39] But to obtain a feasible foundation for 'all' mathematics Zermelo adds two more, namely a version of the axiom of choice, and a version of the axiom of infinity, thus:[40]

6. $\forall xy(x \in a \land y \in a \land x \neq y \rightarrow \neg\exists z(z \in x \land z \in y))$

    $\rightarrow \exists b(b \subseteq a \land \forall x(x \in a \land x \neq \emptyset \rightarrow \exists_1 y(y \in x \land y \in b)))$ (Choice)

7. $\exists y(\emptyset \in y \land \forall x(x \in y \rightarrow \{x\} \in y))$ (Infinity)

As we have seen (p. 108) the theory of infinite cardinal numbers must be very unsatisfactory without an axiom of choice, and without an axiom of infinity we cannot deduce the existence of any infinite set, which would rule out the deduction of almost any area of mathematics other than the elementary theory of the natural numbers.[41]

---

[38] If individuals are included, then axiom 2 holds also where a and b are individuals.

[39] The existence of the unit set is a special case of the pair set axiom. The existence of the null set can be deduced from the subset axiom, given any set whatever to start with. The subset axiom, which is also called the axiom of separation (*Aussonderung*), is really an axiom-schema. It generates infinitely many axioms, one for each formula containing x free (but not y free) in place of 'Fx'. '$x \subseteq a$' abbreviates 'x is a subset of a', i.e. 'every member of x is a member of a'.

[40] Here and hereafter I use '$\emptyset$' for the null set, and '$\{x,y,\ldots\}$' for the set which has just $x, y, \ldots$ as members. ('$\exists_1 y$' is the familiar numerical quantifier 'there is exactly one thing y such that . . .'.)

[41] Given Zermelo's own identifications, which I come to shortly, his axiom of infinity says that there is a set which contains all the natural numbers.

These are the seven axioms proposed in Zermelo's seminal paper of 1908. In that paper he gave almost no reasons for selecting just these axioms rather than others, apart from a demonstration that they are adequate for the deduction of several familiar results. But an important later development was the realization that there is an intuitive model for these axioms, i.e. a way of thinking of sets that would explain why they should be reckoned as true of sets, while other axioms that might be proposed should not be. The idea is essentially due to Mirimanoff (1917), but it first receives a full exposition in Zermelo (1930). It is nowadays called 'the iterative conception' of sets, and it goes like this.

Sets can be thought of as constructed in stages. At stage 0 we have no sets (but, perhaps, a number of 'individuals' which are not sets). At stage 1 we form all possible sets out of these (and if there are no individuals this yields only the null set). At stage 2 we form all possible sets out of the items existing at stages 0 and 1. And so on. At stage n we form all possible sets out of the items existing at all stages before n. Moreover, when we have exhausted all the finite stages, we proceed to an infinite stage. At stage ω we form all possible sets out of the sets existing at any finite stage; at stage $\omega + 1$ all possible sets out of what exists at all previous stages, including stage ω; and so on indefinitely. In what is now the received terminology, the stage at which a set is first formed is called its *rank*, and a set exists if and only if it is of some definite rank, i.e. if and only if there is a rank such that all its members have ranks less than this. The ranks (or stages) are to be well-ordered, and so are indexed by the ordinal numbers.

It is easy to see that, on this picture, all the axioms just listed will hold true. One must of course add that this is only a picture, a metaphor, and it is not to be understood literally. For example, I have spoken as if it is our activity that *forms* new sets at each stage, but this is not to be taken seriously. The suggestion is that sets *exist* at any stage after all their members do; it is not that *we* have to *form* them; that happens, as it were, 'all by itself'. (Consequently there may well exist sets which *we* cannot define, or specify in any other way. Indeed, the point of the axiom of choice is to assure us of the existence of sets which we cannot specify, and that is not in any way contrary to the 'iterative' conception.) Another feature of my explanation that is not to be taken literally is the suggestion that the various stages of 'set-construction' are ordered *in time*, with stage 2 occurring *after* stage 1, and so on, as if one could sensibly ask whether the whole process is still continuing or has now been completed. The iterative conception is not intended to give sense to that question. Its idea is indeed that any stage *depends on* the previous stages, but this is not to be construed as a causal dependence, as if 'previous' meant 'previous in time, i.e. occurring at an earlier time'. I confess that it is not easy to see how to cash this metaphor in literal terms, but still it does provide a useful guide on what can or cannot be reasonably assumed about sets.

One suggestion which arises immediately from this iterative conception is for a new axiom, the axiom of foundation (sometimes called the axiom of regularity). If the stages of set-construction form a well-ordered series, as this conception requires, then it must follow that the membership relation is a well-founded relation, i.e. that there are no infinite descents to be got by descending from any set to one of its members, and from that to one of *its* members, and so on. A simple way of getting the effect of this axiom is

8. $\forall x(x \neq \emptyset \rightarrow \exists y(y \in x \land \forall z(z \in x \rightarrow z \notin y)))$ (Foundation)

The point of this axiom is to ensure that all models do accord with the iterative conception. It does allow one to simplify some definitions and some proofs, but (so far as I am aware) it does not yield any new theorems of mathematical interest that could not have been obtained without it.

Another suggestion for a further axiom, which is only somewhat tangentially related to the iterative conception, is the axiom of replacement. This is due to Fraenkel (1922), and in effect it enforces the idea that 'limitation of size' is a crucial factor when deciding whether a proposed set does or does not exist. The thought is that if the proposed set would be no larger than one that we already know to exist, then it does exist. It may be formulated as an axiom-schema, holding for any 2-place well-formed formula in place of 'R', like this:

9. $\forall xyz(Rxy \land Rxz \rightarrow y=z) \rightarrow \exists y\forall x(x \in y \leftrightarrow \exists z(z \in a \land Rzx))$
                                                        (Replacement)

One might think that this axiom was implicit in the iterative conception of sets, since it is clear that a purported set which fails to exist according to that conception, i.e. because it cannot be assigned to any definite rank, would also be very large, and one might say '*too* large' to be a set. But the new axiom in fact goes further than what we have had so far, in two respects.

(i) The original Zermelo theory made no assumption about what individuals (*urelemente*) there are, i.e. things that are not themselves sets but can be members of sets. However, it would seem to be part of the iterative conception that, however many individuals there are, there will be a set of all the individuals, and it will be a set of lowest rank. Yet if we accept this, and also accept the axiom of replacement, there must be some bound upon the possible number of individuals. For example, one might wish to claim that each ordinal number should be counted as an individual. But Zermelo (as we shall see) provides certain sets within his theory to be surrogates of the ordinal numbers, and he cannot accept that there is a *set* of all these sets without falling into the Burali-Forti paradox. Hence, if the axiom of replacement is adopted, we cannot allow that there are as many individuals as there are ordinal numbers, for that would generate a contradiction. *Nowadays* the mathematician's usual response to this thought will be that the set theory which he favours dispenses with individuals altogether. He can do whatever he wants to do within a set theory in which *only* sets are concerned, and there are no individuals at all. (This is called the theory of *pure* sets. There is only one set of lowest rank, namely the null set, and everything else is constructed from that one starting-point.) That is no doubt a fair response, but it does indicate that the new axiom of replacement is making a difference to what counts as a model of the theory.

(ii) There is another difference that it makes, but in this case one that is generally recognized and generally welcomed: it prevents the models from being too small. Zermelo's original theory permitted a model in which the stages of set-construction did not even reach stage $\omega + \omega$, but the axiom of replacement rules that out. It requires the stages to exist for a long way beyond that, but I shall come back later to the question of just how far they do extend.

The theory given by axioms 1–9 is that generally used today by working mathematicians.[42] It is generally called ZF set theory (for Zermelo–Fraenkel), or ZFC if one wishes to emphasize that the axiom of choice (i.e. C) is included. Axiom 8 of foundation is usually included, though more because it is suited to the intuitive model given by the iterative conception of sets than because it is needed to obtain results of mathematical interest.

It is worth comparing this theory with Russell's (simple) theory of types. Both use the idea of a *hierarchy* of sets, though Russell's hierarchy is strict, whereas the ZF hierarchy is cumulative. But in each case it is the

---

[42] If Fraenkel's axiom 9 of replacement is included then Zermelo's axiom 3 of subsets becomes superfluous, as it can be deduced.

hierarchical structure that prevents the known paradoxes from arising. Russell has to keep his hierarchy strict if he is to be able, *via* the no-classes theory, to convert it to a hierarchy of propositional functions, for in their case a cumulative hierarchy would make no sense. (For example, there could not be a monadic propositional function of second level which can take as arguments *both* propositional functions of first level *and* names of individuals. Grammar will not permit that.[43]) But Zermelo has no such motive, since he takes a realistic attitude to these objects called 'sets', and on this view a cumulative hierarchy is both available and more generous in what it allows. Moreover, it quite naturally extends into the transfinite, and Zermelo does so extend it, whereas Russell came to think that such an extension was impossible. But the chief difference between the theories is this. Zermelo's sets are all objects, of the same logical type as one another, even though they may have different ranks. Consequently there is no reason to be suspicious of a style of variable which ranges over the whole domain of all sets, whatever their rank, and ZF theory is standardly presented in this way. It is therefore a theory of first level, assuming the familiar logic of the truthfunctors and of the first-level quantifiers, but nothing more by way of an underlying logic. So the formulae of the theory contain just the one style of variable ('x', 'y', . . .) and the one two-place predicate '$\in$', and that is all. As we shall see in the following section, this simplicity also carries with it a limitation on what the theory can express, but first let us see how it can represent number theory.

The theory of the natural numbers may be represented within ZF set theory in many ways, but the way chosen by Frege (and initially by Russell) is not among them. Frege had identified the number n with the set of all sets which have just n members, but (apart from the set which would be 0) this identification is not available in ZF, since these sets would be too big to exist. Zermelo himself picked a progression of sets starting with the null set and counting the unit set of any set as its successor. So we have:

$$0 = \emptyset$$
$$1 = \{0\}$$
$$2 = \{1\}$$

---

[43]   Hazen (1983, p. 347) appears not to see the difficulty.

and generally

$n' = \{n\}$

However, it is now customary to use the scheme proposed by von Neumann (1925), which takes each number to be the set of all its predecessors. Thus:

$0 = \emptyset$
$1 = \{0\}$
$2 = \{0,1\}$

and generally

$n' = n \cup \{n\}$

This has the advantage that the same idea is easily extended to the infinite ordinal numbers, so that

$\omega = \{0,1,2,3,\ldots\}$
$\omega + 1 = \{0,1,2,3,\ldots; \omega\}$

and so on. It also simplifies the translation of some familiar arithmetical concepts into the notation of set theory. For example, we have:

$n < m \leftrightarrow n \in m$

Again to say that a set x has n members is just to say that it has as many members as n does. It is quite easy to see how other concepts of elementary arithmetic may be represented in set theory upon this scheme.

To obtain Peano's postulates for the natural numbers we need only observe that zero is not the successor of anything, that every set has a successor distinct from it, and that the successor-relation is a one-one relation. Then one can define 'natural number' in terms of zero and successor in the usual way, and Peano's postulates are at once forthcoming. So the ordinary theory of natural numbers is at hand. Moreover, the ZF theory also legitimizes the usual way of constructing other varieties of number, e.g.: the signed integers as ordered pairs, or sets of ordered pairs, of natural numbers; the rational numbers as ordered pairs, or sets of ordered pairs, of signed integers; the real numbers as infinite sets of rational numbers (using Dedekind's method), or perhaps as sets of infinite sequences of rational numbers (using Cantor's method); and so on. These constructions do presuppose some assumptions about the existence of sets, but the ZF theory allows one to prove those assumptions.

Turning to infinite numbers, I have already indicated how the infinite ordinals are construed on the same plan as the finite numbers, i.e. as each the set of all its predecessors. To ensure that a set a does contain all its predecessors, we simply lay down

(i) $\forall xy(x \in y \land y \in a \ \rightarrow \ x \in a)$

To ensure that it contains no more than its predecessors we need the idea that the $\in$ relation is well-ordered on a, which is what ensures that the principle of transfinite induction holds for all ordinals up to and including a. To spell this out in full, the $\in$ relation well-orders a if and only if it totally orders a (i.e. is transitive, connected, and irreflexive on a) and is also well-founded on a. That is:

(ii) $\forall xyz(\{x,y,z\} \subseteq a \ \rightarrow \ (x \in y \land y \in z \rightarrow x \in z))$ (Transitive)
(iii) $\forall xy(\{x,y\} \subseteq a \ \rightarrow \ (x{=}y \lor x \in y \lor y \in x))$ (Connected)
(iv) $\forall x(\{x\} \subseteq a \ \rightarrow \ x \notin x)$ (Irreflexive)
(v) $\forall x(x \subseteq a \land x{\neq}\emptyset \ \rightarrow \ \exists y(y \in x \land \forall z(z \in x \rightarrow (y{=}z \lor y \in z))))$
                                                      (Well-founded)

Then a is an ordinal number if and only if a satisfies these five conditions. The conditions as stated here are not all independent. For example (v) by itself implies (iii). But to obtain a minimal set of conditions we may invoke the axiom of foundation (or regularity) namely:

$\forall x(x{\neq}\emptyset \ \rightarrow \ \exists y(y \in x \land \forall z(z \in x \rightarrow z \notin y)))$

This axiom forbids infinite descents in the membership relation, and so by itself implies

$\neg(x \in x)$
$\neg(x \in y \land y \in x)$
$\neg(x \in y \land y \in z \land z \in x)$
etc

In its presence, therefore, condition (iv) is at once superfluous, condition (v) now follows simply from condition (iii), and condition (ii) follows from (v) together with the results just noted. So the defining conditions reduce just to (i) and (iii). This is an agreeable simplification, though it does of course depend on the axiom of foundation. But, with or without this axiom, we can in ZF produce a definition of 'ordinal number' which will provide all of Cantor's theory of infinite ordinals.

As for the infinite cardinal numbers, these are simply identified with certain ordinal numbers, i.e. those ordinal numbers that have more members than any of their predecessors. So the finite cardinal numbers are the same as the finite ordinal numbers, the first infinite cardinal $\aleph_0$ is the same as the first infinite ordinal ω, the next cardinal $\aleph_1$ is the first ordinal to have more than $\aleph_0$ members, and so on. I shall not give any details of how Cantor's theory of infinite numbers can be developed from this basis, for the necessary details may be found in any textbook of modern set theory.[44] I just say that the ZF theory is entirely adequate as a background theory for almost all mathematical purposes.

Despite this, one must note that the theory is an incomplete theory, in the sense that there are clear mathematical questions which it does not settle. The best-known example is Cantor's continuum hypothesis, i.e. the hypothesis that the number of the real numbers is the *next* infinite number after the number of the natural numbers. It was shown by Gödel (1944) that this hypothesis is consistent with the axioms of ZF, and by Cohen (1966) that its negation is also consistent with those axioms. So they do not decide whether that hypothesis is true or not. This has been the spur to a search for further axioms, still in tune with the iterative conception of sets, but also strong enough to decide this question (among others). The search began with the idea that the usual ZF axioms do not settle how far the stages of construction go. We can describe cardinal numbers which are too large to be proved to exist on the basis of those axioms, but could add new axioms stating that they do exist. (These are called 'large cardinal' axioms.) I shall not give any details of this approach, but I merely remark that so far it has been unsuccessful in its primary aim, i.e. the proposed new axioms have not been shown to have any important consequences for the familiar ('small') cardinals that we began with. There have also been other suggestions for new axioms which might resolve this, or other

[44] For example, Enderton (1977).

problems that are currently unsolved, but I think that it is fair to say that none has yet found general acceptance. Gödel (1947) urged us to seek for new axioms that could be added to the existing ZF theory, and would improve its power. But so far the search has not borne the fruits that he hoped for.

Another weakness in the usual ZF theory is just that it is a *first-level* theory, formulated with only first-level quantifiers ranging over these objects called sets. Some will claim this as an advantage of the theory, but in the next section I shall show why I disagree. Meanwhile I simply note here how this weakness shows itself. What is usually called the axiom of subsets (or of separation, '*Aussonderung*') is strictly not a single axiom but an axiom-schema, generating infinitely many proper axioms, one for each well-formed formula of the theory in place of the schematic '*Fx*' on p. 142. (The same applies to the schematic '*Rxy*' used in the axiom of replacement on p. 144.) There are only denumerably many well-formed formulae in this theory (or in any other comprehensible theory), so this axiom schema can be used to prove the existence of at most denumerably many subsets of a given set. But, from an intuitive point of view (based on Cantor's theorem), there must be more than denumerably many subsets of any infinite set, such as the set of natural numbers. In ZF we can prove Cantor's theorem, that is, we can prove that there is no relation – i.e. no set of ordered pairs[45] – that correlates the natural numbers one-to-one with the subsets of the set of natural numbers. However, a model of the theory can accommodate this theorem even if it provides only denumerably many subsets, but at the same time no set of ordered pairs which correlates these with the natural numbers themselves, because of a paucity of correlating sets within the model.

This observation is simply an application of the Löwenheim-Skolem theorem, which says that *every* first-level theory which has an infinite (normal[46]) model, will also have (normal) models of every infinite cardinality whatever. So although our theory seems to say, as one of its theorems, 'there are non-denumerable sets', still it has a model in which

[45] In the usual set theory, ordered pairs are somewhat artificially construed as sets, e.g. by putting

$$\langle x,y \rangle \quad \text{for} \quad \{\{x\},\{x,y\}\}$$

(This definition is due to Kuratowski, 1920). Then any set which has only ordered pairs as members will count as a relation.

[46] A 'normal' model is one in which the identity-sign is interpreted as meaning identity.

everything called a 'set' in the theory is actually interpreted as a set with no more than denumerably many members. (This point is known as 'Skolem's paradox', after Skolem 1923.) No first-level theory can avoid this result.

Some cautious moves towards improving the power of the subset axiom are obtained by adding what are called 'proper classes' to the ZF theory. This keeps the idea that the theory is a first-level theory, with just the one style of variable ranging over a single domain of objects, but it will now include more objects than before. For the thought is that the domain will consist of what are called 'classes', and some of these classes will be the same as the sets that we had before, while some are not sets at all. The latter are known as 'proper classes', and their distinguishing feature is that a proper class cannot be a member of anything, i.e. neither of a set nor of another proper class. The sets remain as intended in the ZF theory, and they can be members both of other sets and of proper classes. But the idea is that since the variables now range more widely, over proper classes as well as sets, the formulae that can be used in the axiom-schema for subsets, or replacement, will now have a wider interpretation.[47] One enlargement of this kind is known as NBG (for von Neumann–Bernays–Gödel[48]), but it is somewhat timid. For on this theory the formulae which may take the place of 'Fx' in the axiom of subsets (or of 'Rxy' in the axiom of replacement) are restricted to those in which the quantified variables are confined to range only over those classes which are sets. A more enterprising theory known as MK (for Morse-Kelley) has no such restriction. But in both cases the theory is still construed as a theory of *first* level, with just the one style of variable ranging over both sets and proper classes. I think that this is a mistake. For if proper classes are not allowed to be members of anything then they certainly are not ordinary objects, and one will do better to think of quantification over them as really a use of quantified variables in place of predicates, making the theory into a theory of second level. I shall take up in the next section the relative merits of theories of first level *versus* theories of second level.

Meanwhile I end this section with a footnote. There are other theories of sets beside ZF, and its enlargements to NBG or MK. They have quite

---

[47] A useful description of these theories may be found in Fraenkel, Bar-Hillel and Levy (1973).

[48] The basic idea of a proper class is due to von Neumann (1925), but his formulation of set theory was idiosyncratic. First Bernays and then Gödel presented the idea in a more familiar setting.

different sets of axioms. One of them is Quine's theory NF (for 'New Foundations'), and its later version ML which adds proper classes. (ML is for 'Mathematical Logic', the book in which this version first appeared.) Another is due to Aczel (1988), in which there are sets which disobey the axiom of foundation. I shall not discuss these, since I do not think that they are worth taking seriously. It is true that, so far as is known at present, these two rival theories seem each to be consistent, but they have no intuitive models to match the model for ZF that is provided by the iterative conception. Consequently they are largely ignored by most practising mathematicians.

That completes my account of what has happened to logicism as originally conceived by Frege, i.e. as the attempt to show that mathematics is 'really' just logic plus definitions. But now I shall step back a little, and look more critically at these two central notions. What are we to count as logic, and what are we to count as a correct definition?

5. Logic

There is no general agreement on the first question. Logic certainly begins (with Aristotle) as the study of valid arguments, and this is an aspect of the subject which never disappears, though nowadays there are other aspects too.

In the days before Frege one might have said that the purpose of logic is to explain the notion of validity and to give rules which can be applied to any argument to test whether it is or is not logically valid. That would have been a fair description of all the logical systems proposed before Frege, but it no longer fits Frege's logic, even if for the time being we set aside his use of second-level quantifiers and his inconsistent set theory. This is because even the now familiar logic of the truthfunctors and the first-level quantifiers, which Frege introduced, is not decidable. That is, there is no test which can be applied to any arbitrary formula of that logic and which will tell one whether the formula is or is not valid. (This was proved by Church in 1936.) What one can do is to provide a set of rules which can be used to prove, of all and only the valid formulae, that they are valid. That is, the rules are both sound and complete for the familiar first-level logic, but they do not provide a decision procedure. There are some who suppose that this feature, i.e. soundness and completeness, should count as definitive of what logic is,[49] and even if it is going too far to make this a part of the definition, still there are many who think that a logical system with this feature is highly desirable. But that would rule out a logic of the second-level quantifiers, which was essential to the logicist programme as both Frege and Russell conceived it.

From the fact that the usual rules for first-level logic are complete, as was first proved by Gödel (1930), it follows that this logic is also 'compact', which means that any formula which follows validly from an infinite set of premises must also follow just from a finite subset of those premises. (For, by completeness, whatever is valid can be proved, but no proof can have more than finitely many premises.) This result is disturbing, for it is easy to cite examples of arguments in which an infinite number of premises is essential. For instance, let the premises be all of

x is not a parent of y
x is not a parent of a parent of y
x is not a parent of a parent of a parent of y
etc.

All these premises together imply the conclusion

x is not an ancestor of y,

but no finite subset of them does. Such arguments cannot be presented as valid in a logic of first level.[50] A consequence of this is that no theory that can be stated in this logic can both have an infinite model and be 'categorical', i.e. can have only one kind of model, in the sense that all its models are isomorphic to one another. There will always be what are called 'non-standard' models, i.e. models of a different structure from the one intended.

I illustrate this with the case of the natural numbers, taking as axioms the Peano postulates listed on pp. 102–3, and assuming a suitable definition of 'less than'. We can force a non-standard model in this way. Consider *adding* a new constant symbol τ, and infinitely many new axioms:

$0<\tau,\ 1<\tau,\ 2<\tau,\ 3<\tau,\ \ldots,\ n<\tau,\ \ldots$

---

[49] E.g. Kneale and Kneale (1962, pp. 724, 741–2).

[50] As Frege showed in his *Begriffsschrift* (1879), in a logic of *second* level 'ancestor of' can be defined from 'parent of', and this argument may be certified as valid. (Cf. p. 75, n. 13)

If one adds any finite subset of these axioms the system obviously remains consistent, for τ can be interpreted as any number greater than those mentioned in the axioms added. Consequently the first-level theory which contains *all* of these axioms is still consistent. For, by compactness, if a contradiction could be deduced from the system with all the new axioms added, then it could also be deduced from a system which adds only finitely many of these axioms, and we have observed that this cannot be done. Since the system with all of the new axioms is consistent, by the completeness theorem it has a model, i.e. there is an interpretation which makes all the axioms true. Of course, this interpretation must also be one that yields a model for the original Peano axioms, with no additions, but it cannot be the model originally intended. For it must include an item denoted by 'τ', which cannot be the same as any of the standard natural numbers.

For the curious, I here add a description of such a non-standard model. After all the intended numbers there comes this extra 'number' called τ. It is easily shown from the Peano postulates that every number has a successor and that every number other than 0 has a predecessor, so this will also hold for τ. Let us write 'τ'' for the first and '`τ' for the second. Then our non-standard model must at least include

$$0, 1, 2, 3, \ldots.$$
$$\ldots, ```τ, ``τ, `τ, τ, τ', τ'', τ''', \ldots.$$

Let us call these added extras the τ-numbers. (It is obvious that none of them can be identified with any of the intended natural numbers, for there must be an 'infinite distance' between any numbers of the two kinds.) From now on I assume that our axioms include the usual stipulations for addition and multiplication.[51] τ must be either even or odd, so let us suppose that it is even. (If not, then choose τ' instead.) It is easily shown from the Peano postulates that for any number that is even there is another number which is half that number. Let us call it τ/2. Then again it is easy to see that this cannot be one of the intended natural numbers, nor one of the τ-numbers indicated above, but must be intermediate between them. Similarly there must be a number which results from adding τ to itself,

[51]   If we are given just the first-level Peano axioms stated on p. 102, with no consideration of addition and multiplication, then predecessors and successors of τ are all that are required to form a non-standard model.

which we may call 2τ, and which must come after all of the τ-numbers just indicated. Moreover, what applies to halves and doubles will also apply to thirds and triples, and so on. In addition, each of these new 'numbers' will have their own predecessors and successors, so the non-standard model will include all of these too, in the structure

$$0, \quad 1, \quad 2, \quad 3, \quad \ldots$$
$$\vdots$$

$\ldots,$	$'''τ/3,$	$''τ/3,$	$'τ/3,$	$τ/3,$	$τ/3',$	$τ/3'',$	$τ/3''',$	$\ldots$
$\ldots,$	$'''τ/2,$	$''τ/2,$	$'τ/2,$	$τ/2,$	$τ/2',$	$τ/2'',$	$τ/2''',$	$\ldots$
$\ldots,$	$'''τ,$	$''τ,$	$'τ,$	$τ,$	$τ',$	$τ'',$	$τ''',$	$\ldots$
$\ldots,$	$'''2τ,$	$''2τ,$	$'2τ,$	$2τ,$	$2τ',$	$2τ'',$	$2τ''',$	$\ldots$
$\ldots,$	$'''3τ,$	$''3τ,$	$'3τ,$	$3τ,$	$3τ',$	$3τ'',$	$3τ''',$	$\ldots$

$$\vdots$$

But also, and finally, we must interpose between τ/3 and τ/2 a new number 2τ/3, and between 2τ and 3τ a new number 3τ/2, and similarly for every fraction of τ, proper or improper. (If the fraction has denominator n, and τ itself is not divisible by n, then choose the next τ-number after τ that is divisible by n, for there must be one between τ and the nth successor of τ. This makes no difference to the overall structure.) So we end up with a structure like this. After all the standard numbers come the non-standard ones. They are arranged in an array of non-overlapping (horizontal) rows, each of which has the order type of the signed integers. And the rows themselves are arranged (vertically) in an array which has the order type of the rational numbers. Any non-standard model for Peano arithmetic must include at least this much (and of course it may also include yet more, for it need not all be based on just the *one* non-standard number τ). It is completely obvious that such models are *not* what was originally intended.

I add as a footnote that when the positive rational numbers are constructed from the natural numbers in the usual way, and the positive real numbers from the rational numbers, these non-standard natural numbers will yield non-standard rational and real numbers. The non-standard real numbers will include some that are 'infinitely large', i.e. infinitely distant from 0, as just illustrated, and therefore their reciprocals will be 'infinitely small', i.e. closer to 0 than any standard real number. Abraham Robinson has observed that these latter may be regarded as filling the role that 'infinitesimals' were once supposed to fulfil, and therefore the notion of an infinitesimal is not actually a contradiction.[52] But of course those who believed in infinitesimals did not also believe in all the other non-standard 'numbers' which such a non-standard model must contain as well.

Let us now return to our main question, 'what counts as logic?'. All theories that are of interest to mathematicians have infinite models. A first-level theory which admits infinite models cannot be categorical, i.e. it can never describe an infinite structure uniquely. Only a second-level theory can do this. When Dedekind argued that his postulates for natural numbers were categorical, he was implicitly reasoning within a second-level logic. The difference in this case comes with the axiom or postulate of mathematical induction. In a first-level theory this can only be taken as an axiom-schema, yielding a specific axiom only for each well-formed formula '$(—x—)$' in place of the schematic 'Fx', i.e. only for those particular instances that can be expressed in the vocabulary of the first-level theory. But in the context of a second-level theory this axiom generalizes over *all* ways of interpreting the letter 'F', whether or not we can express them in a first-level formula. That is why the second-level formulation *is* categorical, while the first-level formulation cannot be.

The case is similar with an axiomatic theory of the real numbers, as first propounded by Cantor. This theory contains an axiom of continuity which states that, for *every* way of interpreting 'F', if the real numbers which satisfy 'F' have an upper bound then they also have a least upper bound. With the axiom understood in this way, the theory is a categorical theory, as Cantor showed. But if the relevant interpretations of 'F' are restricted to those that we can express in some suitable first-level language then again categoricity is lost. The case of set theory is partially, but only partially, similar. It is again true that a first-level version has to interpret the axiom of subsets (or the axiom of replacement) as an axiom-schema, applying only to such conditions as we can formulate in a first-level vocabulary, and in that case non-standard models cannot be excluded. But if we think in terms of a second-level theory, so that this axiom applies to absolutely every way of forming subsets from a given set, then the theory is halfway towards being categorical. It does not go all the way, for even the second-level formulation does not settle questions about how far the stages of

---

[52] The full treatment is in Robinson (1966), but there is a short and useful summary of his position in his (1967).

set-construction go, i.e. about the existence of large cardinals.[53] But it does settle questions that arise at lower stages, e.g. the truth or falsehood of Cantor's continuum hypothesis. As I said earlier, when the theory is construed as a first-level theory then both that hypothesis and its negation are consistent with the usual axioms. But with a second-level formulation that is no longer so, and it can be shown that either the continuum hypothesis is true in all models of the theory or it is false in all models, so it is 'decided' one way or the other. (Cf. Kreisel 1967.) However, we still do not know what the decision is.

Second-level theories therefore have the advantage that they can be categorical, i.e. they can describe the intended structure uniquely, even when that structure is infinite. First-level theories cannot do this. Second-level theories also have the advantage that they can provide explicit definitions of many important concepts, which a first-level theory cannot define at all. The simplest example of this is the case of identity. Given the usual second-level definition, i.e.

$$a=b \leftrightarrow \forall F(Fa \leftrightarrow Fb)$$

the interpretation of '=' is fixed uniquely, but no set of axioms in a first-level logic can achieve this. There are plenty of other examples, e.g. Frege's definition of the ancestral of a relation, Cantor's definition of 'as many as', and many other definitions connected with his theory of cardinal numbers, e.g. 'infinitely many', 'countably many', 'continuum-many', '$\aleph_0$-many', '$\aleph_1$-many', and so on. None of these can be adequately defined in a first-level theory. (Definitions are offered in ordinary set theory, but if the set theory is itself a first-level theory then the Lowenheim-Skolem theorems ensure that they can always be given an unintended interpretation.) It is plausible to say that mathematicians do in practice think in a way which can be adequately represented only in a theory of second level.[54]

The cost of all these desirable features of second-level theories is that in their case it is impossible to provide a set of rules and axioms which is complete, i.e. which is sufficient to prove the validity of every formula that

---

[53] I add that if the theory permits individuals (*urelemente*) then a second-level formulation will of course not tell us how many of them there are.

[54] For an example see e.g. Potter (2004, pp. 42–3, and 312–13).

Philosophy of Mathematics

is valid.[55] As noted already (n. 49) there are some who say that this fact by itself is enough to show that second-level logic does not qualify for the title 'logic'. I have no sympathy with this view, for you might just as well claim that only a decidable system can count as logic. There are others who do not put their objection in quite this way, but still do object that second-level logic requires an understanding of infinity that we do not have.[56] This is mainly because, if the domain of objects is denumerably infinite, then by Cantor's theorem the ways of interpreting a letter 'F' on that domain are non-denumerably infinite, and the claim is that we do not understand non-denumerable infinity. This is allegedly shown by our (present) inability to decide Cantor's continuum hypothesis. But this line of thought is vulnerable to the response that even *first-level* logic requires such an understanding. For in that logic a formula containing a schematic letter 'F' is counted as valid if and only if it is interpreted as true for *every* way of interpreting 'F' on any domain. If this idea is comprehensible, then so is the second-level quantifier '$\forall F$', for its explanation is exactly the same.

To my mind, this shows that if what is ordinarily recognized as first-level logic does deserve to be called 'logic', then so also does second-level logic, for the central ideas are no different. But this claim is certainly disputed. I also admit that the second-level logic so far defended is only a partial one, since it does not yet contain schematic expressions (such as Frege's '$\mathcal{M}_x(\text{---}x\text{---})$') to stand in place of any arbitrary second-level predicate. This is required before we can progress further to a logic of third level, and so on up. But in my view the upwards move does not involve any new principles at any stage, and so I would defend the view that all of Russell's simple theory of types deserves to be accepted as 'logic' (though in practice one seldom needs more than three or four levels above the first).[57] Others will not agree.

As for the further addition of what I call 'type-neutral predicates', all that one can say is that these predicates have always been recognized to be predicates of a logical nature. But at the same time the attempt to give them

[55] This is a corollary of Gödel's incompleteness theorem, which I shall discuss on pp. 175–80. It assumes a very natural condition on what is to count as a rule or axiom, namely that we – or a machine – can always recognize when a formula is an axiom, or when a rule is correctly applied.

[56] Putnam's early (1967b), sections 1–2, is a vigorous statement of this view.

[57] As Shapiro remarks, little is gained by going beyond second level, for 'there is a sense in which nth-order logic, for $n \geq 3$, is reducible to second-order logic' (1991, p. 134). But the reduction is complex, so I do not give it. (There is something similar in Rayo, 2002.)

a suitable recognition, and to work out a logical system that incorporates them, has (so far) proved rather difficult. The systems that I have proposed turn out to be unexpectedly complex in their detail, and to run so close to unexpected paradox that one cannot have much confidence in them. I hope that others will be able to do better, but until they do the claim of such systems to qualify as 'logic' must be viewed with some suspicion.

If we turn to the alternative approach of set theory, especially as based on the ZF tradition, I think that there will be no one nowadays willing to defend the claim that it should be counted as 'pure logic'. This is mainly because pure logic is by tradition neutral on the question of how many objects there are, for a logical truth should remain true in *every* domain of objects (save possibly the empty domain[58]). But evidently the usual ZF set theory makes huge demands on the number of objects, i.e. sets, that there are. I add that the basic principles of logic are traditionally supposed to be in some way self-evident, but that obviously does not apply to the ZF assumptions on set existence. (Consider, for example, the axiom of replacement.[59]) To clarify this, I would not want to say that pure logic can never make any existential claims, for this would not hold of higher-level logics. For example, if there is at least one object in the domain then there must be at least two ways of interpreting a predicate-letter on that domain, i.e. as true of that object or as false of it. (Generally, for n objects in the domain, there will be 2^n ways of interpreting a predicate-letter on that domain.) But these are not assumptions about the number of the objects at lowest level, which is just what one does find in set theory. I point to a consequence of this line of thought: if logic is neutral on the number of objects that there are, then a logicist cannot construe numbers as objects, for arithmetic is certainly not neutral on the number of numbers that there are. They must be construed in some other way, perhaps as numerical quantifiers, if the logicist thesis is to be maintained. But we have seen that the logic of the numerical quantifiers is not altogether simple, and this brings me to a further objection to the logicist programme.

---

[58] I have discussed this possible exception at some length in my (1997, chapter 8). In the present context it is of no importance.

[59] It seems to me that, if Fraenkel's axiom of replacement is to be believed at all, then that can only be for the kind of reason that Russell held to justify his axiom of reducibility (which I discuss in chapter 8), namely that the axiom is fruitful. It allows us to prove many other results which we want to prove, but cannot see any other way of proving. Gödel in his (1947) defended this way of justifying an axiom.

The thesis is that mathematics – or anyway arithmetic (Frege) – is 'really' just a matter of logic (plus definitions). This thesis was certainly considered, both by Frege and by Russell, as providing an explanation of our knowledge of mathematics. But it has now become clear that if the thesis is to succeed then what is to count as 'logic' must be unexpectedly complex. It is obvious that most ordinary people, who know quite well how to use numbers for counting, how to add them and multiply them and so on, did not learn these skills by first learning a complicated logical theory in which the relevant truths can be derived. (This holds whether the theory in question is a logic of type-neutral numerical quantifiers, or a theory of sets, or any other way of providing a deduction of arithmetic within some more inclusive theory.) It follows that, even if the logicist thesis can be defended as technically correct, it still does not answer the epistemological question that it was supposed to answer, namely: how do we know anything about the numbers? For it is quite clear that our actual knowledge was not in fact derived by the complex route that logicism requires. But it does not follow that the knowledge *could not* be attained by such a route. As space is limited, I must here leave that question open. I further observe that questions of ontology are still open, as the next section will indicate.

6. Definition

The objection just raised to the logicist's account of 'logic' also applies to his use of 'definition', for the central definitions that he offers are evidently controversial, and not in any way familiar to ordinary people. Frege defined numbers as being certain sets (i.e. extensions of concepts), and no ordinary person thinks of them in that way. Besides, the sets that Frege chose do not even exist according to the set theory that is now generally accepted. Russell in effect construed the theory of numbers as a theory of numerical quantifiers, and again no ordinary person thinks of the numbers in that way, though all of us do constantly *use* numerical quantifiers when applying our knowledge of numbers. Zermelo returned to the Fregean idea that numbers should be identified with sets of a certain kind, but he chose quite different sets. Moreover, it is obvious that this choice was simply arbitrary. Von Neumann made a different choice, and clearly there are many further ways of choosing from the sets of ZF set theory a progression which may represent the numbers. For purely mathematical purposes, any choice is as good as any other.

This point is stressed in an important article by Benacerraf, entitled 'What Numbers Could Not Be' (1965). He begins by imagining a dispute between two characters called 'Ernie' and 'Johnny', who have each been brought up to believe in a different way of identifying numbers with sets. (Curiously, his 'Ernie' believes the identification due to John von Neumann, and his 'Johnny' believes that due to Ernst Zermelo.) Naturally, they differ over questions arising from their different identifications. For example, according to Johnny, the number 2 is a set with just one member, namely {1}, but according to Ernie this same number is a set with just two members, namely {0,1}. Benacerraf argues that there is no principled way of choosing between these identifications, or many many others that could perfectly well be proposed. For each is perfectly good for purely mathematical purposes, and there are no other purposes that would favour one rather than another. So each has an equally good claim to be true. But it cannot be the case that all of them *are* true, since {1} and {0,1} are not the same set as one another, and therefore the only rational conclusion to draw is that all of them are false. Numbers cannot actually *be* sets, for there would be no way of telling *which* set any given number is.

This is a powerful line of argument, and (as I shall show later[60]) it can also be turned against several other ways of trying to 'reduce' the ordinary theory of natural numbers to something simpler. I further postpone to the next chapter a discussion of the moral that Benacerraf himself drew, namely that since *any* progression of objects could equally well 'play the role of' the natural numbers, those numbers cannot actually be objects of any kind. What I wish to note here is that the same line of argument can also be applied to the usual logicist ways of constructing all the *other* kinds of number from the natural numbers. I will illustrate this point just for the rational numbers and the real numbers, though obviously it applies more widely.

In effect Russell defines the rational number n/m (where $m \neq 0$) as the set of all ordered pairs $<p,q>$ of natural numbers such that $n \cdot q = m \cdot p$ and $q \neq 0$.[61] This is a definition which works perfectly well for mathematical purposes, but so also would many others. One which is quite appealing is to identify the rational number n/m with the single pair $<p,q>$ of natural numbers that expresses that fraction in its lowest terms, i.e. the pair $<p,q>$

---

[60] Chapter 9, section 4.
[61] Actually Russell speaks of 'relations' *rather than* 'sets of ordered pairs', but that is of no significance in the present context.

such that p and q are mutually prime and $n \cdot q = m \cdot p$. Another identification, which is evidently entirely arbitrary, is to take n/m to be the set of all pairs $<p,q>$ such that p is greater than 100 and $n \cdot q = m \cdot p$. There is no good mathematical reason for choosing between these different identifications, and so Benacerraf's argument applies once more: all are equally good, and hence none of them can be correct.

The case is entirely similar with the real numbers. As we know, they can be identified with 'cuts' in the series of rational numbers, following Dedekind's plan. Or they can be identified with sets of convergent series of rationals which all converge to the same limit, following Cantor's plan. It is easy to think of further identifications. For example, a real number can be taken to be a set of all sets of rational numbers that share all their (rational) upper bounds. Once again we have the same situation. All of these identifications work equally well for mathematical purposes, but they cannot all be correct, so they must all be incorrect.

In his *Introduction to Mathematical Philosophy* (1919), Russell has some remarks on 'postulation' versus 'construction' which are not irrelevant here. The logicist approach is to 'construct' these further kinds of number, which is what Russell himself does. The alternative is to 'postulate' them, and this better describes Dedekind's own attitude to the real numbers, and Cantor's own attitude to the infinite numbers. On these alternatives Russell famously commented (p. 71) that 'postulation has many advantages; they are the advantages of theft over honest toil'. Let us pause to consider the advantages of each (and again I shall illustrate just with the nonnegative rational numbers and real numbers).

If one is to postulate the rational numbers, the simplest way to proceed is this. One postulates that, for any natural numbers n and m (except where $m = 0$), there is to be a number n/m which is the result of dividing n by m. Since division is the inverse of multiplication, the basic thought here is

$$m \cdot (n/m) = n$$

That is, in more general terms

$$m \neq 0 \rightarrow \exists x(m \cdot x = n)$$

The idea is that one postulates new numbers to be intercalated between the natural numbers that we began with, and to fill out that original series.

Given suitable auxiliary postulates (which are easy to devise) we therefore prove such things as

$$1 = {}^2/_2, \quad 2 = {}^4/_2$$

and hence

$$1 < {}^3/_2 < 2$$

On this approach each natural number *is* also a rational number, but we are postulating more numbers to be added to those that we began with.

It is similar with the real numbers. In the case of the rationals we postulated that the familiar operation of multiplication has an inverse which is (nearly) always performable, and with the reals one may begin with the thought that the familiar operation of exponentiation should also have an inverse. So, for example, there should be a number which is $\sqrt{2}$. But it soon becomes clear that this postulate is not general enough, and we are led on to the algebraic numbers which are needed to provide roots for all suitable algebraic equations. Then it turns out that the algebraic numbers will not satisfy all our needs – for example the number π is not algebraic – and so today's full series of real numbers is required. There are many ways of framing such a postulate, as we have seen, but the natural approach is once more to think of the real but irrational numbers as *added* to the familiar series of rational numbers, to fill the 'gaps' in that series. So our postulates will easily lead to such results as

$$1.4 < \sqrt{2} < 1.5$$
$$1.41 < \sqrt{2} < 1.42$$
etc.

The thought is that each rational number *is* also a real number, belonging to the series of real numbers, but there are also other real numbers in that series which are not rational. Nevertheless the same 'less than' relation orders them all.

I claim that this is an advantage of the approach by postulation, because it allows us to say that the natural number 2 is exactly the same number as the rational number 2, and as the real number 2, and similarly for other varieties of number. That is, after all, the way that we all think of these things when we first learn, at school, about the several different kinds of number. In a university course in mathematics it is nowadays the opposite view that will be taught, i.e. the view that the natural number 2 is not the same entity as the positive integer 2, and that the rational number 2 is different again, and so is the real number 2. This must be the case if each new kind of number is to be 'constructed' from those that we have already. But it strikes me as very unnatural, and the fact that 'postulation' can avoid it gives it an advantage.

The other main advantage is just that postulation is not subject to the Benacerraf objection mentioned earlier. There are no doubt many different ways of postulating the new kinds of number, each equally good from the mathematical point of view, but we do not have to say that each different set of postulates is postulating a different kind of number. For example, there may be just one set of entities, the real numbers, which satisfies all the different ways of postulating the real numbers. Different constructions will construct different entities, and then there is no rational way of choosing between them, though they cannot all be right. But different sets of postulates may perfectly well be equivalent to one another, and if so then all of them may be correct.

What, then, are the 'honest' advantages of construction over postulation? It is obvious that the point which Russell had in mind was mainly this. One who simply *postulates* a new kind of number is evidently vulnerable to the question: why should we suppose that your postulate is true, i.e. that there really are numbers of this new kind? By contrast, one who *constructs* the new numbers from material already available has thereby shown that they do exist. But this thought brings me on to the topic of my next chapter, namely formalism. For the formalist replies that *truth* has nothing to do with the matter, and is simply irrelevant. In mathematics we study various sets of postulates because we are interested in what follows from them. But we are simply not concerned with whether those postulates are in some sense 'true', and from our perspective that question makes no sense, for we consider the postulates just as defining a formal system to which no interpretation is assigned. Admittedly, it does make sense to talk about truth when some definite interpretation is in question, i.e. when the formal system is to be *applied* in some way to the real world. For example, you can ask whether it is actually true that between any two points of the space we live in there is another point. But that is a question for the physicist, not for the (pure) mathematician. According to the formalist, the sets of postulates studied in mathematics are treated as *uninterpreted* formal systems, so the notion of truth simply does not apply.

Suggestions for further reading

It is essential to read the positive part of Frege (1884), i.e. sections 55–91 and 106–9. These sections are reprinted, in a different translation by M.S. Mahony, in Benacerraf and Putnam (1983). By far the best commentary on Frege is Dummett (1991). As for Russell, it is best to begin with his (1919), and one could start by concentrating just on chapters 1–3, 12–13, 18. Chapters 1–2 and 18 are reprinted in Benacerraf and Putnam (1983). A useful discussion of his simple theory of types is in Copi (1971, chapter 2).[62] My own account of this theory is more fully elaborated in my (2009b).

The suggestion of adding to Russell's type theory some type-neutral elements is pursued in my (1980). That work is somewhat complex, and I suggest reading only pp. 211–17 and 233–6 in the first instance.

The suggestion of replacing Russell's theory by a set theory of the ZF kind has proved much more popular. A useful study of the basic ideas in such a theory is Boolos (1971); one might also look at sections I–II of Parsons (1977). Its application to the natural numbers is explained in a somewhat unusual way in Quine (1963), chapters 11–12. An unorthodox and very interesting full-length treatment of set theory is Potter (2004). This pays full attention to its philosophical aspects.[63] The book is not intended as an introduction, but you should be able to cope with its part I and appendix A. Then you might look at the summary on pp. 284–7 for some idea of the contents of what you have not yet read.

The claim that second-level logic should be admitted as logic has consistently been opposed by Quine; see e.g. section 38 of his (1950) and chapter 5 of his (1970). A different objection is vigorously stated in sections 1-2 of Putnam's early (1967b), and there is a more nuanced review of several different objections in Wagner (1987). On the other side

---

[62] Copi (1971) contains some minor inaccuracies, which do not obscure the main lines of his exposition, but which should be corrected. The needed corrections are clearly and succinctly given in the review by Grover (1974).

[63] There is a review of this book which may be helpful in Uzquiano (2005).

second-level logic is defended in one way by Boolos (1975) and in another way by Boolos (1984).[64] I have offered my own view in my (1998). But the fullest defence is Shapiro (1991), which is a whole book devoted to the topic. I suggest that the main arguments come in his chapter 5, and attention might reasonably be focused on that. (Alternatively, much of the argument reappears in his 2005.)

For the central question of whether numbers may be identified with *anything* that is already known in some other way, the crucial reading is Benacerraf (1965). No one can ignore his argument.

---

[64]   Boolos (1984) is criticized by Rayo and Yablo (2001).

Chapter 6

Formalism

There are varieties of formalism, and the idea that in mathematics we study systems of postulates that are treated as uninterpreted is only one of the central ideas. Another is the negative idea that in mathematics we avoid contentious metaphysical claims, such as the claims that there really are such things as natural numbers, or real numbers, or geometrical points, and so on. But this negative idea can (in principle) be combined with the view that mathematical claims are interpreted, and are evidently true, because they are to be understood not as claims about such things as numbers but rather as claims about symbols, or formulae, or formal systems, or whatever. Here is a pronouncement from the early formalist E. Heine, writing in 1872:

> I do not answer the question 'What is a number?' by defining number conceptually . . . I define from the standpoint of the pure formalist, and *call certain tangible signs numbers.* Thus the existence of these numbers is not in question.[1]

On the surface this affirms that there are numbers, but that they are just signs, so what the mathematician says about numbers should be understood simply as talk about signs. I would rather say that its moral is that there are no such things as numbers are commonly supposed to be, or anyway that the mathematician need not suppose that there are, because what he says can be interpreted as talk about signs and not about numbers. But, as Frege's criticism at once goes on to say, it is not at all clear how this

[1] The passage is cited in Frege (1903, §87). I presume that Heine means it to apply to all kinds of number, but in fact this particular passage is concerned with real numbers, which – as Frege goes on to point out in §§124ff. – is a particularly difficult case. For there are more real numbers than there are signs (unless you count an infinite sequence of signs as itself 'a sign').

interpretation is supposed to work, i.e. not clear just what (according to Heine) the mathematician is supposed to be saying about his 'signs'.

In any case, I set aside the early attempts at formalism by E. Heine and J. Thomae, whom Frege criticizes at some (unnecessary) length in §§86–137 of his (1903). This is mainly because they were writing with no knowledge of what *we* call a formal system, and so their views seem to us to be somewhat primitive. Instead, I begin with Hilbert, who was quite deliberately concerned with formal systems, i.e. systems which are presented with a clear specification of what is to count as a formula, as an axiom, as a rule of proof, and hence as a theorem.

1. Hilbert

We start with the idea that the mathematician may study any formal system he likes, whatever the axioms and whatever the rules of proof. But Hilbert always added a further condition: the system must be consistent. In early days, during his controversy with Frege on the nature of geometry, he was apt to say both that consistency was all that was required and that – from a mathematical point of view – it by itself ensured the existence of the objects posited by the theory.[2] (Frege could not agree with this.) But later he changed his mind and adopted a more complex view of the existence of mathematical objects. This change of mind was largely due to his thought that, in order to establish mathematics on a sure foundation, one had to *prove* the consistency of its various branches, and this became his goal.

At the time the only branch of mathematics in which consistency was seriously in doubt was Cantor's theory of infinite numbers. Cantor's way of presenting this theory, with its reliance on a set theory that was not yet clearly formulated, did seem to give rise to contradictions, and some reform was called for. But Hilbert was determined that this theory had to be rescued from the threat of inconsistency, because it was so very attractive. As he said, 'no one shall drive us out of the paradise which Cantor has created for us' (1925, p. 191). So his ultimate aim was to find a consistency proof for Cantor's theory. But in fact he never did any serious work on this, for there was too much else to be done before that problem was reached. His leading idea was simple: we have run into difficulties with

---

[2] The correspondence is translated in Frege (1980). There are useful commentaries in Resnik (1980, chapter 3), and in Shapiro (2000, pp. 154–7).

Cantor's theory because the human mind has no intuitive grasp of infinity. But it is soon obvious that almost all other areas of mathematics also involve infinity in one way or another, so apparently what is needed is a consistency-proof not just for Cantor's theory but for all the rest of mathematics too. By this date the notion of the infinitely small could be set aside, for Cauchy and Weierstrass had shown how to reformulate the so-called 'infinitesimal calculus' in a way which eliminated all mention of 'infinitesimals'. But the elimination did, of course, make use of the notion of an infinite series, which is one kind of totality that is infinitely large. And there are many other infinite totalities in ordinary and conventional mathematics, so it appears that a great many different consistency-proofs will be needed. As a matter of fact some reduction of the proofs needed is quickly available, by making use of the logicist 'constructions' already discussed.

For example, we can show the consistency of geometry (whether Euclidean or non-Euclidean) by using the theory of sets of real numbers to provide a model for it. The geometry will originally be formulated with certain notions taken as primitive and undefined – e.g. points, lines, betweenness, congruence – and axioms laid down for them. We then show that when a 'point' is interpreted as a pair (or triple) of real numbers, as given by its Cartesian coordinates, and when a 'line' is interpreted as a set of points of a certain sort, and so on, then the axioms are all interpreted as true. This proves the consistency of the axioms in a model-theoretic way, i.e. by showing that they do have a model, whether or not it is the 'intended' model. The proof is a 'relative' consistency proof, because it assumes the usual theory of sets of real numbers, and it only shows that *if* that theory is consistent then so also is the geometry that we began by considering.[3] In the same way the usual logicist constructions provide proofs (i) that if the theory of sets of rational numbers is consistent then so too is the theory of real numbers, (ii) that if the theory of (sets of) pairs of signed integers is consistent then so too is the theory of rational numbers, and (iii) that if the theory of (sets of) pairs of natural numbers is consistent then so too is the theory of signed integers. In fact the question of the consistency of almost all areas of traditional mathematics can be reduced in this way to the two questions of the consistency of some quite elementary parts of set theory and the consistency of the usual theory of the natural

---

[3] We need only assume the consistency of the theory of sets of real numbers, rather than its truth, because the model shows us that if there is a contradiction in the geometry then it will transfer to a contradiction in the theory which provides a model for that geometry.

numbers. Cantor's theory of infinite numbers is a more serious question, for its use of set theory is altogether more demanding. But in these other areas the set theory used is really very simple, and we may reasonably begin by setting aside the question of its consistency. So the first problem is just to show the consistency of the usual theory of the natural numbers. On Hilbert's general approach this does need to be shown, for we do take it that the natural numbers form an infinite totality, and Hilbert thinks that ordinary human reasoning is no longer reliable in such a context. As things then turned out, Hilbert's serious work never got any further than this first problem.

In this case there would seem to be no possibility of a model-theoretic proof, assuming the consistency of some simpler theory which would yield a model. For what theory of any infinite totality do we have that can fairly be regarded as 'simpler' than the theory of the natural numbers that we all learn at school? Some new kind of consistency-proof is therefore needed, and Hilbert proposed one. He invented what today is called 'proof theory' (as opposed to 'model theory'). The basic idea is this. You begin by setting out a formal theory with an explicit vocabulary, a set of formation rules saying what counts as a formula of the theory, and a set of axioms and rules of proof saying what sequences of formulae count as proofs in this theory. Then you hope to show what can or cannot be proved in this theory by an argument based on the structure of what is counted as a proof. In particular, Hilbert's first problem was to show that in the elementary theory of natural numbers, as specified by Peano's postulates or in some other equivalent way, there could not be a proof of a contradiction, i.e. a proof both of 'P' and of '$\neg P$' for some formula 'P'. (Given a classical logic, which we have in this case, from any one contradiction you can prove any other, so it is sufficient to consider a particular case. The usual case selected is the question whether the formula '$0 = 1$' can be proved, since we already know that its negation '$0 \neq 1$' can certainly be proved.)

At once we are faced by this question: what techniques of argument may be employed in this desired consistency-proof? It is immediately evident that this is a relevant question, because Hilbert began with a suspicion of all infinite totalities, but what we are now trying to prove is a generalization about an infinite totality. For there are infinitely many proofs in elementary arithmetic, and the object is to show that *none* of them ends in a contradiction. Further, the desired argument will surely have to use methods that are familiar from the theory of natural numbers, in particular the method of mathematical induction. For example, one might hope to argue by induction on the length of a proof, i.e. (i) no one-line proof ends in a contradiction, and (ii) if no n-line proof ends in a contradiction then nor does any $(n + 1)$-line proof. Hence, since no proof has more than a finite number of lines, no proof ends in a contradiction. But here we are already assuming the correctness of a method of arguing about an infinite totality when trying to prove the consistency of our simplest theory about such a totality, which itself uses the same method. And what use is that? If you make use of a particular theory in order to prove the consistency of that very theory itself, then clearly you have achieved nothing at all. For even an obviously inconsistent theory can be 'proved' to be consistent in this way, i.e. by using in the 'proof' the methods of that theory itself.

This problem led Hilbert to restrict the methods to be used in 'metamathematics', i.e. in proving things about the systems that result from formalizing ordinary mathematics, to what he regarded as 'finitary' methods. He never did give any clear characterization of just what does count as 'finitary', though he accepted the suggestion put forward by Skolem (1923) that what we now call 'primitive recursive arithmetic' (henceforth 'PRA') would qualify. This theory does not require the explicit use of quantifiers which range over all the infinitely many natural numbers, so it may also be described as 'quantifier-free arithmetic'. Nevertheless it does allow us some unrestricted generalizations, namely those that can be expressed simply by using free variables which are tacitly understood as being universally quantified. (It is a familiar convention both in logic and in mathematics that *initial* universal quantifiers may be omitted; PRA takes this convention to be reflecting a significant distinction.) Its method of argument is to introduce functions (and relations) on the natural numbers by a pair of recursive equations, to assume without more ado that such equations do indeed define just one function (or relation), and then to prove generalizations which involve them by the method of mathematical induction. This cannot be stated explicitly as an axiom, but is used as a rule of proof which can be applied to any formulae already obtained. It may help if I give a couple of examples of what can and cannot be done in this allegedly 'finitary' theory of PRA.

First, consider what is perhaps the simplest two-place recursive function, namely addition. This is introduced by the usual recursive equations:

$$n + 0 = n$$
$$n + m' = (n + m)'$$

It is taken for granted that this is a successful way of introducing a function from numbers to numbers, so (e.g.) it will be *true* that, for all numbers *n* and *m*

$$\exists x(n + m = x)$$

But although this is assumed to be true, in PRA we cannot say it, since the unrestricted quantifier $\exists$ is not available. Nor can we reproduce anything like Dedekind's recursion theorem, which *shows* that it is true, for that theorem requires methods of argument which go way beyond any that are available in PRA. Instead, it is simply built into the rules of inference that, when a function-sign such as '+' is introduced in the usual way, then any expression of the form '*n* + *m*' does designate one and only one number.

Given such an introduction for '+', and given suitable definitions of the usual Arabic numerals, we can of course prove all specific equations of the kind '7 + 5 = 12'. Further, by using mathematical induction we can prove generalizations which can be expressed just by using free variables, as in

$$n + m = m + n$$

This is understood as a thesis which holds universally, though in PRA we do not have the unrestricted quantifier $\forall$ that would be needed to state this explicitly. Consequently there is no way of negating such a thesis, i.e. of saying that for *some* *n,m* the above thesis is not true. For the formula

$$\neg(n + m = m + n)$$

can only be understood as saying that the thesis is *never* true.

Let us turn to a different example. Supposing that multiplication has been introduced in the usual way, one can define what it is for a number to be prime, e.g. in this way:

$$Pr(n) \quad \text{for} \quad (\forall x{:}x{\leq}n)(\forall y{:}y{\leq}n)(x{\cdot}y = n \to (x = 1 \lor y = 1))$$

This is permitted, because the quantifier '$\forall x{:}x{\leq}n$' is a restricted quantifier, restricted to range only over the numbers less than or equal to *n*, and that is a finite totality. (Similarly for '$\forall y{:}y{\leq}n$'.) When the quantifiers $\forall$ and $\exists$

are so restricted, they can be thought of as simply a short way of writing what could otherwise be expressed by a long conjunction or disjunction. In PRA we cannot state that there are infinitely many prime numbers in what might seem the natural way, i.e.

$$\forall n(Pr(n) \to \exists x(Pr(x) \wedge n{<}x)),$$

for the quantifier '$\exists x$' is here unrestricted. But, if we think of how Euclid actually proved this theorem, this enables us to supply a suitable restriction. That is, we can state, and indeed prove,

$$Pr(n) \to (\exists x{:}x{\leq}(n! + 1))(Pr(x) \wedge n{<}x)$$

Quantifiers which are bounded in this way are permissible, and can indeed be introduced by recursion. But unbounded quantification is permissible only when it can be expressed without any explicit use of quantifiers, by exploiting a familiar use of free variables.

Let this suffice as a description of PRA.[4] The theory was proposed by Skolem (1923) and accepted by Hilbert as a 'finitary' theory. While he never did say that *only* this theory counts as 'finitary', he did not suggest anything more that might be added to it, so it is fair to assume that this is the needed elucidation of 'finitary methods'. We can now come back to Hilbert's general position.

Hilbert has to maintain that what he calls 'finitary' methods of argument are ones that we do understand and do see to be correct. So he claims that they deal in notions that have a genuine content, and that what we can say by means of them does have a genuine truthvalue. He has to make this claim if he is himself to make use of these methods in order to prove, in metamathematics, the consistency of various mathematical theories. For otherwise these 'proofs' would be worthless. But he adopts the usual formalist attitude to other areas of mathematics, mainly because of his thought that they involve the infinite in ways that we do not really understand. What he claims here is that we do not need to understand them, for we can treat the formal systems in question as lacking in any genuine content, and – so long as they are consistent – this is entirely adequate for the purposes of pure mathematics. As he says:

---

[4] A classic exposition of PRA is Kleene (1952, chapter IX). For a more modern treatment I suggest Takeuti (1987).

We conceive mathematics to be a stock of two kinds of formulas: first, those to which the meaningful communications of finitary statements correspond; and secondly, other formulas which signify nothing and which are the *ideal structures of our theory*. (Hilbert, 1925, p. 196)

A familiar complaint, raised by Frege in his discussion of earlier formalists, is that if these non-finitary statements 'signify nothing' then it becomes mysterious that they should have useful applications in science and in daily life. But that is not a matter that Hilbert was concerned with.

If finitary arithmetic does have a genuine content, then what is that content? What is it about? Hilbert's somewhat unexpected reply is a return to the view of Heine with which we began, namely that it is about *numerals*, and in particular the so-called 'stroke numerals' that are written simply as a series of strokes. For example, what is really meant by an equation such as '2 + 3 = 5' is that the result of writing the numeral '| |' followed by the numeral '| | |' is the numeral '| | | | |'. Similarly for more complex cases. I observe incidentally that, in the more complex cases, we have to understand the numerals in question as abstract objects, i.e. as types and not as tokens, and as existing quite independently of whether there are any suitable tokens. For example, it is probably the case that there never has been and never will be a string of 1,234 strokes making the stroke-numeral for the number 1,234. But Hilbert would not conclude from this that the number 1,234 does not and will not exist. On the contrary, he would insist that there is, and always has been, such a numeral, whether or not it has ever been written down. So this is a Platonist attitude of a sort, even if Hilbert's numerals are in a way 'less abstract' than the kind of numbers that Plato himself believed in.

Similarly the arithmetical generalizations which Hilbert accepts as still 'finitary' are presumably to be understood as generalizations about these stroke numerals. For example, the simple thesis

$$n + m = m + n$$

is to be understood as saying something like this: the result of concatenating any two stroke numerals is always the same, whichever of them is written first. More complex formulae of 'finitary' arithmetic will need more complex translations, if they are to be understood as concerned with numerals of this pattern, but presumably the required translations could always be provided. For example, the equation '$3^3 = 27$' is presumably taken to mean something like this: if you begin with the numeral '| | |', and write down that numeral as many times as there are strokes in '| | |', you will form a new numeral (i.e. for the number 3 + 3 + 3); if you then write down that new numeral as many times as there are strokes in '| | |' you will form (a numeral with 27 strokes, i.e.) '| | | | | | | | | | | | | | | | | | | | | | | | | | |'. One can hardly say that this is a useful way of thinking of arithmetical equations, but presumably it will always be possible.

From a philosophical point of view there is obviously a Benacerraf-style objection: why choose just this style of numerals rather than any other? More importantly there is the obvious objection that, since 'there are' (from Hilbert's point of view) infinitely many such numerals, it cannot be reasonable to count generalizations about all of them as 'finitary', even if (by a familiar convention) they can be expressed without the explicit use of an unrestricted quantifier '∀'. But Hilbert was never persuaded to respond to any such objections. We can see that, if his programme is to be possible, he simply must accept some way of generalizing over infinite totalities, and the method allowed in PRA is certainly more restricted than what is available in the full theory of Peano arithmetic. So if the programme could be completed that would anyway be a significant result, even if the means employed do not really deserve to be called 'finitary'. As for the point about which numerals are to be used, and why it is fair to assume their existence, this is quite unimportant. For Hilbert was not interested in finitary arithmetic (e.g. PRA) for its own sake, but as a source of *methods* that could be applied in other areas, in particular in proofs of the consistency of a formal system. When working upon such a proof one naturally takes for granted the existence of the formulae of the system, and of those series of formulae that constitute proofs in the system, whether or not any actual tokens of them have ever been written.

As I have said, Hilbert's own work never got beyond his first problem, namely to show the consistency of the ordinary theory of natural numbers, with its unrestricted quantifiers. He and his followers did achieve some partial successes, but I shall not describe either their methods or what progress was made. This is because Gödel quite soon showed that full success was impossible.

2. Gödel

Gödel's two incompleteness theorems of 1931 showed conclusively that what Hilbert was trying to do cannot be done. The first theorem proves that any (reasonable) system which aims to formalize elementary arithmetic must be incomplete, i.e. must fail to prove some true statement of arithmetic, and the second goes on to infer from this that there can be no proof of the consistency of arithmetic that uses only the methods of argument that are available within arithmetic. It is the second theorem that most directly addresses Hilbert's aims, but I think that in fact it is the first that has the greater significance from a philosophical point of view. I therefore give a quick sketch of how it is proved.

We assume that we are dealing with a formal system for arithmetic that is in fact consistent (for an inconsistent system is automatically 'complete', since it can prove every formula), and that is also a 'reasonable' system in this sense: it is specified what is to count as a symbol of the system, what combinations of symbols are to count as formulae of the system, and what combinations of formulae are to count as proofs of the system. Moreover this is specified in such a way that we (or a machine) can recognize in some finite number of steps whether or not some string of symbols is a formula, and whether or not some string of formulae is a proof. That is, these questions are effectively decidable.[5] We also assume that the system is at least as strong as PRA, and so can represent any arithmetical function (or relation) that is recursive, i.e. is such that we (or a machine) can decide in a finite number of steps whether a closed formula without quantifiers that involves that function (or relation) is or is not correct.

The point of these assumptions is that they ensure that we can *encode* statements about this system, formulated in its 'metalanguage', into statements of the arithmetical system itself. We can assign a number to each primitive symbol of the system, a number to each string of symbols that counts as a formula of the system, and a number to each string of formulae that counts as a proof in the system. Moreover, the statements that such and such a number is the number of a symbol of the system, or a formula of the system, or a proof of the system, can themselves be represented within the vocabulary of arithmetic. So for each n there is an arithmetical statement of the form 'Fn' which is true (and hence provable) if and only if n is the number of a formula of the system, and for each n, m there is an arithmetical statement which I shall write as 'n proves m' which is true

---

[5] All formal systems that anyone ever uses in practice are in this sense 'reasonable'. An 'unreasonable' one would be (e.g.) one that says that *any* true statement of arithmetic is to count as an axiom, for then we have no way of deciding whether a given formula is or is not an axiom.

(and hence provable) if and only if n is a number which encodes a string of formulae that is a proof of the formula with number m. This is the one that is important. We can now tackle Gödel's argument.

There is an open sentence of arithmetic which we abbreviate to

$$\neg\, \exists x(x \text{ proves } m)$$

This becomes a definite sentence for each name of a particular number in place of 'm', and for each such choice there is a corresponding statement in the metalanguage which is itself encoded as a number by the numbering technique. Hence (by a diagonal argument) there will be some number G such that the arithmetical statement '$\neg\, \exists x(x \text{ proves } G)$' is itself encoded by the number G. That is, we have:

$$G \text{ is the number of } `\neg\, \exists x(x \text{ proves } G)\text{'}$$

So, in a way, the sentence with number G says *of itself* that it cannot be proved. That is the construction which the argument depends upon.

(i) Suppose that for some specific N we could prove

$$(*) \; N \text{ proves } G$$

Then there would be a proof of the statement G, and hence G itself would be false, so we could prove a falsehood. But, worse, the system would be inconsistent. For if $(*)$ is provable then of course G is provable, but so also is its negation, which arises just by existentially quantifying the number N. Hence, if our initial system is consistent (as we have assumed) there is no number which is the number of a proof of G.

(ii) For a given N, the statement $(*)$ is decidable. So, if we cannot prove it then we can disprove it. Thus the argument so far shows that our system contains a proof of each of:

$$\neg\, (1 \text{ proves } G)$$
$$\neg\, (2 \text{ proves } G)$$
$$\neg\, (3 \text{ proves } G)$$
$$\vdots$$
$$\neg\, (n \text{ proves } G)$$
$$\vdots$$

If we could *also* prove the negation of G, that would be to prove

$\exists x(x$ proves $G)$

The system would therefore be what Gödel calls 'ω-inconsistent'. Assuming that the formal system cannot be like this, we have therefore shown that $\neg G$ is not provable, and we have earlier shown that G is not provable. So there is an arithmetical statement, namely the one with number G, which can neither be proved nor disproved in our formal system.

In his original paper of (1931) Gödel's argument simply assumed, as we have just done, that the formal system would not be ω-inconsistent. This is a perfectly reasonable assumption, for no one would want a formal system for arithmetic that is ω-inconsistent. It would be a system which could not be given the intended model, of just the natural numbers and nothing else. But later Rosser (1936) showed that by complicating the formula G one could rework this second phase of the argument so that it relied only on the simple consistency of the system, and not on the more complex condition of ω-consistency. I shall not describe Rosser's complication, but I shall rely on his result in the discussion that follows. The result, then, is that any formal system for arithmetic which is both consistent and 'reasonable' in the sense given earlier will also be incomplete. There will be a statement in the language of the theory, namely G, such that neither it nor its negation is provable in the theory. And I add that anyone who can follow the proof can see that it must be G, rather than $\neg G$, that is actually true in the intended interpretation of the system. This means that the system cannot represent *all* the reasoning that an ordinary mathematician will accept as correct. So not even the *first* stage of Hilbert's programme can be carried out, i.e. the construction of a formal system for arithmetic which will capture all correct arithmetical reasoning. Some is bound to be left out.

I append a quick account of Gödel's second theorem on this topic, which draws a corollary that directly attacks the conclusion that Hilbert aimed for. The argument that we have just rehearsed shows that

If the system is consistent, G cannot be proved in it.

This conditional can *itself* be encoded as a statement of arithmetic, and can be proved in arithmetic by a proof that mirrors what we have just said in the metalanguage. It follows that if, in arithmetic, you could prove the encoded version of 'this system is consistent' then also in arithmetic you could prove the encoded version of 'in this system G is not provable'. That is, you could prove in the system that what G says is true, and our argument showed that in fact this could not be proved. The moral is that you cannot prove, within arithmetic, the arithmetical statement that encodes 'this system is consistent'. It follows that you cannot prove the original statement 'this system is consistent' by methods of argument which are themselves available in the arithmetical system in question. So, *a fortiori*, you cannot prove the consistency of the full system of arithmetic (with unrestricted quantifiers) by using only the methods that are available in the subsystem that is PRA. Hilbert's hopes vanish.[6]

I end this section on Gödel by pointing to a couple of morals. The first is that, whatever formal system we have started with, we can always add its Gödel-statement G as an extra axiom, and then the system can prove more. Moreover, the strengthened system must be satisfactory if the previous one was, for if the previous system was consistent then the addition must be true. But then the new strengthened system will also have *its* Gödel-statement G', which again we can see to be both true and unprovable within the system. So we can strengthen the system yet further by adding G' as another axiom, and so on for ever. One can in fact describe an infinite sequence of new axioms G, G', G'', . . . , and one can consider the system which results when all are added. (This is still a 'reasonable' system in the sense used earlier.) But this system too will still be incomplete, and will have its own Gödel-statement G∞, which again we can see to be true but not provable within the system. There is here a weakness in formal systems which cannot be escaped.

The second observation is that this is a weakness in what can be *proved* in the system, not in what can be *said*. I have explained earlier (pp. 153–7) that, although a first-level theory for arithmetic must have non-standard models, a second-level theory can be categorical. Indeed the usual Peano axioms are categorical when taken in the second-level way (as Dedekind himself took them). Technically speaking this means that all models of the

---

[6] There are complications here which I am ignoring, for there are several ways of representing within the system the statement that it itself is consistent, and this can make a difference to the argument. A useful account of these complications is given in Giaquinto (2002, part V, chapter 2). But Giaquinto himself concludes that, whatever way we choose, Hilbert's hopes must be frustrated. In any case, as I said earlier, Gödel's *first* incompleteness theorem already destroys Hilbert's programme, because it shows that we cannot actually formulate the formal systems which were to have been his starting point.

theory are isomorphic to one another, and hence that any arithmetical statement is either true in all models that satisfy Peano's axioms or false in all such models. As we have seen G must be true in all of them. That is, G is a *semantic consequence* of the axioms, though not *provable* from them. In a first-level logic we know that validity and provability must coincide, but at higher levels this is no longer so. We have as true

Axioms $\models G$

And as false

Axioms $\vdash G$

The difference between validity (semantic consequence) and provability (called syntactic consequence) is thus clearly revealed. One must bear in mind that Gödel's theorem concerns the latter, i.e. '$\vdash$' and not '$\models$'.

3. Pure formalism

Hilbert was only partially a formalist, for he held that the simplest area of mathematics, i.e. PRA, does have a genuine content which we all understand, and which we can all see to be true. This position was apparently forced on him by his metamathematical programme, but Gödel's result showed that this programme could not be carried out, and since then no one has seriously pursued it.[7] However, we could apparently abandon Hilbert's metamathematical programme without abandoning his overall philosophical position, namely that PRA has genuine content and genuine truthvalue, whereas other areas of mathematics do not. In these other areas all that is required is a consistent set of axioms, and mathematicians can happily explore the consequences of these axioms without even asking whether they count as 'true'. Gödel's results do not obviously show that there is anything wrong with this philosophical position.

[7] Gentzen (1936) provided a proof of the consistency of arithmetic by using methods of argument which go beyond those available in ordinary arithmetic, in particular by using transfinite induction on the infinite ordinal numbers up to the ordinal e_0. It is not clear that this kind of consistency-proof achieves anything worthwhile. (For the curious, a readable account of a slightly modernized version of Gentzen's proof may be found in Mendelson, 1964, Appendix.)

They do perhaps show that it lacks an adequate motivation. For what would be the point of taking an attitude to PRA which contrasts with one's attitude to all the rest of mathematics? Why not say about *all* mathematics what Hilbert did say about all mathematics other than PRA, namely that the (pure) mathematician does not have to suppose that it *means* anything at all. His job is simply to explore the consequences of axioms that can be regarded as arbitrarily given. Of course it may be expected that some of these formal systems will find useful applications, and in several cases the axioms will have been deliberately chosen with a view to some desired application, but that is of no importance to the pure mathematician. For there are also other cases where the axioms are proposed just because their study has a purely mathematical interest, independently of any practical applications. (Before Einstein, this was the case with all non-Euclidean geometries, and it is still the case with the more distant parts of 'Cantor's paradise'.) The pure formalist, then, makes no distinction between one area of mathematics and another. There are sometimes applications of his theories, but that is a question to be explored by others. In his own enquiries these theories are not assigned any definite interpretation, and so the question of their truth cannot arise.

A variant on this position says that there is a notion of truth that is applicable, but it is simply that a mathematical statement counts as 'true' if and only if it does follow from whatever axioms are in question. In this sense the axioms themselves are trivially 'true', for of course each axiom follows from itself. A further variation is to say that the axioms count as 'definitions', for they simply define what the topic is. In several areas of mathematics, most obviously in what is called 'abstract algebra', this seems exactly the right thing to say. For example, you cannot seriously ask whether the axioms of group theory count as 'true', for they simply define what is to count as a 'group'. The same applies to the theory of rings, fields, and so on. It is common to all these variations that there is no serious question of the truth of claims in pure mathematics. Surrogates for truth can be provided, and it is a real question whether some formula does or does not follow from a given set of axioms by the rules of proof that are specified, but that is not the question of whether what this formula states is *true*. For, on the formalist approach, the formula is not to be regarded as *stating* anything at all.

Haskell Curry (1951) takes such a view. He claims that mathematics simply is 'the study of formal systems', and that any formal system is as good as any other as a topic for the mathematician. He is clear that one need not suppose that such formal systems mean anything at all, for they are suitable objects for mathematical study whether or not any particular interpretation is intended. This holds whatever the rules of inference may be, and whether or not they have any connection with the normal rules for logical inference. (Curry does not even require consistency – e.g. in the sense of Post, i.e. that not every formula is a theorem – though he does of course recognize that there is little interest in a system that is *known* to be inconsistent.) There is an extremely obvious objection to this position, namely that the mathematician (on this account) will be trying to prove things about these formal systems, and his proofs will use language that does have meaning, and reasoning that aims to be correct reasoning, taking one only to conclusions that do follow from the premises. So mathematics still has genuine content. It is not clear what is gained by the proposal to shift that content from what is ordinarily counted as mathematics to what is now called metamathematics. How has that avoided the original problems? The idea must presumably be that we have made an ontological gain, if not an epistemological one. For the *existence* of formal systems is (perhaps) less problematic than the *existence* of such things as numbers, functions, and so on, even if our *knowledge* is equally problematic in either case.

As I have said, the initial idea behind the formalist approach is surely right for certain areas of mathematics, and the 'pure formalist' is one who applies it to all. Apart from the point that the only interesting formal systems are those that use logically correct rules of inference, the areas where this approach is commonly resisted are principally in the theory of natural numbers, the theory of real numbers, and the theory of sets. We find it very difficult to believe that (e.g.) '$7 + 5 = 12$' can be said to be 'true' only in the surrogate sense that it follows from axioms which define what is meant by 'natural number', or again that '$\sqrt{2}$ is irrational' is also 'true' only in this surrogate sense. But what good reason do we have for rejecting the formalist claim in such cases?

Before Gödel, formalists thought mainly in terms of proof from axioms ($\vdash$) rather than in terms of being a valid consequence of those axioms ($\vDash$). Gödel's argument showed quite conclusively that this way of thinking cannot be maintained.[8] For there are cases in which '$\vdash$' and '$\vDash$' fail to coincide, namely in any logic above the first level, and in such a case all mathematicians think that what matters is '$\vDash$' rather than '$\vdash$'. But the formalist position, as I have expressed it so far, is neutral between these notions,

---

[8] Curry (1951) does try to maintain it.

and we can easily switch it from the second to the first. The result is this: in mathematics we deal with formal systems which have axioms and rules of inference, and we are concerned with what follows from these axioms in the sense of '⊨', i.e. with what has to be true in any model that makes the axioms true. There is no particular model that is the 'intended' one. For example, there may perhaps be such things as Plato imagined the numbers to be, but even if there are they are not what arithmetic is about. Equally, there may perhaps be some mental entities called 'ideas' which satisfy the usual axioms for elementary arithmetic, but the subject does not specially concern them. For Peano arithmetic is about *any* system of entities that satisfies the Peano axioms, i.e. any system that has a first member, for each member a next, and no others. There are no doubt many systems of objects that exemplify this structure, and arithmetic is equally about all of them.

This is a version of what nowadays is called 'structuralism'. As I see it, it is what old-fashioned formalism naturally becomes when the change is made from '⊢' to '⊨'. In either case a leading thought is that we do not have to suppose that mathematics is about some peculiar things called 'numbers', and that the mathematician enjoys a special kind of 'insight' into these things. Each philosophy supplies some other subject for mathematics to be about, either formal systems (for the old-fashioned formalist) or the 'structures' which model these systems (for the structuralist). I think that no contemporary philosopher would admit to being a formalist in the old sense, whereas structuralism has plenty of adherents today. My suggestion that structuralism is really just a modernized version of old-fashioned formalism, resulting from substituting '⊨' for '⊢', will surely be controversial. But I do believe that it contains a lot of truth.

4. Structuralism

As a theory about the nature of mathematics, structuralism may be traced back to Dedekind (1888), if not before, but the modern discussion starts with Benacerraf's important article 'What Numbers Could Not Be' (1965). This considers only the case of the natural numbers, and I too shall begin by limiting attention to this topic. Benacerraf's discussion begins (as already noted, p. 161) with the claim that numbers cannot be sets, because there are many different ways of identifying numbers with some arbitrarily chosen sets, and they are each equally good for all mathematical purposes. So they all stand or fall together, and therefore they all fall. This, it seems to me, is a very convincing argument. But Benacerraf then goes on to generalize his claim to the conclusion that numbers are not any other kind of objects either, and the truth is that 'there are no such things as numbers' (final sentence[9]). His argument is basically this: any progression at all (i.e. any ω-sequence) can 'play the role' of the numbers,[10] and it does not in the end matter to arithmetic what the objects are that constitute that progression. 'Arithmetic is the science that elaborates the abstract structure that all progressions have in common merely in virtue of being progressions' (p. 291). The role of our different numerals for the different numbers is simply to mark the different positions in such a structure. As he puts it: 'to *be* the number 3 is no more and no less than to be preceded by 2, 1, and possibly 0, and to be followed by 4, 5, and so forth. . . . *Any* object can *play the role* of 3; that is, any object can be the third element in some progression. What is peculiar to 3 is that it defines that role . . . by representing the relation that any third member of a progression bears to the rest of that progression' (p. 291).

Benacerraf is taking it for granted that there are progressions (both of sets and of other things), and the particular case that he takes to be most relevant is the usual progression of *numerals* in whatever language is in question. This is because in practice we learn how to count by learning how to construct this series of numerals. He is happy to accept that *numerals* exist, though he does not wish to say that numbers exist, in the sense of being the objects denoted by numerals. For numerals do not denote objects, though they can be regarded as representing something other than just themselves, namely the various positions in a progression. For example, '$7 + 5 = 12$' should be understood as a short way of saying something like: 'whatever progression you take, the fifth place after the seventh place is the twelfth place'. This does not construe the numeral '7' as an expression that is the *name* of a place, though it does get the overall effect that Benacerraf desires.

The general claim of structuralism is that mathematics is the study of structures, and in this particular case that elementary arithmetic is the study of the structure of a progression (or ω-sequence). The claims of arithmetic

---

[9] The sentence continues: 'which is not to say that there are not at least two prime numbers between 15 and 20'. This will become clearer in what follows.

[10] Benacerraf adds that the progression must be recursively generated. I shall take this qualification for granted in what follows. Later, in his (1996), he retracted the qualification, but I think it better to retain it.

are therefore claims about what is true of any progression. To prevent such claims being vacuous, it must be assumed that *there are* progressions, for if in fact there are none then it will be equally true that 'in any progression, $7 + 5 = 100$', which is surely not what is wanted. Benacerraf himself saw no problem over this assumption, but others have done. This is because structuralism does not assume the existence of numbers as objects, and so appeals to the nominalist who wishes to deny the existence of *all* abstract objects. But in that case the existence of progressions must become problematic. For there is no good reason for supposing that there are infinitely many concrete objects, or that – even if there are – there is any relation which arranges (some of) them into a progression. At this point some structuralists (e.g. Hellman, 1989) will reply that all that is needed is the *possibility* of a progression, and that that is not controversial.

I pause to put this in more detail. In the present case we do have an explicit definition of the relevant structure, namely a progression. The crucial idea is in Frege (1879) and was spelt out in more detail by Dedekind (1888). So where S is any two-place relation we can say just what is required if S is to be a successor-relation that generates a progression. First we confine all quantifiers to the field of S, and define 0 as the sole entity in that field which is not the successor of anything in it. Then we add axioms stating that 0 exists and is unique, that the relation S is one-one, and that everything in its field has a successor which is in its field. Finally, and most importantly, we add the induction axiom, formulated with a second-level quantifier, so that (all and) *only* genuine progressions will satisfy those axioms. Let us abbreviate all these conditions to '$Prog(S)$'. Next, where 'A' is any ordinary statement of arithmetic, we have seen that it can be viewed as abbreviating a statement with the relation 'S' as its only non-logical constant. Let this be '$A(S)$'. Then Benacerraf's idea is that any statement 'A' of arithmetic should be understood as abbreviating

$$\forall S(Prog(S) \rightarrow A(S))$$

Here 'S' is treated as a variable, ranging impartially over all two-place relations.[11] In this way all statements of arithmetic are turned into generalized conditional statements, an approach sometimes called 'if-then-ism', which

---

[11] To avoid what would be a premature invocation of set theory, we may here take the notion of a two-place relation as primitive, and not as a set of ordered pairs which themselves are sets of a further kind.

is intended to retain the truth-values of these statements while removing the apparent commitment to such abstract objects as numbers are often supposed to be. But it does not give the right results unless it is assumed that *there are* some progressions.[12]

Unlike Benacerraf, Hellman (1989) aims to avoid this assumption, and so adds a necessity-operator to the generalized conditional:[13]

$$\Box \forall S(Prog(S) \to A(S))$$

Given this revision, the only assumption needed is that it is possible for there to be a progression, i.e.

$$\Diamond \exists S(Prog(S))$$

But what entitles us to make even that assumption? I think that everyone believes it. Even if one rejects all abstract objects, one is likely to hold that there *could* be an infinity of concrete objects, and a relation S that orders (some of) them into a progression. (For example, there *could* have been infinitely many stars, as many people believed before Einstein adopted the idea of a non-Euclidean space.[14] If so, then there *could* have been a progression generated by taking 'S' as 'is the next nearest star to here'.) But it seems odd to take this as a presupposition of arithmetic. Is it not more natural to suppose that what arithmetic presupposes is just the progression which the structuralist wishes to deny, namely the progression of the numbers themselves?

The case with other varieties of numbers is similar. For example, one can define the structure of the real numbers in various ways, and this definition too is categorical – i.e. it does define a unique structure – provided that it is formulated within a second-level logic. For brevity, let us simply call the structure in question 'a continuum'. Then propositions apparently about the real numbers can be reformulated as propositions about all continua. This includes not just propositions that concern only the real

---

[12] Note incidentally that the 'if-then' interpretation is available only if the structure in question can be characterized by a finite set of axioms, but this is not a problem so long as second-level quantifiers are available.

[13] The idea of exchanging an invocation of modality for an ontological commitment to abstract entities is introduced in Putnam (1967a). But Hellman's version is more straightforward.

[14] As notable examples I mention Democritus, Giordano Bruno and Isaac Newton.

numbers, e.g. that there are more of them that are transcendental than there are that are algebraic, but also comparisons with other numerical structures. For example, given also the definition of '*Prog(S)*', we can say that there are no more algebraic real numbers than there are natural numbers. Once more, to avoid triviality one must assume that a continuum is at least possible. This is obviously no problem if one is already assuming the existence of abstract objects such as sets, but even without this I would expect general agreement on the claim that a continuum is *possible*. After all, it has been universally believed by almost all scientists from Aristotle onwards that space and time are actually continuous. Of course it may turn out that that belief is a mistake, but it is surely *possible*.

The case with set theory is more difficult. When structuralism began, Benacerraf's idea was to apply it to numbers, but not to sets. His argument was that numbers, as ordinarily conceived, have no properties other than those that are given by their position in a structure, whereas any genuine *objects* that exemplify some structure must have other properties too, so that they can be identified independently of that structure. So he inferred that numbers are not objects. Would the same apply to sets? Should we say that sets have no identity other than what is given to them by their relation to other sets, as specified by the relation '∈' that we think of as membership? In this case the initial motivation is surely weaker, for we can say some things about a set which do not involve its relation to all other sets. The most obvious is just that a set is supposed to be a *collection* of those things that we call its members. This is not to be taken in too naive a way, as requiring the members to be somehow 'gathered together' before they form a set, but still it is important to the modern conception that a set *depends* upon its members, and not conversely. That is why sets count as the same if and only if their members are the same, and it is why the theory which is based on the so-called 'iterative conception' of sets has been generally adopted. It is no doubt a metaphor to say that a set cannot exist until '*after*' all its members do, but still the metaphor represents a very basic thought about sets, and this militates against the idea that any relation whatever will do in place of '∈', so long as it obeys the same axioms. Nevertheless it is the idea that Hellman supports.

As it happens, we do know of one way of interpreting the usual set theory which does not involve what we naturally regard as sets, namely by taking each so-called 'set' to be an ordinal number (Takeuti, 1987). On this approach ordinal numbers are taken to be objects which exist in their own right, and which can be used to provide a model for set theory, even if it is not the intended model. If mathematics is really about structures, then presumably the mathematician should be just as happy with the one interpretation as with the other, but that seems to me to be improbable. For example, I have mentioned (pp. 149–50) that for some time there has been a search for new axioms which might be added to set theory to improve its power. But one who is engaged on such a search surely would not feel that his new axioms had to be equally compatible with *all* possible ways of interpreting the relation '∈' as that figures in the usual set theory. On the contrary, his hope would be that the new axioms will do something to pin down the relation '∈' so that it can *only* be understood in the intended way, i.e. as a relation between sets. The ambition would be to make some progress on ruling out the unintended models.

It is relevant here that we do not (yet) have in the case of sets, what we do have for natural numbers and for real numbers, namely an explicit definition of the structure intended. There is so far no agreed set of axioms for set theory which is categorical, i.e. which does pick just one structure. This makes it rather difficult to believe that present-day mathematicians are not particularly concerned with sets themselves, but only with the structure that they are supposed to exemplify. For we do not know what structure that is.[15]

It is also relevant that the appropriate assumption of possibility is in this case wholly unpersuasive. According to what is nowadays the *usual* account of possibility, something is possible if and only if there is a 'possible world' in which it is true. That world may be our world (the actual world) or it may be some other. There is a problem here for any philosopher who has nominalist inclinations: should he accept that there are all these other 'worlds', and – even if there are – that we can get to know something about them? No nominalist will be happy with the very literal realism about possible worlds proposed by David Lewis (1986), and will seek for some alternative. All the same, it is a helpful way of thinking about possibilities, so let us apply it here, to Hellman's claim that all that is needed

---

[15]   A comprehensible outline of the situation may be found in Maddy (1990, pp. 125–49). She contrasts two very different approaches. One adds the axiom that all sets are constructible (i.e. $V = L$), and this makes the universe of sets as small as is possible (compatibly with the axioms). The other adds the axiom that there is a supercompact cardinal (i.e. SC), and this makes the universe as large as we now know how to. But should either of them be believed? (I remark that *most* contemporary mathematicians reject the first, and keep an open mind about the second. But that situation may change.)

for mathematical purposes is the *possibility* that our present axioms of set theory should be true. I cannot see any rationale for that. If the sets in question do not exist in this, the actual, world, how could they exist in some other possible world? What comprehensible difference could there be between that world and this, which brings it about that the sets there are different from, or are differently arranged from, the sets here? I see no answer to this question.[16]

I conclude that in the case of set theory the structuralist must maintain outright that *there is in this world* an exemplification of this structure, for otherwise his analysis of set-theoretical claims must collapse. In that case he is of course committed to the view that *there is* an exemplification of the structure of the natural numbers, and of the real numbers, and of other systems too that interest the mathematician. For set theory itself provides examples of such structures. Modality, as introduced by Hellman, does not help. I add that in another way it hinders, namely in our understanding of how mathematics is applied by scientists in their theories of how *this* world works. For one cannot easily see how the exemplification of a certain structure in some other possible world might be in any way relevant to that project. (Hellman's chapter 3 is a discussion of the applications of mathematics, and one sees how roundabout his account has to be.) But there is another structuralist ploy, followed by Resnik (1997) and by Shapiro (1997), which is to insist that the *structures* do exist, whether or not they have any exemplifications in the world familiar to us.

First, I wish to dissociate this thought from another which is held by both Resnik and Shapiro, namely that a number may be *identified* with a position in a structure. The idea would be that the number 3, for example, may be taken to *be* the third position in the structure of a progression. The obvious objection is that this idea could work only when we are dealing with a mathematical structure (such as a progression) which does have some number singled out from all the others. For example, it would not work in the case of the signed integers, where every member has both a successor and a predecessor. So there is no first member, and no last member, and (consequently) no middle member. How could one possibly decide which position in that structure was to be identified with the signed integer +3?

---

[16] Recall that for mathematical purposes all that is needed is the theory of the *pure* sets, i.e. of sets which exist quite independently of any 'individuals' (i.e. non-sets) that there may happen to be.

Such cases are frequent, and one might say that they are the rule. For example, the structure of the positive rationals has no first member, no last member, and (consequently) no middle member. So which position in that structure is to be identified with the rational number $^3/_1$? As Benacerraf pointed out, when discussing the idea that numbers really are sets, it should then be possible to determine just *which* set any given number is. By parity of reasoning, if numbers really are positions in structures, it should be possible to determine just *which* position any given number is. But we cannot do this in either case. It may perhaps be that the only properties which numbers have are their relations to other numbers, and certainly Benacerraf makes that claim. It may also be that the only properties which positions in structures have are their relations to other positions. But this is no ground for saying that a number *is* a position, if the truth is that there is nothing to determine which number is which position.

Even in the case of the natural numbers, where one particular identification does seem natural, the idea leads to some pretty odd ways of talking. The same position in a progression may be occupied by various different items in different exemplifications. For example, in the progression 0, 1, 2, 3, . . . the items are themselves numbers, and the number 2 occupies the third position. But it is certainly odd to report this by saying that here the number 2 'occupies' the number 3, or that the number 3 'is filled by' the number 2. Again, although these authors allow that the same position may be filled by different items in different cases, they think that it will *always* be filled by *itself*. For the position itself is both an example of something that fills a position and is a position. Consequently the structure is *both* a universal with many particular exemplifications *and* a perfect exemplification of itself. (In Platonic terms, it is both a 'form' and an 'intermediate'.) On this view, unexemplified structures turn out to be impossible. These points are of less importance. It is more important that there is no way of making the required identifications, and therefore we should drop the unfortunate idea that a number *is* a position.

What is left is the claim that these structures, which mathematics investigates, exist (as it were) in their own Platonic world, whether or not they happen to be exemplified 'in this world'. So, of course, it gives rise to the same problem as other alleged Platonic entities: how are we supposed to know about such structures, if (as these authors agree) they are abstract entities, which cannot enter into any causal relations, either with our minds or with anything else?

Both Resnik and Shapiro provide discussions of this problem, but it seems to me that they do not really meet the issue.[17] They begin with the idea that human beings are capable of pattern-recognition, at least when the patterns are small enough to be perceived as wholes. Then the thought is that we can extend the idea of such a pattern beyond what can be perceived, first to an arbitrarily large finite pattern, and then to an infinite pattern.[18] But why should we be entitled to assume that *there are* such infinite patterns, in any sense other than that we human beings can conceive them? And do we need to assume that they have any more existence than this? That is a theme of the next two chapters, and I will come back to it once more at the end of chapter 9.

5.   Some comments

There are differences between old-fashioned formalism, which claims that mathematics is the study of formal systems, and its more modern descendant, which claims that mathematics is the study of structures. One is that for the formalist absolutely *any* formal system would qualify as a legitimate object of study, including systems in which the rules of inference bore no relation to the usual meanings of the logical constants. So far as I am aware no formalist ever did become interested in such a system, possibly because it would be very unlikely to have any useful applications. But his overall position would not rule this out, for it assigns no meanings to *any* of the symbols that may be used. By contrast, the structuralist cannot take such a view, for the idea of a 'structure', exemplified by every model for his axioms, already assumes that the logical constants in those axioms do have their intended meaning. This applies not only to the truthfunctors and the first-level quantifiers '$\forall x$', but also to the second-level quantifiers '$\forall F$' if they are used. (And, as I have said, they *must* be used if the axioms are to describe a structure uniquely.) For the structuralist it is only the non-logical symbols, such as 'S' (for 'succeeds') or '$<$' or '$\in$' that are not assigned any definite meaning. In their case any interpretation which yields a model of the axioms is equally good, for any such interpretation

---

[17]   Resnik (1997, chapter 9); Shapiro (1997, chapter 4).

[18]   As is emphasized by Shapiro (1997, chapter 4, section 7) our grasp of an infinite structure has to be mediated by language. We can describe it, even though we cannot visualize it.

will yield a 'structure' of the kind desired. But even this much lack of interpretation still leaves us with a *kind* of formalist position, as Benacerraf's seminal paper avows. Describing his own position, he says: 'One can be this sort of formalist without denying that there is such a thing as arithmetical truth, other than derivability within some given system' (pp. 293–4).

Another difference between formalism and structuralism is related to this. The formalist holds that mathematics is the study of uninterpreted formal systems, and he wishes to prove results *about* these formal systems. He can therefore fairly be asked what methods of argument may be used in this 'metamathematical' reasoning. As we have seen, when Hilbert was faced with this question he came up with a (fairly) definite answer, namely that in reasoning about formal systems the methods that he called 'finitary' could be employed, but no others. However, we now know that what Hilbert was hoping to do with these 'finitary' methods cannot be done, and that (for all interesting systems) the best kind of consistency proof that can be hoped for is a relative proof, i.e. one showing that *if* one system is consistent then so is another. So Hilbert's goal is no longer seriously pursued, but the old question still remains. The old-fashioned formalist holds that his formal systems need have no meaning, so he must presumably be debarred from reasoning *within* those systems when he is aiming for a result that does have meaning, and is indeed true. So what methods of argument are available to him?

This problem does not affect the structuralist, for he does not take his axioms (which define the structure in question) to be lacking in logical meaning. Moreover, he is not debarred from working within one particular exemplification of a structure in order to obtain significant results, for he never intended to deny that structures do have particular exemplifications. And if by working in one structure he reaches a result in another, say in number theory, he will no doubt be able to claim that this result is a truth that applies to all progressions. Moreover, when he aims to reason not *within* a given axiom-system but *about* it (for example, to show that it is or is not categorical), there is again no bar on his using any methods of argument that seem helpful, including those licensed by the axioms themselves. So he is free from this problem faced by his formalist ancestor. But he certainly faces others.

The most important is that Benacerraf's structuralism begins from a false premise, namely that numbers have no properties other than their relations to other numbers. A simple way of bringing out this point is to go back to our primitive tribe (pp. 47–8) which uses only the finite system of numerals

1, 2, 3, 4, 5, 6, 7, 8, 9, 10, Many.

This system obviously contrasts with our own infinite system, and one must presumably accept that the structures are not the same. But does it follow that the numbers are not the same? For example, their numeral '3' may perfectly well function exactly as ours does in the numerical quantifier 'there are 3 . . .', in the ordinal adjective 'the 3rd . . .', in the numerical comparison '3 times as much as . . .', and so on. Is this not enough to justify the claim that their number 3 is the same as ours, even though the 'structures' in question are evidently different? If so, then there must be more to being the number 3 than occupying a certain place in a specified structure. I add that both Resnik and Shapiro do consider the question of whether a number in one structure could be the same as a number in another. They are thinking of such issues as whether the natural number 3 should or should not be identified with the rational number 3, the real number 3, and so on. Resnik simply says 'there is no fact of the matter', and Shapiro similarly says that it is a question for 'decision, based on convenience', rather than a matter of discovery. But both of them are taking for granted that all that matters is the relation that the one number bears to all the others in a given 'structure'. Applications of the number in ordinary affairs, which have nothing to do with its role in a particular 'structure', are simply being ignored.

This illustrates a general point that applies to both formalists and structuralists: for the formalist, formal systems come first, and applications may happen to follow. Similarly, for the structuralist it is structures that come first and applications may happen to follow. The application, if it comes, will be due to the discovery that the whole of the structure, or some significant part of it, applies to some area of physics, or biology, or whatever. It is admitted that there is likely to be some idealization in such a discovery, for the mathematical structures considered will always be infinite structures, whereas the physical systems are very probably finite. But I do not complain about this idealization. I do complain that the whole account is geared to the complex applications that one finds in the advanced sciences, and not to the familiar applications of numbers that are so frequent in everyday life. These have almost nothing to do with the structures of the orderings of the whole series of natural numbers, of rational numbers, of real numbers, and so on. Expressions for particular numbers have a well-established meaning long before there is any thought of the general structures that these numbers fall into.

This leads me on to a final objection, which one might reasonably call the objection 'from history'. How are we to understand the claims either that mathematics *just is* the study of formal systems, or that it *just is* the study of structures? Is it supposed to be a claim about mathematics as it happens to be *today*, or about how it always has been (and always will be)? On the face of it, the claim must be limited to mathematics *today*. Formal systems are a modern invention. In the sense in which we now understand them, they did not exist until the end of the nineteenth century. A defendant might say that people were 'implicitly' working within a formal system for many years before that system was explicitly written down, but one can only understand this as implying an 'implicit' recognition of what we now think of as the central principles. The principle of mathematical induction is now known to be of central importance to the theory of the natural numbers, and yet it was never even formulated before the late sixteenth century, and it took some time for it to become an explicitly acknowledged method of proof.[19] One can hardly suppose that it was even 'implicitly' recognized in the times of ancient Greece, ancient Egypt, or ancient Babylon. But surely those people did make contributions to the theory of natural numbers.

A similar point can be raised against the structuralist, especially if the relevant structures are supposed to be given by suitable axioms. Again, no one knew how to do this until the late nineteenth century. Here one has more sympathy with the idea that structures can be recognized, and made the subject of enquiry, in advance of any axiomatic definition of them. For example, it may plausibly be claimed that the reason why everyone at once recognized as correct Dedekind's (quasi-)axiomatic treatment of the real numbers (1872) and then of the natural numbers (1888) was just that everyone already understood the structure being aimed at. But, even so, this idea of a structure is still a modern invention.

When Euclid proved that there are infinitely many prime numbers, he surely thought of himself as proving something about *the numbers*; he did not think in terms of 'all progressions', and could not have done, since he had no general concept of a progression. Similarly, when Archimedes proved that the volume of a sphere was (what we call) $^4/_3.\pi r^3$, he was surely

[19] I take the information from Kline (1972, p. 272).

thinking of *spheres*, and not of any other entities that happen to form a 'structure' matching that of a sphere (e.g. a certain kind of set of triples of real numbers). It is really no more plausible to say that mathematics has always been concerned with 'structures' than it is to say that it has always been concerned with formal systems. History refutes both of these claims. What present-day structuralists are recommending can only be understood as the view that that is what mathematics is *now*.[20] But (i) we surely cannot suppose that ordinary elementary mathematics as taught in schools today is conceived as a general theory of structures, and (ii) we cannot suppose that mathematicians of past ages, e.g. Euclid or Archimedes, thought of their work in this way. What reason is there to say that the topic has *changed* between ancient days and now, or between today's school and university? Surely it is still *numbers* that are being studied?

That completes my remarks on formalism and on structuralism. Many philosophers now adhere to the latter doctrine, and I think almost none to the former. I do agree that there is *more* truth in structuralism than in formalism, and I will come back to this point at the end of chapter 9. But now it is time to move on to our third main '-ism'.

Suggestions for further reading

Frege's attack on the earlier formalisms of E. Heine and J. Thomae is translated in Frege (1903, pp. 182–233). Those who have plenty of time at their disposal might be interested to look at this, but I would expect most readers to skip it. Essential reading from Hilbert is his (1925), and there is also something of interest in his (1927). For a commentary on Hilbert, and an exposition of Gödel's incompleteness theorems, I recommend Kneebone (1963, chapters 7 and 8). I mention that Detlefsen (1986) attempts to minimize the destructive force of Gödel's theorems on Hilbert's programme, but (in my view) without much success.

Pure formalism is best represented by Curry (1951); I suggest chapters 1–3, 6, 9–12. A readable evaluation both of Frege's attack on Heine and Thomae, and of Curry's subsequent defence, is in sections 1–2 of chapter 2 of Resnik (1980). (The chapter continues with an account of Hilbert's programme and of Gödel's refutation, but I would prefer Kneebone on this subject.)

---

[20] This limitation is explicit in Shapiro (1997, p. 143).

Key reading on structuralism is Benacerraf (1965). I recommend Reck and Price (2000) for a useful review of the different kinds of structuralism. Important advocates are Hellman (1989), Resnik (1997, part III), and Shapiro (1997, part II). There is an assessment of the basic ideas in Parsons (1990). Another author whose position is near to structuralism, as ordinarily construed, is Chihara (2004).

Chapter 7

Intuitionism

One usually thinks of the approach to the philosophy of mathematics that is called 'intuitionism' as beginning with Brouwer. This is perhaps not entirely fair, for a central feature of this approach is its emphasis on 'constructive procedures', and there were others before or contemporary with Brouwer who shared this view. I mention Kronecker (1899), Poincaré (1908), and Weyl (1918). But I shall follow the usual course and begin my discussion with Brouwer. However, I shall deal with his views in a very cursory way, since I can find nothing in them that looks like an *argument* for the approach that he adopts. But I shall then give a rather fuller discussion of what is called 'intuitionist logic', as proposed by Brouwer's pupil Heyting, and provisionally accepted by Brouwer as at least on the right lines. The main burden of my subsequent discussion will then focus on Michael Dummett, whose advocacy of this approach certainly does provide arguments to which a philosopher must pay serious attention.

I add as a final introductory remark that the title 'intuitionism' is entirely misleading. It is used because Brouwer claimed to be continuing the Kantian doctrine of the importance of 'intuition' in mathematics. But this supposed debt to Kant is not to be taken very seriously, even in Brouwer's own writings, and certainly not in later developments of this general approach.

1. Brouwer

There is some common ground between Brouwer and Hilbert. Both were reacting to the paradoxes that affected Cantor's set theory, and both held that the problems had arisen because human beings have no adequate grasp of infinity. But whereas Hilbert wished to rescue Cantor's theory, by proving that it was at least consistent, Brouwer claimed that it should simply be abandoned. In his view, the whole topic was one which was beyond the scope of proper mathematics. Hilbert wished to keep all of traditional mathematics, and Cantor's theory too, though at the price of saying that the ordinary notions of meaning and truth did not apply. Brouwer insisted on genuine meaning and genuine truthvalue, and as a result was led to reject not only Cantor's theory but also many of the usual claims of traditional mathematics. For he thought that it had lost touch with its proper foundations.

For Brouwer, as for the earlier philosophers mentioned in chapter 2, mathematics is about what our minds can construct. On his account it begins with 'an intuition of the bare two-one-ness' that comes when one reflects upon the temporal nature of experience (Brouwer, 1912, p. 80).[1] This basic 'intuition' can then be repeated indefinitely, which yields the theory of the natural numbers, taken to be our own 'mental mathematical constructions'. Moreover, proofs about the natural numbers are again regarded as further 'mental mathematical constructions' of ours, so that the whole of this theory is a human 'construction'. The same applies not only to the natural numbers but also to all other areas of mathematics too, e.g. to whatever can be rescued from the ordinary theory of the real numbers. Brouwer regards as somehow 'wicked' the idea that mathematics can be *applied* to a non-mental subject matter, the physical world, and that it might develop in response to the needs which that application reveals. In his view, proper mathematics concerns only the creations of the human mind.[2] For example, he said (somewhat later): 'Classical analysis, however appropriate it be for technique and science, has less mathematical truth than intuitionistic analysis' (1948, p. 90). To paraphrase: classical mathematics may be useful, but only intuitionistic mathematics is true.

As I have said, Brouwer gives no reasonable argument for these claims, and when no argument is offered then no philosophical response is in place. But these claims do (in Brouwer's view) have certain consequences, which others have pursued without necessarily subscribing to Brouwer's overall position. The best known is the insistence on 'constructive' proofs in mathematics, and hence the rejection of ordinary logic, and in particular its Law of Excluded Middle (LEM). Here is an example of a proof

---

[1]   Brouwer adopts what is a rather tenuous thread in Kant's writing, namely that arithmetic somehow reflects our experience of time. But he rejects the Kantian idea that geometry is connected with our experience of space. (The contrast with Frege may be noted.)

[2]   Brouwer (1905, 1907) as cited in van Stigt (1998, p. 6). This article is a sympathetic but still conveniently brief summary of Brouwer's general views. It abridges themes that are more fully developed in van Stigt (1990).

which relies on LEM, and would clearly be rejected by Brouwer for that reason.[3]

To prove: that there are irrational numbers, x, y such that x^y is rational.

Proof. We know that $\sqrt{2}$ is irrational. If $\left(\sqrt{2}\right)^{\sqrt{2}}$ is rational, then we have our result. But if it is irrational then $\left(\left(\sqrt{2}\right)^{\sqrt{2}}\right)^{\sqrt{2}}$ will yield the result, for this is certainly rational, since

$$\left(\left(\sqrt{2}\right)^{\sqrt{2}}\right)^{\sqrt{2}} = \left(\sqrt{2}\right)^{\left(\sqrt{2}\,\times\,\sqrt{2}\right)} = \left(\sqrt{2}\right)^2 = 2$$

Either way, the result follows.

The proof evidently relies on LEM, since it assumes that $\left(\sqrt{2}\right)^{\sqrt{2}}$ must be either rational or irrational, but it is clearly 'non-constructive', since it gives no way of finding out which of these alternatives is correct.

Brouwer will accept that in ordinary everyday reasoning, when we are speaking of *finite* collections, LEM is acceptable. That is, if the predicate 'F' is a decidable predicate, and the totality involved is finite, then '$Fx \vee \neg Fx$' may properly be assumed for any x, since in this case one can (in principle) run through all the examples and will thereby reach a definite decision one way or the other. But, where the relevant totality is infinite, a decision may be in principle impossible, and that is the situation here. For there are infinitely many rational numbers which might, for all we know, happen to be equal to $\left(\sqrt{2}\right)^{\sqrt{2}}$. We can (in principle) calculate the value of this expression to any desired *finite* number of decimal places, but no such calculation can tell us that the value is irrational. So the question is not (finitely) decidable, and for that reason Brouwer would reject this use of LEM.

Moreover, his own understanding of what counts as 'truth' and as 'falsehood' in mathematics provides a rationale. Though he does quite often speak simply of 'truth' in mathematics, he always equates a true proposition with one that can be proved. As it happens he very seldom speaks simply of 'falsehood' in mathematics, but instead talks always of 'absurdity', and his notion of an absurd proposition is in effect of one that

---

[3]   I take the example from Dummett (1977, p. 10).

can be disproved.[4] He therefore thinks of LEM (in mathematics) as claiming that every mathematical proposition can be either proved or disproved, and it is entirely understandable that he should reject *that* claim. Hilbert had optimistically affirmed that 'all mathematicians' believe it, i.e. they all believe that every mathematical problem is soluble (Hilbert, 1925, p. 200). But I think that on this point most of today's mathematicians would side with Brouwer: it really is too optimistic to suppose that *all* propositions of mathematics can be either proved or disproved.[5]

A third thread in Brouwer's thought is the rather extraordinary claim that mathematics is a mental activity which *uses no language*. I am sure that what he mainly has in mind in this claim is intended as a rejection of Hilbert's views. His thought is that mathematics is *not* a matter of working within any formally specified language, with its vocabulary and rules of inference explicitly laid down (e.g. Brouwer, 1912, pp. 78–9). We now know that there is truth in this. But Brouwer's own pronouncements do not limit the claim in this way, and he apparently accepts with equanimity the strange consequence that genuine (i.e. intuitionist) mathematicians cannot communicate to one another their results and their proofs.[6] I regard this as simply absurd, and obviously falsified by the well-established fact that even intuitionist mathematicians very evidently do communicate with one another. Moreover, they *agree* with one another on what counts as a proof, apparently because their private 'mental constructions' do always match, though on Brouwer's principles one cannot see how this inter-subjectivity is to be explained. This is an aspect of his thought which is quite prominent in the early writings, though it became more muted as time went on, and in my own opinion it is best ignored altogether. So I shall simply set it aside.

There are two other aspects that one cannot reasonably set aside. One is the claim that numbers – and indeed *all* the things that mathematics is concerned with – are our own 'mental mathematical constructions'. We invent them, and they have no kind of existence outside our own minds. The other is the claim that classical logic does not apply in mathematics, and

---

[4] As the next section will show, this is not quite how 'absurdity' figures in the usual treatment of intuitionist logic.

[5] We nowadays have many 'unprovability' results of the general form: *within such-and-such a specified formal system* this proposition cannot be either proved or disproved. But, as we have seen, it does not follow from this that the proposition cannot be proved or disproved *at all*.

[6] Brouwer (1905, p. 37), cited in van Stigt (1998, pp. 6, 9).

the idea behind this is that in mathematics 'true' means (something like) 'provable', and similarly 'false' means (something like) 'disprovable'. As I shall argue in section 3 of this chapter, there is actually no close relationship between these two claims. But in the next section I consider just intuitionist logic, and how that reflects the idea that truth (in mathematics) *is* provability.

2. Intuitionist logic

What is now the accepted system of intuitionist logic was proposed by Brouwer's pupil Heyting in his (1930). He there put forward a set of axioms for the propositional connectives, for use with the usual rule of detachment as sole rule of inference. I do not reproduce precisely his axioms, for they are less convenient than some slightly altered versions that have emerged since. The version that I do give is due to Kleene (1952, pp. 82 and 101). It will be seen that the first two axioms concern → alone, the next three → and ∧, the next three → and ∨, and the final two → and ¬. The implication sign → occurs in all of them, because that is what the sole rule of inference requires.

Axiom-schemas (for any formulae in place of φ, ψ, χ)[7]
1. $\varphi \rightarrow (\psi \rightarrow \varphi)$
2. $(\varphi \rightarrow (\psi \rightarrow \chi)) \rightarrow ((\varphi \rightarrow \psi) \rightarrow (\varphi \rightarrow \chi))$
3. $\varphi \wedge \psi \rightarrow \varphi$
4. $\varphi \wedge \psi \rightarrow \psi$
5. $\varphi \rightarrow (\psi \rightarrow \varphi \wedge \psi)$
6. $\varphi \rightarrow \varphi \vee \psi$
7. $\psi \rightarrow \varphi \vee \psi$
8. $(\varphi \rightarrow \chi) \rightarrow ((\psi \rightarrow \chi) \rightarrow (\varphi \vee \psi \rightarrow \chi))$
9. $(\varphi \rightarrow \psi) \rightarrow ((\varphi \rightarrow \neg \psi) \rightarrow \neg \varphi)$
10. $\varphi \rightarrow (\neg \varphi \rightarrow \psi)$

Rule of Detachment: if both φ and $\varphi \rightarrow \psi$ are theorems, so is ψ.

---

[7] I remark that, to emphasize the symmetry between axioms 5 and 8, axiom 5 may be rewritten without loss as

5. $(\chi \rightarrow \varphi) \rightarrow ((\chi \rightarrow \psi) \rightarrow (\chi \rightarrow \varphi \wedge \psi))$

Although Heyting (1930) is silent on axioms for the quantifiers, it is convenient to add them here. They are:[8]

11　　$\forall\xi(\varphi) \rightarrow \varphi(\xi/t)$
12　　$(\psi \rightarrow \varphi) \rightarrow (\psi \rightarrow \forall\xi(\varphi(t/\xi)))$, provided that t is not in ψ.
13　　$\varphi(\xi/t) \rightarrow \exists\xi(\varphi)$
14　　$(\varphi \rightarrow \psi) \rightarrow (\exists\xi(\varphi(t/\xi)) \rightarrow \psi)$, provided that t is not in ψ.

Readers who know some classical logic should find all these axioms familiar, for they are all classically correct.

I at once turn to a different way of formulating the system, namely as a system for natural deduction, as I imagine that this will be more familiar to many readers. The transformation from either presentation to the other is entirely obvious in the case of axioms 3–10 and 11–14. It is less obvious in the case of axioms 1–2, because they are the axioms used in proving 'the deduction theorem', which is what licenses a system of 'natural deduction' in the first place.

Natural Deduction Rules (for Γ,Δ,Θ any sets of formulae)
1*　　　If $\Gamma,\varphi \vdash \psi$ then $\Gamma \vdash \varphi \rightarrow \psi$
2*　　　If $\Gamma \vdash \varphi$ and $\Delta \vdash \varphi \rightarrow \psi$ then $\Gamma,\Delta \vdash \psi$
3*,4*　　If $\Gamma \vdash \varphi \wedge \psi$ then $\Gamma \vdash \varphi$ and $\Gamma \vdash \psi$
5*　　　If $\Gamma \vdash \varphi$ and $\Delta \vdash \psi$ then $\Gamma,\Delta \vdash \varphi \wedge \psi$
6*,7*　　If $\Gamma \vdash \varphi$ or $\Gamma \vdash \psi$ then $\Gamma \vdash \varphi \vee \psi$
8*　　　If $\Gamma \vdash \varphi \vee \psi$ and $\Delta,\varphi \vdash \chi$ and $\Theta,\psi \vdash \chi$ then $\Gamma,\Delta,\Theta \vdash \chi$
9*　　　If $\Gamma,\varphi \vdash \psi$ and $\Delta,\varphi \vdash \neg\psi$ then $\Gamma,\Delta \vdash \neg\varphi$
10*　　If $\Gamma \vdash \varphi$ and $\Delta \vdash \neg\varphi$ then $\Gamma,\Delta \vdash \psi$
11*　　If $\Gamma \vdash \forall\xi(\varphi)$ then $\Gamma \vdash \varphi(\xi/t)$
12*　　If $\Gamma \vdash \varphi$ then $\Gamma \vdash \forall\xi(\varphi(t/\xi))$, provided that t is not in Γ
13*　　If $\Gamma \vdash \varphi(\xi/t)$ then $\Gamma \vdash \exists\xi(\varphi)$
14*　　If $\Gamma,\varphi \vdash \psi$ then $\Gamma,\exists\xi(\varphi(t/\xi)) \vdash \psi$, provided that t is not in Γ or in ψ

The rules are formulated here in such a way that they may be interpreted as rules for constructing tree proofs, or for constructing linear proofs which

---

[8]　Here 'ξ' represents any variable, and 't' any term. I use '$\varphi(\alpha/\beta)$' to represent the result of substituting 'β' for all free occurrences of 'α' in φ.

keep note of the assumptions on which each line rests. The reader may think of them in whichever way is the more familiar.[9]

As I have said, every axiom or rule presented here is correct from the point of view of the classical logician. Hence whatever can be proved in intuitionist logic can also be proved in classical logic. The converse is not true, for from the classical viewpoint the axioms and rules here given are not complete, i.e. they do not allow the proof of every thesis that is classically correct. In the formulations given here the difference appears only in the axioms or rules for negation, for in a classical logic these would be stronger.[10] The best known divergence is that in intuitionist logic one cannot prove the law of excluded middle, $\varphi \vee \neg\varphi$, and it follows that one cannot prove various other theses that would imply this. Here is some indication of theses involving negation which are all classically correct but of which only some are accepted in intuitionist logic, while some are rejected.

Rejected	Accepted
$\vdash \varphi \vee \neg\varphi$	$\vdash \neg\neg(\varphi \vee \neg\varphi)$
$\neg\neg\varphi \vdash \varphi$	$\varphi \vdash \neg\neg\varphi$
$(\neg\varphi \to \varphi) \vdash \varphi$	$(\varphi \to \neg\varphi) \vdash \neg\varphi$
$(\neg\varphi \to \psi) \vdash (\neg\psi \to \varphi)$	$(\varphi \to \neg\psi) \vdash (\psi \to \neg\varphi)$
$\neg\forall x(\varphi x) \vdash \exists x\neg(\varphi x)$	$\exists x\neg(\varphi x) \vdash \neg\forall x(\varphi x)$

But I add that, at least in the *usual* formulations of classical logic (e.g. as given in n. 10), there are theorems which do not themselves contain the negation sign though they can only be proved by using the classical rules for negation. Some well-known examples are:

$(P \to Q) \to P \vdash P$

$\vdash (P \to Q) \vee (Q \to P)$

$\forall x(P \vee Fx) \vdash P \vee \forall x Fx$

---

[9] In my (1997, part II), I have compared various different ways of presenting logical systems. I remark here that the method of semantic tableaux cannot be applied to intuitionist logic, but it can be presented as a Gentzen cut-free sequent calculus, and this is particularly rewarding from a proof-theoretic point of view. See e.g. Scott (1981, pp. 176–81).

[10] As I show in my (1997, section 5.4), there are many different ways of introducing negation, all equally correct from a classical point of view. The method adopted in Kleene (1952) is to replace the axiom 10 given above by $\neg\neg\varphi \to \varphi$.

In consequence, these are *not* theorems of intuitionist logic, even though the axioms or rules for →,∨,∀ are in both cases the same. This makes it clear that sharing of logical rules does not always betoken sameness of meaning for the logical signs involved. In fact, when one looks at how intuitionists usually explain what they mean by the propositional connectives, it soon becomes obvious that in *no* case is the intuitionist meaning the same as the classical meaning, even though it is usual to use the same symbols in each case.

From the point of view of classical logic, the notion of truth is so understood that truth and falsehood exist quite independently of our ability to recognize them. Thus '$P \vee \neg P$' will always be true, even though we cannot say whether it is 'P' or '$\neg P$' that is true, and perhaps we never will be able to say. But, as I have said, intuitionists cannot accept this notion of truth, or anyway not in mathematics. In their view the only notion of truth that makes sense in mathematics is one that equates truth with provability. Consequently the meaning of their logical constants is given in terms of proof, and what they think of as crucial is what would count as *proof* of $P \wedge Q$, $P \vee Q$, $P \rightarrow Q$, and so on. Let us temporarily use '□' to mean something like 'there is a proof of'. (I shall look into this more closely in a moment.) Then the standard intuitionist explanations are:

$\Box(\varphi\wedge\psi)$	iff	$\Box\varphi$ and $\Box\psi$	$\Box(\varphi\rightarrow\psi)$	iff	$\Box$(If $\Box\varphi$ then $\Box\psi$)
$\Box(\varphi\vee\psi)$	iff	$\Box\varphi$ or $\Box\psi$	$\Box\neg\varphi$	iff	$\Box$ not $\Box\varphi$
$\Box\exists x\varphi x$	iff	for some x, $\Box\varphi x$	$\Box\forall x\varphi x$	iff	$\Box$ for all x, $\Box\varphi x$

There are some variations on these patterns of explanation, but they will not affect my subsequent discussion.[11,12] Now let us consider how this symbol '□' should be understood in the present context.

---

[11]   Negation is sometimes explained thus: '$\neg\varphi$' abbreviates '$\varphi\rightarrow\bot$', where '$\bot$' is a sign for what is sometimes called 'the constant false proposition', say '$0=1$'. Then we have

$$\Box\neg\varphi \quad \text{iff} \quad \Box(\varphi\rightarrow\bot), \text{ i.e. iff} \quad \Box(\text{if } \Box\varphi \text{ then } \Box\bot)$$

Assuming that '$\Box\bot$' must in all circumstances be rejected, we reach a result equivalent to that given above.

[12]   Let $\mathcal{D}$ represent whatever domain is in question. Then the explanations of the quantifiers may be given more fully as

$\Box\exists x\varphi x$	iff	for some x, $\Box(x\in \mathcal{D}$ and $\varphi x)$
$\Box\forall x\varphi x$	iff	$\Box$ for all x, if $\Box x\in \mathcal{D}$ then $\Box\varphi x$

The usual explanations of intuitionist logic use English words where I have written 'Ⅱ', and those words tend to differ from one author to another. For example, Dummett's version of the explanations of ∧ and ∨ (in his 1977, p. 12) is:

A proof of A∧B is anything that is a proof of A and of B.
A proof of A∨B is anything that is a proof either of A or of B.

By contrast, the version given in Scott (1981, p. 164), is:

We have grounds for asserting φ∧ψ just in case we have grounds for asserting φ and grounds for asserting ψ
We have grounds for asserting φ∨ψ just in case we have grounds for asserting φ or grounds for asserting ψ.

A first reaction might be that although my Ⅱ is differently worded in each case, nevertheless the two explanations come to essentially the same thing. What might seem more important is that when the Ⅱ occurs twice in the explanation, as in the account of →, the first occurrence is usually rendered differently from the other(s). Thus Dummett's version of → is:

A proof of A→B is a construction of which we can recognize that, applied to any proof of A, it yields a proof of B.

And the Scott version is:

We have grounds for asserting φ→ψ just in case we have a general procedure by which grounds for asserting φ can be transformed into grounds for asserting ψ.

(This latter fails to add, as presumably it should, that we *recognize* that the procedure in question does perform the desired transformation.) But although the wording has changed between the first occurrence of Ⅱ and the others, it still seems correct to say that the thought is the same, for a 'construction' or a 'procedure' which we can recognize as doing so-and-so *is* a proof that so-and-so, and conversely any such proof will be recognized by the authors as a 'construction' or 'procedure'. I regard these variations as of no importance, and shall pay them no more attention. But there is quite an important difference between the Dummett version and the Scott version which I do wish to pay attention to. One perhaps does not notice it at first sight, but it is there. The Dummett version speaks of what a proof *is*, and this is presumably to be understood as applying whether or not *we have* any such proof. By contrast, the Scott version is concerned only with what it is for us *to have* a proof. Which notion is really more central to intuitionist thought?[13]

In a well-known passage Heyting once said:

> A mathematical theorem expresses a purely empirical fact, namely the success of a certain construction. '2 + 2 = 3 + 1' must be read as an abbreviation for the statement: 'I have effected the mental constructions indicated by "2 + 2" and by "3 + 1" and I have found that they lead to the same result'. (Heyting, 1956, p. 8)

As everyone has commented, this cannot possibly be right. The claim '2 + 2 = 3 + 1' means the same whoever makes it, and it is absurd to suppose that anyone who makes it is saying something about what *Heyting* has done. Nor is it really any better to suppose that he is saying something about what he himself has done, for it is obvious that one may advance a mathematical claim – like any other claim – relying on someone else's authority, and not to report one's own activities. If, then, there is any concealed reference to people in a mathematical claim such as this, it can only be a reference to people in general, and not specifically either to Heyting or to the speaker. So let us replace Heyting's thesis by something along these lines: '2 + 2 = 3 + 1' is short for 'it has been proved that 2 + 2 = 3 + 1', where it is not stated who found the proof, or when, but it is still an empirical fact that is being reported: such a proof *has been* discovered.

Even on this revised proposal a mathematical statement has a tense, and consequently its truthvalue may change over time, for it will fail to be true when we do not yet have a proof, but will become true when later we discover one. This seems to most people highly unintuitive, but I shall not rely on this as an argument, for what is called Brouwer's theory of 'the

---

[13] Some other explanations incline more to what I am calling the Dummett version, and some to what I am calling the Scott version. For example, Kleene's version (1952, p. 51) is: 'An implication "if *A* then *B*" expresses that *B* follows from *A* by intuitionistic reasoning, or more explicitly that one possesses a method which, from any proof of *A*, would procure a proof of *B*'. The first clause here is in harmony with Dummett (i.e. '*B* follows from *A*, whether or not we know this'), and the second is in harmony with Scott ('one *possesses* a method'). I add that what I am calling the Scott version was not in fact written by Dana Scott, nor even approved by him. It was written by D.R. Isaacson, and by the time that this part of the work was put together the editorship had in fact passed from Dana Scott to me. But I call it the 'Scott' version because it appears in the list of references as 'Scott *et al.*'.

creative subject' does apparently propose that mathematical statements should be construed as tensed in just this kind of way. Instead, I turn to a different line of argument.

According to the Scott version of disjunction, we have grounds for asserting a disjunction if and only if we have grounds for asserting one or other of its disjuncts. But this is not how intuitionists in practice behave. For example, where '*F*' is a decidable predicate, e.g. '. . . is prime', they will be ready to assert

$Fx \lor \neg Fx$

for *any* specific numeral in place of '*x*'. They take themselves to have grounds for asserting this even in cases where the number *x* is taken to be so large that we do not in practice know which of *Fx* or ¬*Fx* obtains, and their ground is that we do have a method, a procedure, which *would* decide the matter if it *were* applied (e.g. the so-called 'sieve of Eratosthenes'). This indeed is – roughly – what it means to say that *F* is (known to be) decid-*able*. They are not concerned about the fact that the procedure has not actually been applied to the number in question. To put this more strongly, they are not concerned about whether it ever will be applied, or even whether the laws of physics will allow that it *could* be applied. (Human brainpower is evidently limited, and a similar limitation must presumably apply also to computer-power; for even computers need energy, and the stock of available energy is presumably finite.) It is held to be enough that we do have a method which 'in principle' would decide the question, where this 'in principle' transcends from all limitations that may be imposed by non-mathematical factors. This suggests that the Scott version of my '□' might be altered to 'we now have a proof, or at least a method which, if it were applied, would "in principle" yield a proof'. But this alteration still retains the reference to our *present* condition, and for that reason it still will not do.

It is a generally accepted principle in philosophical logic that the very same proposition may both be asserted and occur in other contexts where it is not asserted. The standard argument is due to P.T. Geach (1965), and is extremely simple. Consider a case of Modus Ponens, where the first premise is '*P*' and the second '*P*→*Q*'. In the first premise '*P*' is asserted, but in the second it is not. Yet it must *mean* the same in both premises, for if not the argument is guilty of the fallacy of equivocation and so cannot be accepted as valid. We may apply exactly this argument to intuitionist logic, which of course accepts Modus Ponens. When an intuitionist asserts

$Fx \lor \neg Fx$

then *what* he asserts must be the same as what occurs as antecedent in a conditional such as

$(Fx \lor \neg Fx) \rightarrow Gx$

But it is quite clear that in the *latter* context '$Fx \lor \neg Fx$' does not mean anything like 'There now exists either a proof that Fx or a proof that $\neg Fx$, or anyway a method which would yield a proof of one or the other if it were applied'. For one is entitled to assert the conditional only if one is entitled to assert that *any* proof of its antecedent – whether or not we now happen to know of it – could be modified into a proof of its consequent.[14]

To put the point succinctly, when $\square$ is embedded, as it is in

$\square$(If $\square P$, then $\square Q$),

then we cannot take the embedded clause $\square P$ as speaking just of proofs, or methods of proof, that *we now know*. We get quite the wrong results if we do. For example, the intuitionist conditional cannot be construed as true simply on the ground that we do not *now* have either a proof of its antecedent or a method that would yield such a proof. If it were so construed, then it could be true today but false tomorrow, when a proof of the antecedent is discovered, and that is not what is intended.

I conclude in this way. When the $\square$ that occurs in these explanations is embedded in some longer sentence, as it is in 'If $\square P$, then . . .', then it must be interpreted without reference to persons or to times, as simply meaning (atemporally) 'If *there is* a proof that P, then . . .'. Since this interpretation is required for embedded occurrences, it is evidently simplest to suppose that $\square$ is to receive the same interpretation everywhere, even when it begins a statement that is being asserted. The assertion is still to be interpreted as claiming that *there is* a proof, and not that I or we now have such a proof, or at least a method which would yield a proof if it were applied,

---

[14] Note that Heyting explicitly denies Geach's point, when he says (1931, p. 60): If we symbolize the proposition 'the proposition p is provable' by '$+p$', then '$+$' is a logical function, viz. 'provability'. The assertions '$\vdash p$' and '$\vdash +p$' have exactly the same meaning. For if p is proved, the provability of p is also proved, and if $+p$ is proved, then the intention of a proof of p has been fulfilled, i.e. p has been proved. Nevertheless, the propositions p and $+p$ are not identical, as can best be made clear by an example. . . .

and so on. These are *grounds* for the assertion; but they are not part of the meaning of what is asserted. For, quite generally, when one asserts that *P*, one is not also asserting either what grounds one has for believing it, or even that one has grounds. By a familiar Gricean implicature, if one does assert that *P* then one will (in normal circumstances) be understood as *also* claiming to have grounds for asserting it, but this latter is not part of what one asserts. What one asserts is just that *P*, as is shown by the fact that *what* one asserts may occur in other contexts too, when no one is asserting it. Consequently, where *P* is a mathematical truth (e.g. '2 + 2 = 3 + 1'), if *what* one asserts means anything about proof at all, it can only be construed as meaning that *there is* – impersonally and atemporally – a proof that *P*. So the □ that I have used in my schematic representation of the standard intuitionist explanation of what the intuitionist logical constants mean should be understood in this way throughout. It must be taken to mean '*there is* a proof', and not anything like '*I have* a proof'. It is, apparently, an unbounded existential quantification over an infinite domain. But is this acceptable from an intuitionist point of view?

Dummett (1977) has suggested that we do not have to treat this quantification as unbounded. He accepts that there is indeed no limit to the kind of considerations which may occur in what is ordinarily counted as a mathematical proof (say in a mathematical journal), but he proposes that we call these ordinary proofs 'demonstrations', and distinguish them from what may be called '*canonical* proofs'. His thought is that a demonstration will be something that gives us good reason to suppose that 'there is' a canonical proof, even though it is not a canonical proof itself. Then he hopes to impose a suitable hierarchy on canonical proofs, whereby

> The complexity of the [canonical] proof matches the complexity of the statement to be proved; that is, for any given statement, if it can be proved at all, then it can be proved by a proof whose complexity does not exceed a bound depending on the structure of the statement. (p. 394)

The thought is that we can certainly impose a bound on what is to count as a canonical proof of an atomic statement, for if such a statement can be proved at all then it *can* be proved by an elementary computation – i.e. can 'in principle', for the computation may in practice be too long and complex for anyone ever to be able to carry it out – so we can set a simple bound on what is to count as a canonical proof of these simple statements. Then the idea is to extend this step by step to more complex statements, including those which involve quantifiers, still retaining the idea that if there is a proof at all then there is one which stays within a suitable bound. He admits, of course, that he cannot himself provide a specification of what these bounds are.

It seems to me that this proposal is hopelessly over-optimistic.[15] Even if we stay within a specified formal system, e.g. the usual intuitionist system for elementary arithmetic, still there is no bound on what is to count, within that system, as a 'canonical' proof of a quantified statement. The point is yet more obvious when we note that, taught by Gödel, intuitionists nowadays do not wish to confine the notion of a proof to what can be achieved *within* any formal system specified in advance. For Gödel's argument shows that we can always prove more by stepping outside any specified system, and arguing about what can be proved in it, and intuitionists do not wish to reject such proofs. Or again, there are proofs of theses which can be stated in the vocabulary of elementary number theory but which make use of the resources of quite different theories, a recent example being the proof of what is always called 'Fermat's last theorem'. This thesis is very simple to state, with all quantifiers universal:

$$\forall xyzw \, \neg \, (x^w + y^w = z^w \, \wedge \, w{>}2)$$

But there is no reason to suppose that our present proof indicates the existence of a suitable 'canonical' proof of matching simplicity. Dummett explicitly says that proofs which appeal to other areas of mathematics (as this one does) are not to count as themselves canonical:

> If . . . we had to envisage the possibility of a proof invoking notions from, say, set theory or complex analysis, this [i.e. having a conception of the form which any proof of our elementary thesis might assume] would be an impossible task. We should be reduced to confessing that we could see nothing in common to all possible such proofs save their having [the thesis in question] as their last line. (pp. 395–6)

But, as I have said, we can place *no* bound on what mathematicians do or will recognize as a proof, even of the most simple statements, and it is quixotic to suppose that absolutely *any* such proof will show that there also exists a suitably simple and 'canonical' proof.

---

[15]   Dummett retains the proposal in his (1991, pp. 177–9). Prawitz (2005) contains a similar proposal.

Dummett himself ends by partially admitting this. At least he admits that our progress in the discovery of new kinds of proof may lead us to change our view on the assertibility of 'Any proof that *P* can be transformed into a proof that *Q*' (pp. 401–2), and he infers that 'the meanings of our mathematical statements are always, to some degree, subject to fluctuation' (pp. 402–3). But this is simply to retreat from '*there is* a proof' to a version of '*we have* a proof', namely '*we have* at least a conception (which may improve, as time goes on) of what any proof would have to be'. But it is clear from everything that he says elsewhere that he does not really wish to accept that mathematical truths change over time, and to achieve this one must interpret my □ as saying, atemporally, just 'there is a proof' and not either 'we *now* have a proof' or even 'there is something which we *now* can conceive of as a proof'. An entirely unbounded quantification over proofs is required. Can the intuitionist accept such a quantification?

Well, why not? Someone may object that this kind of quantification must presuppose that proofs 'exist' in a Platonic kind of way, quite independently of human constructions, and that that is anathema to the intuitionist. I do not see why it would be supposed to have this presupposition, and even if it did why should this matter? After all, it is clear that proofs are not *literally* mental constructions, i.e. constructions that people either have in fact performed or will perform, just as the natural numbers themselves are not *literally* such constructions. A simple consideration here is that both the natural numbers and proofs concerning them are infinite in number, whereas the number of constructions that have been or will be performed is surely finite. So, if we wish to stick to the emphasis on mental constructions, it must be in some sense *possible* constructions that are relevant (i.e. those that are possible 'in principle', where we take no account of limitations that may be imposed by the laws of physics, or by any other non-mathematical source). If this move is legitimate in the case of the numbers, then it should be equally legitimate in the case of proofs concerning them.[16]

So the question is just whether the quantifiers implicit in the informal statements 'there is a proof that *P*' or 'Any proof that *P* can be transformed into a proof that *Q*' can themselves be interpreted intuitionistically. And I see no ground for saying that the answer is 'no'. For example, I see no evidence for claiming that intuitionists *do* commit themselves to quantified claims about proofs which would be correct only on a classical interpretation of the quantifiers, such as: for any *P*

---

[16] I say more of this in the following section.

Either every proof that P can be transformed into a proof that Q, or there is one that cannot.

So far as I can see, they do not make such claims, and there is no reason why they should.

It is very natural to (almost) all of us to suppose that when intuitionists explain their logical constants by translating them into English, the English that they use is to be understood in accordance with the ordinary *classical* understanding of these words. Thus the explanations of the intuitionist $\wedge$, $\vee$, $\rightarrow$, $\neg$ make use of the English 'and, or, if, not', and we tend to take these English words classically. I myself once tried to mount an argument on this basis, purporting to show that if the intuitionist can understand his own explanations then he *must* be able to understand the corresponding classical notions, though he usually claims that they make no sense. I was eventually persuaded that this argument was mistaken. The English words used *can* be interpreted intuitionistically, and do not *have* to be interpreted classically. The same applies when the intuitionist explains his $\exists$ and $\forall$ in a way which uses the English words 'some' and 'all'.[17] I am now suggesting that the same applies too to the quantification originally concealed in my $\square$. This does conceal a quantification, namely 'there is a proof', and it is an unbounded quantification over an infinite domain. But I see no reason to say that it must be construed classically, and cannot be construed intuitionistically.

I finish this section with two footnotes. (i) It is not the least bit surprising that the intuitionist explanations should themselves be understood intuitionistically. After all, the corresponding classical explanations, which also use such English words as 'and, or, if, not, some, all', are clearly to be understood by giving the classical interpretation to these words. Yet in both cases the explanations are still helpful, for a circularity of this kind – roughly speaking, between language and metalanguage – is not in practice a bar to understanding. We are naturally inclined to adopt a classical understanding of the metalanguage because classical logic is at least a better approximation to our normal ways of thinking than intuitionist logic is. But I take it that the intuitionist does not wish to deny *that*.

[17] Dummett (1991, pp. 27–30, 56–9) takes it to be *obvious* that when intuitionist logical constants such as $\wedge$, $\neg$, $\forall$ are explained by using the English words 'and', 'not', 'all' these English words are to be given an intuitionistic meaning.

(ii) The domain of 'all proofs' is certainly not as well understood or as nicely circumscribed as is, say, the domain of all natural numbers.[18] I take it that, from the intuitionist point of view, we might describe a proof somewhat like this: it is something which, *if* it were possible to present the thoughts involved in a nice short form, comprehensible by (suitably educated) human beings, they would recognize as a proof. That is, the idea of being *recognized* as a proof is what is really fundamental. But presumably, even from the intuitionist point of view, there may *be* proofs which cannot be presented in such a nice short comprehensible form. Using once more the phrase which one simply cannot avoid when discussing intuitionist thinking, these are proofs because they are 'in principle' recognizable as proofs, even though in practice no one could so recognize them. Of course this description leaves the domain very open-ended, but the same is true of many other domains that we quantify over in our ordinary ways of thinking and talking. In any case, even if it must be admitted that the domain is not very well circumscribed, still I do not see this as preventing the intuitionist (or even the classicist) from quantifying over it.[19]

Appendix: Some further facts about intuitionist logic[20]
There are two well-known ways of 'translating' between intuitionist logic and classical logic. The first uses double-negation. It is quite easy to show that any formula φ is provable in classical logic if and only if its double-negation $\neg\neg\varphi$ is provable in intuitionist logic. There is also a more complicated translation, in which double-negations are sprinkled *into* the formulae where they are needed, which allows us to generalize this result: let Γ^* be the result of adding suitable double-negations to the formulae in Γ, and φ^* be the result of so modifying φ. Then we have

$\Gamma \models \varphi$ is classically correct if and only if
$\Gamma^* \models \varphi^*$ is intuitionistically correct.

---

[18] Putnam (1980, pp. 479–80) has observed that the intuitionist notion of a constructive proof is 'impredicative', e.g. because $P \rightarrow Q$ is proved by a proof of Q from P, and this proof may be of any complexity, even though P and Q are themselves very simple.
[19] There are useful discussions of what is to count as a 'proof', when that notion is not confined to any particular formal system, in Myhill (1960) and Wagner (1987).
[20] I here give only an outline of the relevant facts. For more detailed statements, and for proofs, a convenient source is Scott (1981, chapters 11–12). But if this is difficult to obtain I suggest McCarty (2005, part III, sections 1–2, i.e. pp. 369–75).

I note that this gives us a relative consistency proof: if intuitionist logic is consistent then so is classical logic. For a proof in classical logic of both φ and ¬φ would translate into a proof in intuitionist logic of both φ* and ¬φ*.

This double-negation translation carries with it some further results, for example this: if the only logical constants in φ are ∧ and ¬, then φ is provable in intuitionist logic if and only if it is provable in classical logic. Since classical logic can be presented with ∧ and ¬ as its only undefined symbols, and all others defined in terms of them, this provides a way in which classical logic can be viewed as a subsystem of intuitionist logic, rather than *vice versa*. For classical logic is just that fragment of intuitionist logic that uses only ∧ and ¬. But this is only a *fragment* of intuitionist logic, for it also uses ∨ and →, which are not intuitionistically definable in terms of ∧ and ¬. Indeed *none* of the classical ways of defining one logical constant in terms of others is available in intuitionist logic. (This includes ∀ and ∃, which are not interdefinable in intuitionist logic.)

The double-negation translation is not a 'translation' which preserves the meaning of the symbols in the two systems. A translation which comes nearer to this is in effect given by my paraphrase of the intuitionist constants in terms of '□' and the English words 'and', 'or', 'if', 'not', 'some', 'all'. If we take these English words in their classical meaning, and compare a formula φ of intuitionist logic with its counterpart □φ' of classical logic, where φ' is calculated according to the □-translations given earlier, then it turns out that φ is provable in intuitionist logic if and only if □φ' is provable in the classical modal logic S4, assuming that '□' now means 'it is necessary that' in any sense appropriate to S4.[21]

There are several ways of providing a semantics for intuitionist logic. The best known, and the one that best represents the intended meaning of the intuitionist logical constants, is due to Kripke (1965), and is usually referred to as the method of 'Kripke trees'. We consider a tree structure, with an initial node α, and branches descending from it. Each further point of branching is also a node, and formulae are evaluated as true or not true at each node. There may well be nodes at which neither φ nor ¬φ is evaluated as true. But there is this constraint: a formula which is evaluated as true at any node must also be evaluated as true at all subsequent nodes.

---

[21] The main principle distinguishing S4 from other modal logics is that it includes the thesis '□φ ↔ □□φ'. (A simple source for modal logics, including S4, is Scott, 1981, but the classical and full-length treatment is Hughes and Cresswell, 1968.)

One may think of the semantics in this way. The initial node α represents our current state of knowledge, and subsequent nodes descending from α represent possible ways in which that knowledge may develop. The constraint just mentioned is then the idea that knowledge is never lost: what is known to be true at one time is also known to be true at every later time. Similarly, in the case of the quantifiers, items known to be in the domain at one time will continue to be counted as in the domain at all later times. But as time goes on our knowledge may increase, with more items being added to the domain and more formulae becoming known to be true.

I add finally that the intuitionist theory of the natural numbers results simply from adding the familiar Peano postulates, with the principle of mathematical induction treated as an axiom-schema. In this theory the intended domain is the domain of the natural numbers, and the quantifiers are just the first-level quantifiers. As we shall see in the next section, this theory is not so very different from the classical theory of the natural numbers (and the double-negation translation continues to hold). Higher-level quantifiers are needed to develop the intuitionist theory of the real numbers, and the interpretation of these quantifiers leads to a theory which is quite unlike the classical theory. I shall not discuss it here, though I remark that it does have some affinities with the 'predicative' theories that are the subject of the next chapter.

3. The irrelevance of ontology

Intuitionists from Brouwer onwards have often spoken as if their resistance to classical mathematics results from their ontology. The idea is that the (alleged) fact that numbers are our own creation, our own 'mental mathematical constructions', is what requires a non-classical logic in reasoning about them. I argue in this section that that cannot be right. My argument is much indebted to what Michael Dummett has said on this topic.[22]

Here is a line of thought that might seem tempting (and I guess that it did tempt Brouwer). Most of our ordinary discourse speaks about objects, i.e. physical objects, which exist independently of us. So in this case it is reasonable to say that 'truth consists in correspondence with reality', or something similar, which allows us to see how a proposition might be true even though we do not now recognize it to be true, and perhaps never will. But

---

[22] Dummett (1977, pp. 383–9). Cf. his (1973b, pp. 228–47).

mathematics is different. According to a formalist view, we do not need to suppose that there are any 'mathematical objects' at all, so naturally truth in mathematics cannot consist in some kind of 'correspondence' with the facts concerning such objects. It can only be understood as provability within whatever formal system is in question. This is not what intuitionists believe, for they think that there are indeed 'mathematical objects', but what is special about them is that they are our own creation. This leads them to a similar conclusion. For the thought is that human creations cannot have any more properties than we humans put into them at their creation, and a property that we have put into an object we can also recognize as being in that object. That is to say, in the mathematical case, that we can prove that the object has that property. So once again truth and provability must be equated.

Now fictional characters are human creations, and it is a well-known point about fictional characters that their creation almost never determines all the properties that they do or do not have. For example, Hamlet is a human – i.e. Shakespearian – creation, but Shakespeare did not determine whether or not Hamlet had a moustache, and consequently no one else can determine this either. So we have a counterexample to the law of excluded middle, for it is not true that Hamlet did have a moustache nor true that he did not. Perhaps, then, the same would apply to the natural numbers if they are also, as intuitionists hold, simply human creations?

But this analogy does not work. If we do 'create' the natural numbers, then we do so by saying or thinking something like this: '0 is a natural number which is not the successor of any natural number; every natural number has one and only one successor; every natural number except 0 has one and only one predecessor; the natural numbers just are the objects that can be reached from 0 by repeated applications of the successor-operation, and consequently mathematical induction holds for them'. (As I have said, intuitionists do adopt Peano's postulates.) This 'creation' does in fact determine all the mathematical properties of the natural numbers. It is true that we still have to add suitable definitions (by recursion) of addition, multiplication, exponentiation, and so forth. But when that has been done then the relationships between the numbers are fully determined. Even if the creation is ours, still it does not provide any ground for rejecting the law of excluded middle in their case.

We can also put the point in this way. The intuitionist and the Platonist do in fact agree on all the *atomic* statements of elementary arithmetic, i.e. those containing no variables but only particular numerals, together with

+, ×, =, <, etc. This is because any atomic statement that is true is also prov-able (by calculation), and *vice versa*. They also agree on all truthfunctions of atomic statements. They give different meanings to these statements, for (as we have seen) the intuitionist explanation is in terms of 'proof-functions' rather than truth-functions, but still the same formulae are accepted by each. They also agree with one another on the domain, at least in this ontologically neutral way: the domain consists of things called num-bers, and there may be a difference of opinion about what kind of things these are, but in any case the domains are isomorphic. For example, each side agrees on what the domain of Arabic *numerals* is, and each agrees that there is one number for each numeral and no number that lacks a numeral. For the classical logician this is enough to settle the truthvalues of all quantifications over the natural numbers, and he need not object to a substitutional interpretation of the quantifiers. That is: '$\forall n(\!-\!n\!-\!)$' is true if and only if '$(\!-\!n\!-\!)$' is true for each numeral in place of 'n'. (Similarly for '$\exists n(\!-\!n\!-\!)$'.)[23] He takes the question of whether a thesis holds for all or for some numerals to be a perfectly clear question, and to have a definite answer, whether we know it or not. This holds whatever the numer-als are taken to stand for. It is simply irrelevant whether or not these things are our own 'mental constructions', for the same logic will apply in either case.

This line of reasoning seems to me to be absolutely right. So far as the debate between the intuitionist and the classical logician is concerned, it does not matter at all what kind of ontological status the 'objects' in ques-tion have, or whether or not they exist 'independently of us'. For the meaning of the quantifiers (and truthfunctors) is simply not affected by this question. I conclude that one cannot argue in *this* way for the claim that ordinary classical logic does not apply in mathematics.[24] My next section therefore addresses a different line of argument that Dummett thinks does work (or, anyway, has a good chance of working). This concerns what it is possible to *mean* by the logical constants, and it is a completely general argument, not in any way confined to mathematics.

---

[23] Substitutional quantification and quantification understood in the usual 'ontological' way will coincide when every object in the (ontological) domain has a name, as is the case here.

[24] I have argued in this section that the ontological status of the *natural* numbers does not affect this debate. The position is rather different with the real numbers, as my next chapter will show.

4. The attack on classical logic

Some who are attracted to the intuitionist way of doing mathematics are liberal-minded. They take the view that the classical approach is entirely legitimate, but they prefer the intuitionist approach, mainly because of its emphasis on 'constructive' proofs, which are mathematically more informative. I shall not discuss this view, which makes intuitionism a matter of taste. But others (from Brouwer onwards) have claimed that classical mathematics is simply *wrong*, that it is based on an important mistake. This charge has generally taken the form of a challenge to classical *logic*, and that is what I discuss in this section. I shall consider only the arguments proposed by Michael Dummett, for in my opinion they are the most interesting of all such arguments (from a philosophical point of view).[25] As I have argued in the last section, the alleged status of mathematical objects as our own mental constructions is here irrelevant. Dummett's argument is that classical logic is wrong *everywhere*, and not just in mathematics. But he also thinks that its wrongness shows in the most blatant way when we are speaking of infinite collections, which is almost always the case in mathematics, though not often elsewhere. However, as I understand his arguments, he is committed to saying that classical logic should not be used in any case in which, *for all we know*, the totality might turn out to be infinite. That includes many ordinary cases, such as 'all men'.[26]

Dummett's opening claim is that one should not argue that a logic (such as classical logic) must be justified simply because it is in common use (e.g. 1977, p. 362). I agree. There may perhaps be something wrong with classical logic, even though it is very widely used, but I do not have space to discuss that question in this book.[27] But in any case we should not endorse the reasoning which Dummett offers for his claim. He first argues against a 'holistic' theory of linguistic meaning, according to which the meaning of any sentence is given by its (inferential) relations to other sentences. On

---

[25] The line of reasoning perhaps comes through more clearly in Dummett's (1973b), pp. 216–27). But the later version in his (1977, pp. 361–80), does add some relevant points.
[26] In later writings, e.g. from the final pages of his (1991) onwards, Dummett has suggested that what causes problems for the classical logician is the 'indefinite extensibility' of an infinite totality. I shall not discuss this, for (so far as I can see) Dummett supposes that *every* infinite totality of abstract objects will be 'indefinitely extensible', including even the totality of the natural numbers. (Cf. his 1991, pp. 318–19.)
[27] My own view remains as it was in my (1990).

this theory, since every sentence of the language is somehow related (perhaps by a very devious chain) to every other, you cannot understand any one sentence of the language without understanding *all* of the others. This 'extreme' holism strikes me as simply absurd.[28] There is room for a much more limited kind of holism about our understanding of language. For example, it is surely to be expected that anyone who can understand the statement '7 + 5 = 12' can also understand the question 'what is 2 + 5?' (Counterexamples may be possible, but only when explained by very unusual circumstances.) More generally, it is reasonable to say that, in the usual case, one does not count as fully understanding a sentence unless one also understands the words in it, and this cannot be done without also understanding these same words in *some* other similar contexts. But the point should not be exaggerated. Obviously a child can grasp the numeral '2' without yet having any understanding of a sentence such as 'Betelgeuse is over 270 light years away'. I do not imagine that one can draw any precise limits to what a limited holism should require, but I – and I hope everyone else – will agree that a quite *unlimited* holism cannot possibly be right.

Dummett's reason for attacking holism is that it threatens to make any established usage automatically justified, simply because it is the established usage which confers meaning. I accept that this must be mistaken. In particular, the fact that classical logic is very commonly used does not by itself amount to a justification of that logic (though it may suggest that a justification is probably available). With so much by way of preamble let us now turn more directly to Dummett's objections to classical logic.

For the sake of this discussion I will accept the (very improbable) idea that someone might understand all the *atomic* statements of elementary arithmetic, without yet having been introduced to the usual logical constants. We may leave it open just how this understanding of atomic statements is acquired. Dummett, of course, thinks that it comes from an understanding of the computation procedures by which we prove or refute such statements. Others may prefer the view that it comes by finding from experiment such facts as that 2 apples plus 3 apples makes 5 apples, and generalizing from there. I here set this question aside. Let us suppose that somehow or other the atomic statements of elementary

---

[28] The holistic theory stems from Quine (1953), though in his (1960) it is modified by the concession that 'observation sentences' have a more direct relation to experience than all others. The theory is endorsed by Davidson (1967 and elsewhere).

arithmetic are already understood, and come to the question of how an understanding of the logical constants might be added.

Dummett starts from the Wittgensteinian slogan 'meaning is use', and he relies on this to claim (a) that one learns the logical constants by learning how to use them, and (b) that one manifests one's understanding by showing that one knows how to use them. He also claims that they could not have anything more to their meaning than could be learnt and manifested in this way. This is already becoming controversial, but I think that we can, for the sake of argument, continue to accept it. What is much more controversial, and what I am sure that we should not accept, is Dummett's next move, namely that the relevant *use* of a statement is *asserting* it – that and nothing more.[29] From this he infers that one understands a logical constant just when one knows when one should and should not *assert* a statement which contains it. For example, the idea is that you understand ∧ when you know that $P \land Q$ can be asserted if and only if P can be asserted and Q can be asserted, and similarly for the other logical constants (in accordance with the scheme given on p. 204 above). In the mathematical case the relevant assertion-condition is that the statement be proved (or anyway, that we have a procedure which we know would lead to a proof of it, and so on). In non-mathematical cases the relevant condition is that the statement should be well verified, or something of the sort, and this is the more general notion, since proof is a special case of verification. It is the kind of verification that one desires in mathematics. Thus, in the teaching situation, the teacher will give as examples of statements that can be asserted only ones that he (the teacher) knows are verifiable (provable). And, when it comes to manifestation, the manifester will give as examples only statements that he knows are verifiable (provable). From this Dummett infers that a grasp of meaning can only consist in a grasp of assertion-conditions, i.e. verification-conditions. It cannot be anything more, because nothing more could be learnt, and nothing more could be manifested.

This does not mean that I understand a statement only when I know whether or not it should be asserted. It is obvious that a statement of any complexity may perfectly well be understood by one who has no idea of

---

[29] As Dummett himself said in a later writing (1991, pp. 47–50), one sometimes needs to explain how the 'assertoric sense' of a sentence may not be enough to determine its 'ingredient sense', i.e. how it contributes to the sense of a longer sentence that contains it. It would seem that his present argument ignores that distinction.

whether it should be asserted (e.g. Goldbach's conjecture).[30] His idea is rather that one's understanding of a complex statement is built up from one's understanding of its simpler components, together with one's understanding of this way of combining them. But both of these kinds of understanding are built up from understanding conditions of assertion, i.e. conditions for asserting simple statements, and rules which allow us to derive from these the conditions for asserting more complex statements built from them.

The classical logician will of course object that there is a difference between being true and being verified (proved), and that meaning is much better understood as a grasp of truth-conditions than as a grasp of verification-conditions. Dummett challenges him to say how a grasp of truth-conditions could be either learnt or manifested if it really is different. The most obvious reply is that we *use* statements not only in asserting them but also in other ways; for example we may *suppose* them, to see what would follow, and surely one would expect grasp of a logical constant to be specially connected with knowledge of what follows from what.[31] To take a very simple example, the teacher can surely by example teach, and the manifester can by example manifest, that if you suppose that $\neg\neg P$ then P will follow. To put the point more generally, we can both teach and manifest our grasp of classical logic, not just in what we assert but also in our other uses of language. For example, we can teach and learn that a disjunction is *true* if and only if one of the disjuncts is, and that hence that $P\vee\neg P$ must always be true, and so may always be asserted. Similarly we may teach, learn, and manifest the knowledge that from $\neg\forall xFx$ there follows $\exists x\neg Fx$, and so on. And why should we not take this as both teaching and manifesting a grasp of the *distinction* between truth and provability?

Dummett's reply is that this move is illegitimate because it simply assumes the correctness of the existing practice, and thereby introduces a kind of holism. Thus the two examples that I have just mentioned make simultaneous use of more than one logical constant (i.e. $\vee$ and $\neg$, or $\forall$ and $\exists$ and $\neg$). Dummett thinks that the basic rules for any one logical constant should mention just it, and not also make use of any other. I accept that it is very natural to cite principles which use more than one constant in demonstrating the difference between classical and intuitionist logic. But

---

[30] Goldbach's conjecture is the thesis that every even number is the sum of two prime numbers. It has still (at the time of writing) been neither proved nor disproved.

[31] Dummett himself makes this point when arguing against Quine in his (1973a, p. 614).

(a) not all natural examples are like this (e.g. $\neg\neg P \vdash P$), and (b) if one is sufficiently knowledgeable and sophisticated about classical logic one can demonstrate the difference using only one logical constant at a time.[32] But of course most teachers, and most manifesters, are not so sophisticated. They do therefore take over just a little of the 'holism' that Dummett so vehemently rejects, but this is not at all unreasonable. It is now more or less orthodox to say about a *physical* theory that it has to be understood as a whole; one cannot test each little bit of it singly, but can only test the whole theory. The same applies to a logical theory, and if it is simplest to use both $\vee$ and $\neg$ together in order to explain the meaning of each, that is not in any way an objection. We still have a whole theory which can be put to a test.

Dummett's claim that we cannot really understand classical logic, as based on a conception of truth which differs from verifiability (provability), seems to me to be evidently wrong. Of course we all *understand* it. And the idea that it is not even a *coherent* theory has no plausibility whatever.[33] But there is still a question over whether it is a *good* theory, though I cannot discuss that here.

Suggestions for further reading

A sympathetic exposition of Brouwer's position is given in van Stigt (1998). (This is largely taken from his fuller and more detailed 1990.) But life is short, and I would suggest skipping Brouwer altogether, and beginning instead with the dialogue that forms chapter 1 of Heyting (1956). This is reprinted in Benacerraf and Putnam (1983), and it presents the basic ideas in an easily digested form. An alternative but fairly brief exposition of Brouwer's point of view is given in part I of Posy (2005), which also continues in an interesting way, though it does overlap with his (1984) to some extent.

In my opinion the author who simply must be read on this topic is Michael Dummett, and the principal work is his (1977). I would suggest starting with his Introduction, and then his sections 1.1, 1.2, 3.1, before tackling the main claims in his chapter 7. The earlier version in his (1973b) might

---

[32] Details are given in my (1997), principally in sections 6.2–6.3.
[33] This sentence is (loosely) quoted from McCarty (2005, pp. 397–8), and McCarty is in general a friend of intuitionism.

also be consulted; it has the advantage of brevity and is in some ways more easily comprehensible.[34]

On intuitionist logic a classic source is Kleene (1952), especially chapter 15. I would rather recommend Scott (1981), chapters 11–12, but that is no doubt because I myself had a hand in its composition. However, it may be difficult to obtain. There are many other sources of basic information on intuitionist logic, for example the exposition in Forbes (1994), but the more succinct account in McCarty (2005), sections 1–2 of part III, contains all that is really needed.

On the intuitionist theory of real numbers, which I have not discussed at all, I suggest that the exposition in Dummett (1977, chapters 2–3), would make a good starting point. But an alternative 'constructive' approach to the real numbers, which is still broadly in the intuitionistic tradition, though presented in a way which classical mathematicians will find more readily intelligible, is given in Bishop (1967). (Bishop's account is discussed in Billinge, 2003.)

[34] I add that there is a long exposition of Dummett's 'manifestation' argument in Tennant (1987, chapters 1–16). But for the most part readers should be able to supply for themselves the extra elucidations that Tennant offers.

Chapter 8

Predicativism

The basic principle behind what is nowadays called the 'predicative' approach to the philosophy of mathematics is what is known as the Vicious Circle Principle (hereafter VCP). This is essentially due to Henri Poincaré (1906), in an explanation of why the logicist approach of Frege and the early Russell had run into contradictions. But Russell then took over the principle himself, and the logic of *Principia Mathematica* (1910) aims to obey it. In chapter 5 I discussed only the 'simple' theory of types, which is entirely independent of the VCP, but the official logic of *Principia* is a more complex theory called the 'ramified' theory of types, in which the VCP is supposed to be fundamental.

In this chapter I shall pass over Poincaré's account and begin with Russell. My discussion is based mainly on his (1908), which is where he first proposed the logic to be adopted in *Principia*. This will occupy my first two sections. Thereafter I turn to more modern developments of the predicative approach. I should add in this preamble that the word 'predicative' is not in itself at all helpful. It comes to be used because both Poincaré and Russell called 'impredicative' those definitions and explanations which infringed the VCP. So 'predicative' has come to mean 'not impredicative'. But, as I say, this is not a helpful title (and, incidentally, Russell himself uses the word 'predicative' in a different sense, as we shall see). What is basic to the topic of this chapter is the Vicious Circle Principle, so we must begin with that.

1. Russell and the VCP

In his (1908) Russell states the principle somewhat vaguely, and in varying terminology, thus:

Whatever *involves*, or *presupposes*, or is *only definable in terms of*, all of a collection cannot itself be one of the collection (e.g. pp. 63, 75; my emphasis)

It is the version in terms of definitions which is mainly important for our purposes, and it is definitions which infringe this principle that are said to be 'impredicative'. I should add here another version, which Russell somewhat oddly regards as the same principle stated 'conversely':

If, provided a certain collection had a total, it would have members only definable in terms of that total, then the said collection has no total. (And he adds in a footnote: when I say that a collection has no total, I mean that statements about *all* its members are nonsense.) (p. 63, Russell's emphasis)

(I shall replace the word 'nonsense' by 'illegitimate', as Russell himself often does.) He offers two reasons for accepting this VCP, first that it seems to be required to solve certain paradoxes, and second that it has 'a certain consonance with common sense' (p. 59). He adds that he finds the first reason more important than the second, though in my opinion it should be the other way round. I begin with the first.

The (1908) article opens by listing a number of paradoxes. Here is a selection of four:[1]

(i)   Russell's own paradox, i.e. the attempt to define a set *w* by the definition

$$(\forall \text{ sets } x)(x \in w \leftrightarrow x \notin x)$$

(ii)   Grelling's paradox, i.e. the attempt to define an adjective 'heterological' by the definition

$$(\forall \text{ adjectives } x)(\text{'het' is true of } x \leftrightarrow x \text{ is not true of } x)$$

(iii)   Berry's paradox: is there or is there not a number named by this name:[2]

---

[1] As it happens, Russell's list in (1908) does not actually include the second stated here, i.e. Grelling's paradox. I regard this as entirely accidental. I include it in order to exhibit its similarity with the first.

[2] I have written 'natural number' where Russell wrote 'integer' (and consequently increased the number of syllables). It is clearly the natural numbers that he is thinking of, and in particular the principle that if any natural number has some property then there must be a least natural number with that property.

The least natural number not nameable in less than 20 syllables.

(The paradox arises because this alleged name itself has less than 20 syllables.) To understand Russell's response it is helpful to paraphrase with an explicit quantification:

The least number n such that ($\forall$ names x)(x names n → x has at least 20 syllables)

(iv)   The Epimenides, which is the oldest one in the book. Epimenides was a Cretan, and he said

All Cretans are liars

We assume that he meant by this

($\forall$ propositions p)(A Cretan says that p → $\neg p$)

To obtain a paradox from this we add the supplementary assumption that everything *else* said by any Cretan is indeed false.

Russell's diagnosis is that in each case there is an illegitimate quantification, i.e. a quantification over a collection that 'has no total'. I give just a couple of examples. First, discussing a simplified version of the Epimenides, Russell says:

> When a man says 'I am lying' we may interpret his statement as: 'There is a proposition which I am affirming and which is false' . . . in other words, 'It is not true for all propositions p that if I affirm p, p is true'. The paradox results from regarding this statement as affirming a proposition, which must therefore come within the scope of the statement. This, however, makes it evident that the notion of 'all propositions' is illegitimate; for otherwise there must be propositions (such as the above) which are about all propositions, and yet can not, without contradiction, be included among the propositions they are about. (pp. 61–2)

Similarly, discussing Berry's paradox, he says:

> 'The least integer not nameable in fewer than nineteen syllables' involves the totality of names. . . . Here we assume, in obtaining the contradiction, that a phrase containing 'all names' is itself a name, though it appears from the contradiction that it cannot be one of the names which were supposed to be all the names there are. Hence 'all names' is an illegitimate notion. (p. 62)

He shows on p. 61 that he is aware of a connection between what is ordinarily called self-reference and these quantifications which may be called 'self-quantifications', i.e. because they allegedly introduce an object by means of a quantification over a collection which includes that object itself. This does not actually lead him to modify his statement of the VCP, no doubt because literal self-reference is not available in the usual logical systems. But I think that we do better explicitly to include it, for several of the paradoxes listed do have an explicitly self-referential version. The best known of these is probably the simpler version of the Epimenides that is known as 'the liar', i.e. the problem of one who says

This statement is false.

Berry's paradox can also be put in this form, i.e. by considering the alleged name:

The least number not named by *this* name.

Let us suppose, then, that the VCP is intended to ban self-reference just as much as self-quantification, as in each case 'viciously circular'.

Attempts to obey the VCP have always led to hierarchies. The items that we are concerned with – e.g. classes, adjectives, names, propositions – are each assigned an order, and these orders are hierarchically arranged. It may not be at once obvious why this should be, but the approach via the paradoxes will easily furnish an explanation. For essentially similar paradoxes may easily be generated not by literal *self*-reference, but by cases where one item refers to another and that other refers back to the first. A simple example is:

The next statement is true.
The previous statement is false.

This too has its overtly quantified version (which is due to Prior, 1961), as when:

The prisoner says 'Everything the policeman says is true'
The policeman says 'Everything the prisoner says is false',

And as it happens neither says anything else. (Or: as it happens everything else that the policeman says is indeed true, or *vice versa*.) These are 'circles' of just two members. It is obvious that longer circles can be generated in the same way. The simplest way to rule out all such vicious circles, in reference or in quantification or in a mixture of both, is indeed to introduce a hierarchy of orders, to insist that all reference or quantification must be restricted to items of some one order, and that what 'involves or presupposes or is only definable in terms of' such a reference or quantification must itself be of a higher order. Russell claims that this ruling will provide a 'solution' to all the paradoxes that he is concerned with.

This claim is disputed, but I shall not enter that dispute.[3] One can at least say that his theory does very often prevent a known contradiction from arising, and I expect that something along these lines could be made to work in all relevant cases, given a small modification here and there. Instead I make two different objections, namely (i) that the supposed solution seems to be extravagant, in that it proscribes many ways of speaking which appear to be harmless, and (ii) that it is certainly not the only solution that is available.

In support of (i) I note that there seem to be many quite harmless examples of self-reference, e.g.

This sentence is in English,

and similarly of self-quantification, e.g.

All English sentences can be translated into French.

Russell might be inclined to reply that these examples are not examples of *definitions* and it is really a certain kind of definition that the VCP aims to rule out. I shall address that topic more directly in what follows. Meanwhile, I add a brief consideration of the objection (ii).

Ramsey (1925, pp. 24ff.) observed that the *simple* theory of types is quite enough to prevent paradoxes that might arise within set theory itself, such as Russell's own paradox, or e.g. the Burali-Forti paradox. Indeed, there are also other methods of handling these, as we have seen, for example the ZF principle of 'limitation of size' (p. 144). What leads Russell into the complications of the ramified theory is largely his wish to provide a

---

[3] For some discussion see e.g. Copi (1971, pp. 89–91 and 107–14), Sainsbury (1979, pp. 320–5), and Giaquinto (2002, pp. 75–9). Giaquinto draws a good distinction between two different kinds of 'semantical paradox'. A surprising moral is drawn from the 'heterological' paradox in Potter (2000, pp. 156–7). (Compare Myhill, 1979.)

solution to many other paradoxes, which Ramsey characterizes as 'semantic' paradoxes, because they all involve semantic notions such as being true, or being true of, or denoting, or something similar. Ramsey proposed that the resolution of these paradoxes be left to semantics, a separate area of study which has no close connection with mathematics. I think that this view is very generally accepted nowadays; there is no strong reason to suppose that one and the same theory should provide solutions both to the paradoxes of set theory and to these semantic paradoxes. Since Russell was writing there has been much further work on the semantic paradoxes, which I do not propose to describe here.[4] But I think that we are entitled to conclude that, even if something like the VCP is helpful in these cases, that is not by itself a good reason for saying that it is needed *in mathematics*. So let us now turn to Russell's other argument for the VCP, namely that it has 'a certain consonance with common sense, which makes it inherently credible'. (In what follows I shall, like Russell himself, ignore self-reference.)

We must begin with some further clarifications of how the VCP is to be understood. First, it applies only to *abstract* objects. Ramsey observed that if we specify a certain man as 'the tallest man in the room' then we are specifying a man by quantifying over a totality (namely all the men in the room) of which he is himself a member. But it would be absurd to infer that such a man cannot exist, or that the quantification was somehow illegitimate (Ramsey, 1925, p. 41). Russell might reply that this is not the *only* way of specifying the man in question, so there is here no counterexample to the VCP, and no doubt we can grant that. But we must also generalize it. Ordinary concrete objects may always be specified in a variety of ways, and in principle simply by a demonstrative which involves no quantification at all. So the VCP does not apply to them, but only to abstract objects, such as sets or adjectives or names or propositions.

A second point is that even such abstract objects may be specified in various ways, e.g. as one might refer to a set as 'the set mentioned on p. 112 of this book'. This is not the kind of specification that is relevant to the VCP, which should be understood as concerned with what one might call 'canonical' specifications of the object, i.e. those that may well be regarded

---

[4] Church (1976) contains a favourable comparison between Russell's way of avoiding the semantical paradoxes and that adopted by Tarski, with his hierarchy of languages. But the similar discussion in Hazen (1983, section 5), reminds us that there are alternatives to Tarski's theory, e.g. that proposed by Kripke (1975). For a general discussion, very clearly expressed, see e.g. Sainsbury (1995, chapter 5).

as definitions. In the case of sets, the canonical specifications are those that cite a membership condition; in the case of propositions, those that cite a that-clause which expresses that proposition; in the case of words or sentences, a quotation of that word or sentence, perhaps with an explanation of its meaning. For simplicity I shall mainly consider sets from now on, but remembering that Russell's 'no-class' theory reduces talk of sets to talk of propositional functions. What the VCP claims in this case is that there does not exist any set such that *all* its canonical specifications are impredicative. The idea is that the same set may well have different canonical specifications, i.e. there may well be different ways of stating its membership condition, and so long as *one* of these is predicative all is well. But if none are, then the supposed set does not exist.

Third, there is an ambiguity over what is to count as a predicative specification of a set. Introducing a set w the specification will take the form

$$\forall x(x{\in}w \leftrightarrow (\text{—}x\text{—})),$$

and it will be predicative or not depending on the quantifiers involved. But *which* quantifiers? Are we to consider only the quantifiers appearing within the clause '(—x—)' on the right-hand side, or do we also include the initial quantifier '$\forall x$'? The usual understanding is that it is only the quantifiers within '(—x—)' that matter, and the constraint is that they must range only over items of an order less than the order of the set w hereby introduced. For the most part Russell too follows this interpretation. But apparently he is committed to saying that the order of the initial quantifier '$\forall x$' must also be taken into account, for he claims that the definition of the Russell set, i.e.

$$\forall x(x{\in}w \leftrightarrow x{\notin}x),$$

is impredicative, and infringes the VCP. But this definition has *no* quantifiers on the right-hand side.

This creates a difficulty, for if the variable 'x' is restricted to items of some definite order, say order n, then the attempted specification of the set w tells us only which items of order n are members of it, and leaves it open whether it also has members of order $n+1$. (This is by no means impossible, as we shall see.) Russell's response might be that the definition of w will also specify *its* order, say as $n+1$, which will imply that it cannot have members of any order greater than n. But one may note that this implication is in effect a quantification over items of *all* orders, and it is not clear that this should be permitted.[5] I would rather say myself that the proposed definition of the Russell class is ruled out by the *simple* theory of types, which distinguishes sets into different *levels*. As we shall see in the next section, this is quite different from the distinction into *orders* which is imposed by the ramified theory. But before we come to that, let us come back to the question of this section: why should Russell think that the VCP has 'a certain consonance with common sense'?

His own discussion is rather reticent upon this point, though it does contain some significant clues. When he is trying to sum up his diagnoses of the paradoxes, he very frequently says things like this:

> Something is said about *all* cases of some kind, and from what is said a new case seems to be generated, which both is and is not of the same kind as the cases of which *all* were concerned in what was said. (p. 61)

Or again, on the Epimenides:

> Whatever we suppose to be the totality of all propositions, statements about this totality generate new propositions which, on pain of contradiction, must lie outside the totality. (p. 62)

And, generalizing,

> All our contradictions have in common the assumption of a totality such that, if it were legitimate, it would at once be enlarged by new members, defined in terms of itself. (p. 63)

The point to notice here is that Russell constantly speaks as though it is *our* statements, or *our* definitions that would (if permitted) *create* the new objects.

He must therefore be adopting a broadly conceptualist position on the existence of abstract objects, according to which they may be said to exist, but their existence depends upon our mental activities. And apparently the crucial activities are our defining activities, i.e. our ability to produce what I have called 'canonical specifications' of such objects, for that is what ensures their existence. Someone of a realist tendency (e.g. a Platonist) will of course disagree with this entirely. He will say that it is absurd to suppose that it

---

[5] The case is relevantly similar to what was noted in chapter 5 as a problem for the simple theory, namely that to explain this theory in English one naturally quantifies over items of all different levels. A similar reply is perhaps appropriate.

is our ability to define or specify concrete objects, say horses, that brings them into existence, and the same applies equally to abstract objects. They exist (or not) quite independently of whether we happen to be able to define them. Russell must be supposing that such a realist approach does not count as 'common sense', whereas conceptualism does.[6]

It is true that Russell's earlier writings show little trace of conceptualism, and this has led some commentators to suppose that he cannot be meaning to invoke it here, when advocating the VCP. For example Hylton (1990, pp. 298–300) plays down the idea of definability, claims that the important thesis is that a propositional function *presupposes* its values, and explains this as meaning that the function's existence depends upon that of all of its values. It is a minor objection that Russell does not himself speak in terms of presupposition in section I of his (1908), where he is recommending the VCP, but brings in this idea only later, at the start of section IV. Moreover his claim there is in fact somewhat ambiguous.[7] It is not until later, i.e. *Principia Mathematica* (1910), that we find the passage which Hylton relies on, namely 'a [propositional] function is not a well-defined function unless *all* its values are already well-defined. It follows from this that no function can have amongst its values *anything* which presupposes the function' (p. 39, my emphasis). However, this claim about presupposition is surely false, and there seems to be no reason why Russell should think that it appeals to common sense. On a different approach, but one which again avoids conceptualism, Demopoulos and Clark (2005) have said: 'the principle underlying ramification is entirely plausible and in no way *ad hoc* if the ramified hierarchy is held to reflect the fact that our epistemic access to [propositional] functions higher in the hierarchy depends on our access to those below it' (p. 157). Once more we have to supply '*all* those below it' to obtain a justification of the VCP,

---

[6] Gödel (1944, esp. pp. 456–9) is often credited with making this observation. But Gödel is not distinguishing between conceptualism and nominalism, i.e. the view that abstract entities do not exist at all. This is particularly clear in the revised (1983) version of his paper (e.g. p. 456, n. 15), and in fact his arguments are more concerned with nominalism than with conceptualism. As we have seen in chapter 5, there are indeed nominalizing tendencies in Russell's thought, but I cannot believe that he would have regarded *that* as 'common sense'.

[7] In his (1908) Russell just says that 'a proposition containing an apparent [i.e. bound] variable presupposes others from which it can be obtained by generalisation' (p. 75). Here 'others' could be interpreted as '*some* others', which yields a very plausible claim, but one that will not justify the VCP.

and then again the principle is not at all plausible. Indeed, Russell himself apparently denies it when he says, shortly after the passage in *Principia Mathematica* just quoted, that 'a function can be apprehended without its being necessary to apprehend its values severally and individually' (pp. 39–40).

I conclude that these reinterpretations fail, and that it is a conceptualist approach that Russell is relying on. Nor is this really surprising if one takes into account the fact that by 1910 he was thinking of propositional functions just as *expressions* of a certain kind, i.e. as linguistic items (cf. p. 132 above). Anyway, I shall henceforth assume that it is a conceptualist approach that lies behind what Russell regards as his second recommendation of the VCP, namely its appeal to 'common sense'. The view is that abstract objects – and in particular propositional functions – exist only because they are definable. It is the definition that would (if permitted) somehow bring them into existence.

However, our explanation cannot stop here. For even if it is our definitions that bring sets (or other entities) into existence, we may still ask why an impredicative definition cannot do this just as well as one that is predicative. To explain this I think that we must suppose that Russell is also subscribing to the kind of 'construction' metaphor that lies behind the standard interpretation of ZF set theory. It is significant that in this discussion (i.e. 1908, pp. 59–64) he constantly speaks in terms of 'totalities', and it is quite easy to think of a totality as 'built out of' its members. So, if we may rely on the usual temporal metaphor we may say that, just as a house cannot exist until after its bricks do, similarly a totality cannot exist until after its members do. We now add that a totality cannot be referred to until it exists, and that an impredicative definition is one that professes to define a member of a totality by reference to that totality. This does indeed introduce what looks like a genuinely *vicious* circle. By the construction principle a totality cannot exist until all its members do; by the conceptualist principle a member cannot exist until it is defined; so if all its definitions are impredicative the member cannot exist until after it exists. This clearly is an impossible situation.

Even so, the argument is not really very convincing. In what I have just said I put in the claim that you cannot *refer* to a totality until it exists, which in turn cannot happen until all of its members exist. That may sound plausible at a first hearing. But now let us recall that this 'reference' to a totality is, in context, just a quantification over all members of that totality, i.e. a use of an expression of the form 'all so-and-so's'. One may note (*a*) that such a use would not normally be thought to presuppose the existence of a singular item called 'the totality' (e.g. set?) of all so-and-so's. (For example, in modern set theory one quantifies over all sets without supposing that there is a set of all sets.) Also (*b*) in ordinary cases one may certainly quantify over all so-and-so's without supposing that they all exist already. (For example, if I say 'All men, past present and future, are mortal', I am surely not supposing that future men somehow exist now.) Why would not this same point apply when the so-and-so's are abstract objects? More generally (*c*) it may reasonably be said that if I am meaningfully to quantify over all so-and-so's then I must understand what it is to be a so-and-so, i.e. (in Frege's language) I must know the sense of this expression. But I do not need to know all the so-and-so's, or even any so-and-so, or even whether there are any so-and-so's. Russell's invocation of 'totalities' makes his argument sound much more convincing than it really is.

Moreover, as an *ad hominem* point against Russell's own theories, one may fairly say that his 'no-class' theory makes the construction metaphor especially inappropriate in his case. We do think of classes as 'defined by' their membership, in the sense that wherever you have the same members you have the same class, and this makes it not unreasonable to think of classes as 'built out of' their members. But in the end Russell intends to eliminate all apparent reference to classes in favour of quantification over propositional functions, and the construction metaphor is quite unreasonable in their case. For propositional functions (as Russell conceived them) may certainly be *different* functions – for example, functions of different orders – even though they are extensionally equivalent. From this it surely follows that they cannot be viewed as 'built out of' the objects which satisfy them. As I have noted, Russell does claim that a propositional function 'is not well-defined unless all its values are *already* well-defined' (*Principia*, p. 39, my emphasis), but since he at once goes on to concede that 'a function can be apprehended without its being necessary to apprehend its values severally and individually' (*ibid*), we are altogether lacking an argument for the 'already'.[8]

---

[8]   Recall that in Russell's terminology the 'values' of a propositional function are the propositions which result from supplying that function with an argument, but the relevant point is that he is taking it for granted that the 'value' is not well defined unless the 'argument' that occurs in it is also well defined. I remark that Giaquinto (2002, pp. 79–84) argues in the opposite direction, that the VCP is more plausible for propositional functions than for classes. But he admits that he cannot supply a plausible 'construction principle' in their case. ('The other underlying thought is that, if an attribute cannot but be defined in terms of some class *c*, that attribute constitutively depends upon every member of *c*. This may be right, but I do not know of a cogent argument for it', p. 82).

In the latter part of this section I have been trying to find a good way of interpreting Russell's claim that the VCP accords with 'common sense'. The search has not yielded a satisfying answer. Later in the chapter (i.e. in section 3) I shall consider some *other* reasons for accepting the VCP. But before we come to that it will be useful to consider the VCP 'in action'.

2. Russell's ramified theory and the axiom of reducibility

The simple theory of types that we considered earlier stratifies propositional functions into *levels*. In the ramified theory that we now come to, propositional functions are distinguished not only into levels but also into *orders*, in obedience to the VCP. Russell explains that a propositional function is of order 1 if its fundamental specification quantifies (if at all) only over individuals, of order 2 if the specification quantifies over functions of order 1, and so on. It follows that some functions of order 2 are of level 1, and some are of level 2. For example, if the variable G_1 ranges over first level functions of order 1, then there will be a first-level function F_2 of order 2 such that

$$\forall x(F_2(x) \leftrightarrow \forall G_1(G_1(x) \leftrightarrow G_1(a))$$

(The function F_2 is true of the individual a, and – presumably?[9] – of nothing else.) But there will also be a second-level function $\mathcal{M}_2$ which is again of order 2, such that[10]

$$\forall G_1(\mathcal{M}_2(G_1) \leftrightarrow \exists x(G_1(x)))$$

(The function $\mathcal{M}_2$ is true of all non-empty first-level functions of order 1.) Thus the classification of propositional functions into *levels*, and into *orders*, is simply a cross-classification, and neither implies the other.

Russell appears to claim that the classification into levels can itself be derived from the VCP, but he gives no such derivation, and I cannot see

---

[9] The 'presumption' is satisfied if the axiom of reducibility is true. (I come to this axiom in a moment.) I add that in the first edition of *Principia* Russell wishes to construe propositional functions intensionally, and in that case the first occurrence of $\leftrightarrow$ in this formula, and the only occurrence in the next, should perhaps be replaced by some notion of identity that is stronger than $\leftrightarrow$.

[10] I abbreviate the full version $\mathcal{M}_x Gx$ simply to $\mathcal{M}(G)$.

how to construct one. A simple consideration which reveals the difficulty is this. Propositional functions of different levels cannot occur in exactly the same contexts, i.e. you cannot ever substitute one for another, for the result is simply ungrammatical. Consequently the hierarchy of levels must be a *strict* hierarchy.[11] But the hierarchy of orders may well be taken as a *cumulative* hierarchy, with variables ranging at once over functions of all orders up to some given order, for that will still satisfy the VCP. So there cannot be a simple way of deriving the first hierarchy from the second. The most that one can say is that Russell apparently assumes that there cannot be a function of level n that is of order less than n, and he pays special attention to the functions of level n that are also of order n, which he calls 'predicative' functions. They play a special role in his theory, as we shall see. (It is quite unclear why he chooses the word 'predicative' to pick out these functions, for there is no proper contrast with his use of the word 'impredicative'.) The general requirement, then, is that quantification is permitted only where the bound variables are restricted *both* to some definite level n *and* to some definite order m, where $m \geq n$. Other quantifications are deemed to be 'illegitimate'.

That is the basic outline of Russell's ramified theory of types. Of course my exposition has simplified by considering only *monadic* propositional functions, and the situation becomes very much more complicated when we also include functions that are dyadic, triadic, and so on. For example, in simple type theory a dyadic relation may be one that has two arguments of the same level as one another, or one which has arguments of different levels. This is retained in the ramified theory, but further complicated by the fact that the orders of the arguments to the relation must now be specified in addition to their levels, and once more these orders may or may not be the same. For our purposes we may continue to set these complications aside. Even so there are matters of detail which are open to dispute, for Russell's own exposition of his theory is often unclear, but I shall not fuss over that.[12] One popular way of thinking of Russell's theory is to do what Russell himself never explicitly did, i.e. to suppose that we *begin* with the simple theory and then *add to it* the 'ramifications' which the VCP

---

[11] This is denied by Hazen (1983, p. 347), but without explanation.

[12] Russell sometimes uses the word 'type' to mean level, sometimes to mean order, and sometimes to mean both. There are expositions of his theory which differ from one another in detail in Copi (1971, chapter 3), Chihara (1973, chapter 1), and more recently Urquhart (2003). I do not regard these variations as important.

demands. In that case one quite naturally thinks of the orders as always beginning in each level with order 1. That makes no important difference.

Now we have already seen that even the simple theory of types makes a difficulty for Russell's attempt to deduce arithmetic from logic, since the restrictions which it introduces necessitate an axiom of infinity, and that axiom does not look like a truth of logic. But the effect of imposing a further ramification into orders is very much more disruptive, for *many* of the crucial definitions in the logicist programme are in fact impredicative.

I start with a very simple example, identity, which is usually defined so that

$$a=b \;\leftrightarrow\; \forall F(Fa \leftrightarrow Fb)$$

Here one understands the quantifier $\forall F$ as ranging over all first-level predicates, so that the definition ensures that if $a = b$ then either may be substituted for the other in any context whatever. But in the ramified theory we are not allowed to do this, since the variable F must now be confined to first-level predicates of some definite order. And whatever order is chosen, the definition cannot rule out the possibility that while a and b share all predicates of that order still they differ on predicates of a higher order. But this must be ruled out if it is genuinely to be *identity* that is defined.

In *this* case there is a straightforward remedy, which is to take identity as a primitive and undefined notion, governed by the usual axioms, namely

$$a = a$$
$$a = b \;\rightarrow\; (Fa \leftrightarrow Fb)$$

Here the letter 'F' is not a bound variable, and we can therefore interpret the second axiom as an axiom-schema, generating infinitely many axioms, one for each first-level predicate in place of 'F'. This covers in one blow all first-level predicates of all orders, and so avoids the problem.

But an essentially similar difficulty keeps cropping up, in one definition after another, and we cannot meet them all in the same way without giving up on logicism altogether. For example, the usual definition of 'natural number' must now be so modified that it no longer yields the desired result. Suppose that '0' is already defined as an item of some definite level and order, and suppose that 'successor' is also defined as a function on such items, and let us use 'x', 'y', ... as variables for those items (whatever is their level and order). Then one standardly defines '$\mathbb{N}$' for 'natural number' by

$$Nx \leftrightarrow \forall F(F0 \wedge \forall y(Fy \rightarrow Fy') \rightarrow Fx)$$

But once more the ramified theory requires the bound variable F to be confined to functions of some definite order, and whatever order that is it will then follow from the definition that mathematical induction holds for the natural numbers for functions of that order, but it will not follow that it also holds for functions of higher order. That is evidently not what is needed.[13]

Once again, we can remedy this situation as the intuitionists do, by abandoning any attempt to *define* 'natural number', and instead taking this notion as primitive and governed by the usual axioms (i.e. Peano's postulates). That is hardly in accordance with logicist aims, though one may feel that in this case it is acceptable.[14] But, as I say, a similar problem *keeps* cropping up. For instance, one standardly defines 'the Fs are as many as the Gs' as 'there is a relation which correlates the Fs and the Gs one-to-one', and one understands this as quantifying over all relations of all orders. But according to the ramified theory some definite order must be specified, and when that is done the consequence is that there may be a correlating relation of a higher order without it following that the Fs are as many as the Gs. This again means that the definition does not adequately define the notion intended. Many further examples could be cited. (I mention by way of example the notions of the ancestral, of a well-ordering, of an isomorphism, of 'infinitely many', and so on.)

Russell's reaction to these difficulties is to introduce an Axiom of Reducibility. This states that to any propositional function of any order, within a given level, there corresponds another which is of the lowest possible order in that level.[15] That is, there corresponds what he calls a

---

[13] The axiom of reducibility would avoid the problem, as we shall see, but the second edition of *Principia* (1925) contains no such axiom. Myhill (1974) argues that there is then no way of showing that full induction holds for the natural numbers, whereas Landini (1996b) argues the opposite. The divergence comes from the fact that each gives a different account of the underlying logic that Russell must be assuming but does not state explicitly.

[14] In this particular case there is another way of circumventing the difficulty, as my next section will show.

[15] This is strictly an axiom-schema for each level, e.g. for the first level

$$\exists F_1 \forall x(F_1(x) \leftrightarrow (\text{---}x\text{---}))$$

for any well-formed formula containing 'x' free, but not 'F_1' free, in place of '$(\text{---}x\text{---})$'.

'predicative' function of that level. The two 'correspond' in the sense that they are equivalent, i.e. true and false of exactly the same things. Hence if we define identity for individuals by quantifying over all predicative functions of individuals, we shall be able to deduce that identical individuals may *always* be substituted for one another in any (extensional) context, since the relevant propositional function will always be *equivalent* to some predicative function. Similarly we may define 'natural number' by quantifying over all predicative functions of natural numbers, and it will follow that induction holds for all extensional functions of natural numbers, whether predicative or not. The same applies to all other cases where one naturally uses a definition that is impredicative.

In particular, it applies to Russell's way of introducing the usual notation for classes while retaining his 'no-classes' theory. In the context of the simple theory of types, the definition for classes of first level was given earlier (p. 130) in this way[16]

$$—\{x{:}Fx\}— \quad \text{for} \quad \exists G(\forall x(Fx \leftrightarrow Gx) \land (—G—))$$

Stated in this way the definition has now become illegitimate, and the bound variable G needs to be confined to some definite order. But if we confine it to functions of order 1, i.e. to the predicative functions of level 1, then the axiom of reducibility assures us that nothing has been left out. For every function of any order will be equivalent to some predicative function, and so will determine some class in accordance with this definition. That is why Russell so closely associates the axiom of reducibility with the ordinary notion of a class, for he thinks that the ordinary notion both supposes that there always is a class for any propositional function, and that classes are not distinguished from one another by being of different orders (e.g. 1908, pp. 81–2; 1919, p. 191).[17] More generally, the effect of the axiom of reducibility is that in the actual deductions of *Principia Mathematica* it is only the simple theory of types that is used. For almost all of the deductions use the notation of classes, and classes are defined by quantifying only over predicative functions, and if we confine attention to these functions then we are in effect working within the simple theory. In fact *Principia*

[16] Strictly speaking, we also need a way of indicating the *scope* of the expression '$\{x{:}Fx\}$' – compare the definition of '$\imath x(Fx)$' – but that is a detail of no importance here.
[17] The ordinary notion *also* supposes that classes are not distinguished from one another by being of different levels. That is a point which Russell here ignores.

does not use quantifiers over any other functions than the predicative functions (as is stated on p. 165). So one naturally asks: what was the point of first introducing the complications of the ramified theory, and then adopting an axiom of reducibility which allows us simply to ignore them?[18]

In Russell's view the point is this. The axiom allows us to ignore the distinction into orders when we are doing *mathematics*, for the propositional functions that occur in mathematics are all extensional functions, i.e. they take the same value when supplied with extensionally equivalent arguments.[19] But when we come to consider the semantic paradoxes the propositional functions involved are not extensional, and so the axiom of reducibility cannot reintroduce the paradox. Here is a simple illustration.[20]

Suppose that I take some object *a* (say a pebble on the beach at Brighton), and assert of it

Whatever is at any time asserted of *a* is false of *a*.

Suppose that nothing else ever is asserted of *a*. Then on the face of it we have a contradiction of the usual type, for what I assert of *a* is true of *a* if and only if it is false of *a*. But it is generated by the use of quantification ('whatever is asserted') which Russell will regard as illegitimate, so his solution is that we must specify a definite order for the propositional functions here generalized over. Let us take them to be first-order functions. Then what I assert of *a* must be more narrowly expressed as

Whatever first-order propositional function is at any time asserted of *a* is false of *a*.

And to assert this of *a* is to assert of *a* not a first-order but a second-order propositional function. Hence what I assert of *a* is vacuously true of *a*, for by hypothesis no first-order propositional function ever is asserted of *a*; and it is not also false of *a* because it is not itself a first-order propositional function. Now by the axiom of reducibility this second-order function is

---

[18] It is clearly explained in Landini (1996a) that *in effect* the logic of *Principia Mathematica* is that of the simple theory of types.

[19] Individuals are deemed to be extensionally equivalent if and only if they are identical. Propositional functions are extensionally equivalent if and only if they are each true of the same arguments and false of the same arguments.

[20] The writings cited in nn. 3–4 above are mainly concerned with the different paradox of heterologicality.

equivalent to some first-order function, so there is a first-order function which is true of *a*. But no contradiction follows from this, for we have no reason to suppose that that first-order function which is true of *a* is also asserted of *a*. The two functions are extensionally equivalent, but it does not follow that the intensional function '. . . is asserted of *a*' must be true of the one if it is true of the other. So in a context such as this, where non-extensional functions are involved, the VCP and its associated distinction into orders, still has an effect which is not destroyed by the axiom of reducibility.

That is why Russell thinks that the ramified theory of types and the axiom of reducibility work nicely together to give us all the desired results. The distinction into orders is needed to solve the semantic paradoxes, and that solution is not affected by the axiom of reducibility. But the distinction into orders is not needed in mathematics – indeed, it upsets much of ordinary mathematics – and in that context the axiom of reducibility is important, for it allows us to ignore those distinctions altogether. Russell does not claim that the axiom of reducibility is a truth of logic, or that it has any of the 'self-evidence' that one might expect in a basic axiom, or even that it has any 'consonance with common sense'. Quite explicitly, in his (1907) he argues that axioms should be judged not by their self-evidence but by whether they yield all the consequences that are desired and none that are not desired, and he evidently thinks that this condition is satisfied in the present case. I shall not discuss this claim, for that would take us further into the semantic paradoxes than is appropriate here. In any case, there is a much more important objection to bring. It seems that, however nice the consequences of the axiom of reducibility may be, still from Russell's own perspective it has to be *false*. That must be a good reason for him not to accept it.

I said earlier that the VCP has some appeal if one takes a broadly conceptualist stance to the existence of abstract objects, but none at all if one takes a more robust and realistic attitude. This is because it is only if abstract objects owe their existence to our mental activity that their existence could depend upon what ways of defining them are available to us. Intuitionists (and others) are concerned with our ability to *construct* such entities, and this is a kind of conceptualism, for the relevant 'constructing' is of course a mental activity. Since the VCP (in its central version) is concerned with defining entities rather than constructing them, its version of conceptualism seems initially to be somewhat different. But actually they are closer than may at first appear, for we saw on pp. 233–4 that a definition must be regarded as a kind of construction if we are to justify the prohibition of impredicative definitions. In any case, my argument will focus on definitions rather than constructions, and on the idea that an abstract entity exists only if it has a definition of a suitable sort. The VCP then requires that impredicative definitions are not to count as suitable.

The argument is now very simple: no learnable language can contain more than denumerably many definitions. This is because any learnable language must have a finite basis – e.g. only finitely many letters, or finitely many phonemes, or finitely many single words, or whatever we wish to take as the starting point. Moreover every expression of the language is of finite length, and so contains only finitely many of the basic symbols. It follows that the expressions of the language are at most denumerable. Hence if an abstract object exists only when there is some suitable way of expressing it, then there are at most denumerably many abstract objects.

This argument is completely general, and it may be useful if I reinforce it by pointing to Russell's own way of understanding his notion of a 'propositional function'. I observed earlier that, even in the context of the simple theory of types, it is simplest to take a propositional function as just an expression (namely an open sentence), and that Russell's later writings are quite explicit on this. But I now add that in the ramified theory this understanding is forced upon us, because in this theory functions are distinguished into orders by the quantifiers that occur in their expressions. It follows that there must *be* a way of expressing the function, for otherwise no order could be assigned to it. So, once again, there cannot be more propositional functions than there are ways of expressing them. I add, incidentally, that there may well be fewer, for we may well wish to say that different expressions which share the same 'meaning' shall count as expressing the same function. But here we cannot identify 'meaning' with 'extension', for Russell presumes that there are different functions with the same extension, and his solution to the semantic paradoxes depends upon this.[21]

[21] On p. 133 I observed that, even if a function is an expression, we do not have to interpret the universal quantifier '$\forall F$' as generalizing only over what we can express. Instead, it should be interpreted as generalizing over all permitted ways of interpreting the letter 'F', where the permitted interpretations simply assign to 'F' an extension. But if, as Russell desires, functions may share the same extension while being different functions (e.g. because they are of different orders), this explanation is no longer available. And in this case I think that the quantifier '$\forall F$' can only be interpreted substitutionally.

Now the axiom of reducibility claims that for every first-level function of any order there is a first-level function of first order that is equivalent to it. Similarly for higher levels, but with 'predicative function' in place of 'first-order function'. We have seen that this means: there is an equivalent function which is expressed by a first order expression (or, more generally, by what Russell would call 'a predicative expression'). This claim is false. Rather than argue the point directly, I choose a roundabout route which is highly relevant to mathematics. Cantor proved that there are more than denumerably many real numbers. In *Principia Mathematica* real numbers are construed as classes, and then reference to classes is analysed away by Russell's 'no-class' theory, so that it is replaced by reference to some or all propositional functions which (as we say) define those classes. Consequently *Principia*'s approach implies that there can be non-denumerably many real numbers only if there are non-denumerably many propositional functions. Moreover *Principia* does contain a proof of Cantor's theorem. Hence *Principia* contains a proof of a falsehood, and so one of its axioms must be false. By reflecting on Cantor's proof, it is quite easy to see that the culprit must be the axiom of reducibility. Here is a quick outline.

Assume (for *reductio ad absurdum*) that all the real numbers between 0 and 1 can be listed in a denumerable list. To state this assumption we must be able to quantify over all those real numbers, so they must all be of some definite order, say order n. Cantor's argument proceeds by 'going down the diagonal', and thereby defining a new real number between 0 and 1 that is not in the original list. This definition quantifies over all the real numbers originally listed, and so is of a higher order, say $n + 1$. *But* the axiom of reducibility is then invoked to show that this function of order $n + 1$ is equivalent to one of order n, and hence that the real number it defines must after all be in the original list. Thus we have our *reductio*, and Cantor's result is proved. Clearly the proof would not go through without the axiom of reducibility. That axiom therefore leads to a result which is incompatible with Russell's adoption of the VCP and of the ramified theory which is based upon it. Russell is therefore committed to its falsehood.[22]

No thinker after Russell who has endorsed the VCP has also endorsed his axiom of reducibility, and they have all accepted that the universe of

---

[22] I am not, of course, claiming that any contradiction is deducible within the system of *Principia Mathematica*, for that system says nothing about expressions. It does not say either that every function can be expressed or that there are at most denumerably many expressions. But the second of these claims is evidently true, and I have argued that Russell's explanations do commit him to the first.

abstract objects is at most denumerable. Their position on the real numbers has been that there is no order which contains all of them, for, by the argument of the last paragraph, whatever order you choose there will always be real numbers of a higher order. Let us proceed, then, to these later predicative theories.

3. Predicative theories after Russell

Nowadays predicative theories are usually presented as theories of sets, and there is no intention of applying anything like Russell's 'no-classes' method of reducing statements about sets to statements about something else. In this way the theories resemble the familiar ZF theory of sets, and in other ways too. The simple theory of types is usually[23] set aside, and all sets are treated as being on the same *level* as one another (and on the same level as individuals, if they too are included). But sets are distinguished into different *orders*. As with ZF, for mathematical purposes one may confine attention to the *pure* sets, i.e. those built up just from the null set, so that the domain of the theory consists of nothing but sets. The natural numbers will be construed as finite sets, e.g. in von Neumann's fashion, so that each natural number is the set of all its predecessors. The same approach is applied to ordinal numbers in general, and not only to those that are finite. Rational numbers may be construed as ordered pairs of natural numbers, and hence as themselves finite pure sets, and real numbers are then construed as infinite sets of rational numbers, e.g. as given by a Dedekind cut in the rationals. (One usually identifies a real number just with the lower half of a Dedekind cut, and one may simplify by confining attention to those cuts where the lower half has no greatest member.) The elementary theory of the natural numbers is intended to be just the same as is the classical theory. It is with the real numbers that the restrictions imposed by predicativity begin to make a real difference.

These restrictions require that every variable shall carry an index (which I write as a subscript on the variable) to show its order, i.e. the order of the sets that it ranges over. The orders are construed as forming a cumulative hierarchy, so that a variable x_n has within its range all sets of orders less than n, as well as the new sets of order n. The basic assumptions are

---

[23] It plays no part in Wang's work, which is what I mainly discuss. It does figure in Feferman, but only in a minimal way.

these. There are axioms of extensionality for each order, which we may put in this way: wherever the index i is the maximum of n and m

$$\forall x_i (x_i \in y_n \leftrightarrow x_i \in z_m) \rightarrow y_n = z_m$$

(Identity is taken as primitive, and governed by the usual axioms, as noted on p. 237) There are also axioms of abstraction, for each order n

$$\exists y_{n+1} \forall x_n (x_n \in y_{n+1} \leftrightarrow (\text{---}x_n\text{---})).$$

Here $(\text{---}x_n\text{---})$ is any formula containing x_n free (but not y_{n+1} free) *provided that* in this formula every bound variable has order less than or equal to n, and every free variable has order less than or equal to $n + 1$. It is this proviso which rules out impredicative cases of the abstraction axiom. Finally, the cumulative structure of the orders is shown by axioms: wherever $n < m$

$$\exists y_m (x_n = y_m),$$

and the restriction imposed by the orders is shown by axioms: for all n, m

$$x_n \in y_{m+1} \rightarrow \exists z_m (x_n = z_m)$$

Since our theory has no individuals, there is no variable indexed 0. To get started we must therefore posit the existence of the null set by axioms

$$\exists y_1 (y_1 = y_1)$$
$$\forall y_1 \neg (x_n \in y_1)$$

This is an outline of the basic principles, which yields a complete theory so long as only finite orders are considered, and therefore only sets that are hereditarily finite (i.e. they are finite, their members are finite, *their* members in turn are finite, and so on). The orders of these finite sets are the same as their *ranks* in the natural model for ZF set theory, but now the distinct orders are built into the symbolism, and there is (so far) no variable which allows us to generalize over all the finite orders at once. But we cannot rest content with this, for our natural numbers are each of a different finite order, and even elementary arithmetic requires us to generalize over all the natural numbers. Consequently, variables of infinite order must be added. I shall start just with the lowest of these, the variables x_ω.

Every set of finite order is also of order ω, as the cumulative structure requires, and there are no other sets of order ω, for the abstraction axiom cannot be applied, since ω is not the successor of any number.[24] (I believe that the idea of infinite orders is due to Wang,[25] but not the use to which I now put it.)

I said earlier (pp. 237–8) that the predicative approach could not accept either the usual definition of identity or the usual definition of the natural numbers, for both of these definitions are impredicative. I see no way of avoiding the first problem except by taking the notion of identity as undefined but governed by the usual axioms. We could do the same with the notion of a natural number, but in this case we do not have to. For there is an alternative to the usual definition which delivers all the right results, and which we can reproduce within the predicative theory, given one extra but apparently very reasonable assumption. So far as I am aware, the idea is due to Quine,[26] though he was not writing with predicative problems in view.

Assuming that 0 and the successor-function are already defined, the usual definition is that the natural numbers are the smallest set containing 0 and closed under the successor-function. That is to say, anything a is a natural number if and only if

$$\forall x(0 \in x \land \forall y(y \in x \rightarrow y' \in x) \rightarrow a \in x)$$

Clearly any set x which satisfies the antecedent to this conditional must be an infinite set, containing all the natural numbers as members. But Quine observed that it will serve just as well if we invert this definition to[27]

$$\forall x(a \in x \land \forall y(y' \in x \rightarrow y \in x) \rightarrow 0 \in x)$$

The point of interest is that in this latter (inverted) version the sets x which are quantified over may all be taken as finite sets, but nevertheless we get all the same consequences.

---

[24] We cannot state as a simple axiom that every set of order ω is also of some finite order, since we cannot quantify over orders within our formal language. Wang proposes a complex way round this deficiency, which I describe in n. 30 below.

[25] See his (1962, chapters 23–4), which record work done in 1953. Given that the ZF theory has infinite ranks, it is hardly surprising to find infinite orders in its predicative variant.

[26] Quine (1963, p. 75).

[27] To see the rationale for this inversion, change '... $\in x$' throughout to '$F(...)$'. Then by substituting '$\neg F$' for 'F' either version can be obtained from the other.

It is easy to see that the inverted definition implies that 0 is a natural number, and that if x is a natural number then so is x'. But it also implies the unrestricted form of mathematical induction in this way. Assume that a is a natural number according to this inverted definition, and suppose also that

(i) $F(0)$
(ii) $\forall y (F(y) \rightarrow F(y'))$.

Then we require to show that $F(a)$, so assume for *reductio ad absurdum*

(iii) $\neg F(a)$.

Now we take some finite set which contains a and all its predecessors. Given von Neumann's way of construing the natural numbers, that is the set $a \cup \{a\}$. Consider the members z of this set such that $\neg F(z)$. By hypothesis (iii), a is one of them. By hypothesis (ii), if y' is one of them then so is y. So, using the inverted definition, it follows that 0 is one of them. Hence $\neg F(0)$. This contradicts assumption (i). So the *reductio* is established, and we have deduced an unrestricted form of mathematical induction, where $F(x)$ may be any condition on x whatever.

Let us now adapt this argument to the predicative context. We begin by subscripting ω to all relevant variables, for all the sets in question are sets of finite order. But then we note that the assumption that there is a subset of $a \cup \{a\}$ containing just those of its members satisfying '$\neg F(\ldots)$' needs some defence. For we are assuming that this subset is of *finite* order, whatever bound variables may occur in '$\neg F(\ldots)$'. Generalizing the assumption, it is this

$$\forall a_\omega \exists y_\omega \forall x_\omega (x_\omega \in y_\omega \leftrightarrow (x_\omega \in a_\omega \wedge F(x_\omega)))$$

This seems to me a perfectly defensible assumption, from the predicativist point of view. Admittedly the given condition 'F', which defines a certain subset of a_ω, may contain quantified variables of arbitrarily high order. But since the subset is a finite subset, however it is defined, we may infer that it *could* (in principle) be specified in a way which involves no higher-order quantifiers at all, i.e. simply by listing its members. This gives us a kind of axiom of reducibility for the special case of the hereditarily *finite* sets, and it seems to me to be entirely in accordance with the overall predicativist position. If so, then there is a suitable predicative definition of the natural numbers, which allows us to deduce that mathematical induction holds for any arbitrary condition on the natural numbers, no matter what the orders of its variables.[28]

We cannot perform with the infinite ordinal numbers the same trick of restricting attention to sets of some one order. The problem with the natural numbers was to obtain an unrestricted principle of ordinary mathematical induction, and the analogous problem for the infinite ordinals is to obtain an unrestricted principle of transfinite induction. In the classical theory this is usually achieved by building into the definition of an ordinal number that it is a set which both contains all ordinals less than it and is well ordered by the relevant 'less-than' relation. But to say that a set is well ordered one must quantify over all its subsets, requiring each to have a least member, and when these subsets may be infinite we cannot put any bound upon their order. Consequently a predicative theory cannot allow such a quantification, and in this case there is no way of finding a substitute quantification which can be confined to some definite order.

But there is another way of doing things, which is available with the von Neumann approach to ordinals, whereby their 'less-than' relation is just $\in$. For an axiom that applies specially to $\in$ is the axiom of foundation (or regularity), which in ZF is formulated thus:

$$\forall x(x \neq \emptyset \ \rightarrow \ \exists y(y \in x \land \forall z(z \in x \rightarrow z \notin y)))$$

We can adapt this to a predicative theory by making it an axiom-schema, with one instance for each order-index subscripted to the variable x. (The orders of y and z cannot be higher than that of x.) Moreover, it is entirely suited to the predicative conception, because its role in ZF is to forbid infinite descents in the membership relation, and that is something which the predicativist evidently endorses. For there can be no infinite descents in the order-indices which he attaches to each variable. Finally, we have already seen (pp. 148–9, above) that in the presence of this axiom the definition of 'a is an ordinal' may be simplified just to

$$\forall xy(x \in y \land y \in a \ \rightarrow \ x \in a)$$
$$\forall xy(x \in a \land y \in a \ \rightarrow \ (x = y \lor x \in y \lor y \in x))$$

---

[28] Compare Feferman and Hellman (1995), who make a similar assumption about the predicativity of all finite sets. (Note in particular their definition of '$\leq$' on p. 6, and their proof of induction on p. 11.)

The predicativist may therefore take over this definition, for each indexed variable a_i, and the problem is resolved.[29]

Before we move on to consider the real numbers, it will be useful to be more explicit on how infinite orders may figure in a predicative theory. This topic has been explored by Wang (1962, chapters 23–5), and I shall base myself on his discussion.

We begin with a change of nomenclature. As we have observed, once it is agreed that a predicative theory may allow infinite orders, it will follow that there is an order which contains all the hereditarily finite pure sets. I have been calling this the order ω, but Wang simplifies by counting this as his first order, the order 0. The variables ranging over all hereditarily finite pure sets are therefore variables of order 0, and the system which employs only these variables is Wang's first system Σ_0. Then his next system Σ_1 contains variables of order 1 in addition, and these range over the sets which have sets of order 0 as members, and which can be defined without quantifying over sets of any higher order. After this come systems Σ_2, Σ_3, ... Σ_ω, $\Sigma_{\omega+1}$, ..., each understood in the same way, i.e. as containing variables of all orders up to and including the index given, ranging over all sets of the corresponding order, i.e. those that can be specified without quantifying over sets of any higher order. So we have a family of systems Σ_α, for each permitted ordinal α, each extending all the systems Σ_β for $\beta < \alpha$. I postpone for the moment the question of just which ordinals count as 'permitted'.

Every system obeys the axioms listed earlier as basic axioms, with the subscripts n and m now understood as taking the place of any (permitted) ordinal indices. But once infinite ordinals are included those basic axioms are no longer sufficient, and Wang's attempt to remedy the deficiency involves axioms of a quite new kind, which he calls 'axioms of limitation'. These cannot be stated simply, so I here leave my account of his ideas undeveloped.[30] But let us come back to the question of which ordinals are permitted.[31]

---

[29] It may be noted that this gives us another way of defining the natural numbers, namely as the ordinal numbers of order ω. This is the approach taken by Wang.

[30] The basic idea is this. For any ordinal α the intended sets of order α will be denumerable, for each can be defined in terms of sets of lower orders, and there are only denumerably many definitions. So, for each α, the axioms actually enumerate the sets of that order, and state that these and only these exist in that order.

[31] I remark that if *all* of the classically accepted ordinals are permitted then what results from the predicative approach is in effect the Gödelian proposition '$V=L$'. But no predicativist would accept the assumption.

Wang's original suggestion was that we might begin with some well-known collection of ordinals, say the finite ordinals, and then extend the series in this way. If we already have a permitted system Σ_α, then we consider all the well-ordered series which can be defined in that system, and we are then entitled to use the ordinals β of these well-ordered series to construct further systems Σ_β. In the new systems Σ_β we may define further well-ordered series, yielding new ordinals γ, which then entitles us to introduce yet further systems Σ_γ. And so on (Wang, 1954). But it was then shown by Spector (1955) that all the ordinals that can be reached in this way are in fact already available in Σ_1, and moreover they are the ordinals of well-orderings of the natural numbers which are given by *recursive* relations on the natural numbers. (A relation is recursive when there is a method, which a machine could apply, for determining whether any two given numbers do or do not stand in that relation.) They are naturally called 'the recursive ordinals', and I think it is now the usual view that the orders permitted in a predicative approach are just these recursive ordinals.

The point is not universally agreed, but I think that for our purposes a further pursuit of this problem is unnecessary.[32] So let us just say that the 'permitted' ordinals are either just the recursive ordinals or some closely related class of ordinals. In any case, they will certainly not include all of Cantor's ordinal numbers, but will be a subclass of Cantor's second number class, i.e. of those ordinals which have no more than denumerably many predecessors. For the predicative approach will never lead us into non-denumerable sets.

This observation naturally brings us on to the topic of the real numbers. In the classical treatment, the real numbers are infinite sets of rational numbers, i.e. the lower halves of a Dedekind cut, and the rational numbers may perfectly well be treated simply as ordered pairs of natural numbers. So, on Wang's approach, the real numbers are infinite sets of sets of order 0, and therefore are themselves sets of at least order 1. But not all real numbers will be of order 1. The usual way of demonstrating this is by reflecting on the classical theorem of the least upper bound, which tells us that for any set of real numbers which is bounded above there exists a real

---

[32]   Wang had originally intended that the permissible ordinals should extend beyond the recursive ordinals, and when he found that his initial characterization did not achieve this he made some tentative proposals, in his (1959), for a revised criterion which would permit more. On the other side Feferman (1964) has proposed to narrow Wang's original characterization in a way that may permit less.

number which is the least upper bound of all the real numbers in that set. The theorem is proved by showing that the set of all rational numbers which are members of *some* real number in the given set is itself a real number, and satisfies the condition for being the least upper bound of that set. But this defines the required least upper bound by quantifying over all the reals in the original set, and so the order of this least upper bound, as thus specified, must be greater than the orders of all the reals in that set. That is, if they are all of order 1, then (so far as this proof goes) we can only assume that their least upper bound is of order 2.

As a matter of fact Weyl showed in his (1918) that a surprisingly large amount of the classical theory of real numbers can be obtained while still restricting attention just to the real numbers of order 1. His methods and his results are described in chapter 13 of Feferman (1998), and I do not reproduce them here.[33] But in any case we cannot suppose that all real numbers are of order 1. On the contrary, take any order α and consider the real numbers of that order. As we have seen, they can all be enumerated by enumerating their defining conditions. Given any such enumeration we can reproduce Cantor's theorem, and show that by 'going down the diagonal' a new real number can be defined that was not in the original enumeration. So there are real numbers not only of order α but also of order $\alpha + 1$, and no order can contain *all* the real numbers. That is a conclusion which the predicative approach cannot avoid. But should we count it as an objection which destroys this approach? Opinions differ.

First, the classical mathematician will think that predicative mathematics must omit several areas of mathematics that are of great interest to him, all of them stemming from Cantor's discovery that the real numbers (as classically construed) are not denumerable. The most obvious example is Cantor's whole theory of infinite numbers, with its claim that there are sets of *huge* cardinalities, going way beyond what is denumerable. There cannot be a predicative version of this theory. In fact, the predicativist cannot even accept the first step of Cantor's theory, which is the idea that two sets have the same number of members if and only if there is a relation which correlates their members one-to-one. For this quantification 'there is a relation which . . .' cannot be confined to relations of some definite order,

---

[33] In the first part of this chapter, i.e. pp. 249–68, Feferman is describing Weyl's own work (though the last few pages do become rather technical). The second part is more Feferman's own way of pursuing what may still be regarded as Weyl's aims, and it is quite difficult to follow.

if the definition is to do the work intended of it. But there are also several other areas of mathematics, which are essentially concerned with non-denumerable totalities, and which are therefore beyond the predicative approach. An example is measure theory, which begins from the principle that only a non-denumerable set can have a positive measure. Other examples could be mentioned.

On the other hand it is fair to add that none of these (comparatively new) areas of mathematics, which explicitly go beyond what is denumerable, have found any practical application in the sciences. There is an empiricist approach to mathematics, which I shall discuss in the next chapter, which claims that the only *justification* for any branch of mathematics is its usefulness for science. In pursuit of this approach it has also been claimed that it is only predicative mathematics that has a justification of this sort, and this claim is in fact quite plausible.[34] On this ground it may be argued that predicative mathematics is all the mathematics that we actually need.

But I think that serious advocates of predicative mathematics have never thought of this as the point that really matters. Their position is much more likely to be that it is only predicative mathematics that makes good sense, and – as I have said – this claim is best understood as stemming from the philosophical view that mathematics is our own creation. It is a theory of such things as numbers, sets, functions, relations, and so on, which exist only because they are objects of our thought. The restrictions of a predicativist theory are consequences of this. For if we accept the initial idea that we can think only of what we ourselves can describe, then something like this theory quite naturally results. My final section in this chapter returns to this question of justifying the underlying conceptualist viewpoint.

Before I come to this I should say that for the sake of simplicity my discussion has focused on just one version of a predicative theory, which is due mainly to Wang. But it is not the only version. Wang's approach follows Russell, insofar as it distinguishes sets into different orders, depending upon the quantifiers in their definitions. A different approach can be based upon Poincaré's original (and not very clear) suggestion that what is wrong with an impredicative definition is that it allows the set defined to alter its composition as more sets are added to the theory. He can therefore be interpreted as requiring a certain kind of 'stability' in the

---

[34]　See e.g. Hellman (1998) and Feferman (1998, chapter 14). A stronger claim is made in Chihara (1973, pp. 200–11), namely that science need not go beyond Wang's system Σ_ω.

definition of a set. This approach is described in Feferman (1964), and compared with the Russellian approach pursued by Wang. Feferman concludes that each of these two approaches does in fact yield the same results, though they reach them in rather different ways. An entirely different theory, which must nevertheless be counted as a predicative theory, is the intuitionist theory of the real numbers. This is much more complex, both because the underlying logic is non-classical and because intuitionists aim to preserve the Aristotelian principle that even a denumerable totality is only 'potentially' infinite. This leads them into a convoluted theory which I shall not describe. (The interested reader may consult Dummett, 1977, chapter 3.) The intuitionist also believes that mathematical objects are our own creation, but he puts some very severe limits on our creative powers. It is only the more moderate predicative proposals that I shall pursue further.

4. Concluding remarks

Our first brush with conceptualism was in chapter 2, with the claim of Locke and Hume that the objects of mathematics are simply 'ideas' in our minds. The intuitionist thesis that these objects are our own 'mental mathematical constructions' is evidently similar. At that stage three simple objections immediately suggested themselves. The first was that we do not ascribe to our 'ideas' such mathematical properties as having 6 sides and 12 edges; the second that our actual 'ideas' or 'mental constructions' are presumably finite, whereas the objects of mathematics are not; the third that we do not think of mathematical truths as being without truthvalue in the days before there were any conscious beings to have 'ideas' or to make 'mental constructions'. The conceptualist whom we are now considering can certainly offer replies to the first two of these objections, though it is not clear quite how he would respond to the third.

First, our present-day conceptualist does not *identify* mathematical objects with any mental objects, though he does claim that their existence depends upon facts about the mind. For example, a perfect cube has 6 equal sides and 12 equal edges because that is how we think of it, which is not to say that any thought of ours itself has 6 sides and 12 edges. As for the second objection, our present conceptualist does not claim that mathematical objects exist only if we do actually think of them, or produce verbal definitions of them. For example, his position on sets is that they exist if and only if *there is* a suitable definition of them; he does not require that anyone has actually formulated this definition, in thought or in speech or in writing or in any other way. What matters is just that such a definition is part of the language that we have adopted for mathematical purposes. The rules of this language are explicitly created by us, and so must be capable of a finite formulation. But these rules will allow for an infinity of (suitable) defining expressions, and in this way there may perfectly well be infinitely many mathematical objects. As for the third objection, that the truths of mathematics should antedate (and postdate) all human creations whatever, it is not obvious how the conceptualist should reply, and I leave this for your consideration.

A fourth objection has emerged in this chapter, namely that the conceptualist approach cannot admit more than denumerably many objects, whereas most of us have been convinced by Cantor's theorem that the real numbers (and the ordinal numbers) are not denumerable. On this point the conceptualist, as I see him, simply refuses to budge. He will say that most of us are just mistaken, for there cannot be more than denumerably many possibilities for our thinking. Cantor's proof that there are more real numbers than natural numbers, has an interpretation which he can accept. This is that, if you consider the real numbers of any specified order, then there will always be some further real numbers of a still higher order, but he will not admit that such an ascent ever introduces a non-denumerable totality of real numbers. And there is no good reason which forces him to this admission, so long as he will not admit a quantification over absolutely all real numbers, whatever their order happens to be.

So much for very general objections to conceptualism. I now come to something more detailed, namely the approach pursued in this chapter, which is based on the Vicious Circle Principle (VCP). This says that abstract objects of the kind that we are concerned with, in particular sets, must each be of some definite order, and can have members only of lower orders. The order of a set is fixed by the orders of the sets quantified over in its defining condition, and the idea is that when these quantifiers range only over sets of orders less than n, then the definition does define a set, which we can take to be of order n. (The same set may be defined in different ways, and a set which can only be assumed to be of order n according to one definition may be shown by another definition to be of some order less than n.) It is the VCP that dictates this stratification of sets into different orders. So let us come to the main question: is there any good reason for believing in the VCP?

I earlier suggested a justification on these lines. (i) According to the conceptualist principle, sets owe their existence to our ability to define them.

But (ii) according to the construction principle sets owe their existence to the existence of their members. Hence (iii) the members of a set must all be definable 'before' that set is itself defined. (Thus no set can have members definable only in terms of itself.) But to get from here to the criterion that is used in practice we must add (iv) that a quantification over all sets of a certain kind is not admissible unless all such sets 'already' exist, i.e. are 'already' definable without using the quantification in question. Claim (i) is the conceptualist claim; claims (ii) and (iii) introduce a usual view about the existence of sets, and draw from it a conclusion which conforms to claim (i). But we also need claim (iv), on how quantifiers are legitimately used, if we are to reach the principle on which predicative theories are built. I have remarked earlier that, when we are concerned with everyday quantifications over physical objects, the analogue of claim (iv) has absolutely no support from common sense. For example, I can quantify over all human beings, past present and future, without supposing that in some sense the past ones 'still exist' and the future ones 'already exist'. It is not obvious why we should not say the same about quantifying over all sets.

This brings me to an argument proposed by Michael Dummett, which he introduces as supporting the VCP, and which is explicitly addressed to legitimate domains of quantification.[35] At this point he is discussing Frege's supposed domain for his quantifiers of first level, which Frege simply describes as 'all objects'. Dummett objects that 'the totality of all objects is, par excellence, an illegitimate totality in Russell's sense, a totality which cannot be taken as a domain of quantification' (p. 529). This, he says, is because it is 'an impredicative totality, one the specification of which offends against Russell's "vicious-circle principle".' His explanation is that for Frege objects are understood simply as what proper names stand for, and Frege will accept as a name an expression which itself includes a quantification over all objects. This is what introduces the 'illegitimacy'. Dummett explains it thus, with the example of sets (or classes):

If there is some determinate totality over which the variable 'x' ranges, and if '$F(\xi)$' is any specific predicate which is well defined over that totality, then of course there will be some definite subset of objects of the totality which satisfy the predicate '$F(\xi)$'; and it is perfectly in order to assume that we can form the term 'the class of all x's such that $F(x)$' . . . and regard this as

---

[35] Dummett (1973, pp. 529–35). There is a later version of the argument in his (1991, pp. 228–9), but this is somewhat toned down.

standing for a specific abstract object. What there is no warrant for is the assumption that the object so denoted must belong to the totality with which we started. (p. 530)

Applied simply to sets, what Dummett is imagining is this. We may start with some nicely circumscribed totality of sets, e.g. the pure sets that are hereditarily finite (i.e. the ZF sets of ranks less than ω.) Then he supposes that we are automatically entitled to assume that there is a set of all these sets, which of course is not itself a finite set. We may also assume the existence of any specifiable subset of this set. Evidently the new set, and most of its subsets, will not themselves be among the finite sets that we began with. Then we can iterate this procedure. We may start now with the totality of all sets that *either* are hereditarily finite pure sets *or* are sets of these in turn. This gives us a new totality, and once more there will be a set of all its members, and subsets of this set given by any well-defined condition on its members. The whole process may be iterated again and again, until we have sets of all the ZF ranks less than $\omega.2$. Then we throw all these together, to form a set of rank $\omega.2 + 1$ which has as members all sets of ranks less than $\omega.2$. Of course, it too will have its own subsets. Then we can do it all again, and again, and again. This is how the familiar ZF hierarchy of sets is generated, and is apparently what Dummett is describing.[36]

He concludes that, since the notion of a set is in this way 'indefinitely extensible', one cannot legitimately quantify over all sets (p. 532). I see no reason to accept this inference. One might just as well argue that the notion of a natural number is also 'indefinitely extensible', on the ground that the totality of all numbers up to and including the number n itself introduces another number, for it has the number $n + 1$. But surely no one will conclude from this that it is somehow 'illegitimate' to quantify over all natural numbers?[37] Why should not the same apply to sets? They too can fairly be said to be generated by an 'indefinitely extensible' procedure, but why should that prevent us from quantifying over them all? So far as I can see, Dummett has only one answer. If one could quantify over all sets, then

[36] There is this difference. Dummett is apparently restricting attention to subsets that we can describe, whereas the usual accounts of ZF are not so restricted. (That is what justifies the addition of the axiom of choice.) But the main argument is not affected by this divergence.

[37] See n. 26 of chapter 7. More accurately, Dummett's later position seems to be that it may be legitimate to quantify over a totality that is 'indefinitely extensible', but only if the quantifiers are understood intuitionistically, and not classically (1991, p. 319). But it is difficult to see any reasonable connection between these claims.

one could produce descriptions of sets which cannot exist, for example 'the set of *all* sets'. So the totality is 'illegitimate' just in the sense that by quantifying over all its members we can produce 'definitions' that fail to define anything. But what is wrong with that? Only, I think, the idea that a 'legitimate' definition must always be successful, i.e. that there must always be an object which it defines. But this can only be understood as an appeal to the conceptualist principle, and those who take a more 'realist' view of the existence of sets will see no reason to accept it.

Dummett then moves from this claim to another, which he explicitly says does not depend upon any view about the ontological status of the objects in question. This claim concerns legitimate ways of *specifying* a domain of quantification. It is no longer being said that certain domains are 'illegitimate', but only that certain ways of specifying them are illegitimate, because 'viciously circular'. This view of the VCP is, he says, 'the weakest possible; we have indeed so weakened it as to put its validity beyond question' (p. 533). And he goes on: 'Understood as we have understood it here, merely as ruling out any circular specification of a totality, the principle becomes indisputable, and independent of any metaphysical views about the ontological character of the elements of the totality' (p. 534). The circularity in question is that of using quantifiers which range over some totality in the course of trying to explain what the totality is.

A first reaction is that in many cases that are of philosophical interest a circular explanation is all that is available, but even so it can succeed in generating understanding. A well-known example is the explanation of a necessary truth as one that is true in all possible situations. As everyone who has taught the subject knows, this description – perhaps with a few examples added – does in practice generate the required understanding, even though we cannot give a non-circular answer to the question 'And when is a situation a possible one?'. There are many other cases. It is difficult to explain the (simple) notion of truth without using a substitutional quantification which in turn is explained by the conditions in which it is true; it is difficult to explain the English 'if' without making use of that word itself; it is difficult to explain the logician's '∀' without using a word such as 'any' or 'every' or 'all', which in many contexts is simply a synonym. And so on. It is nice to be able to avoid a circular explanation, but it is often very difficult to do so, and it cannot always be possible (for otherwise an infinite regress must result). So it would not be surprising if, in interesting cases, a circular definition of a domain of quantification was all that was available.

In fact when a domain can be introduced by a formal definition, which quite often applies to domains of interest to the mathematician – for example the natural numbers, the real numbers, the functions on the reals, and so on – the standard definitions are never circular in the way that Dummett is talking of. His point applies only to cases when we do not attempt a formal definition, as may reasonably be said in the case of 'all sets'. Here the informal explanation of what a set is will cite various conditions which all sets satisfy, and of course one will quantify over all sets when stating these conditions. This is just what happens when one explains what a set is (on the iterative conception) partly by invoking the accepted axioms for sets. For those axioms do quantify over all sets. But there is nothing wrong with this. The same happens in many entirely ordinary cases. For example, to explain what a puppy is one may well say such things as 'all puppies are born of dogs, and (if they live) will grow into dogs'. This quantifies over puppies in order to give a partial explanation of what a puppy is, and hence to say what is meant by 'all puppies'. But what is wrong with that?

I conclude that Dummett's reformed and 'weakened' version of the VCP is indeed not committed either to a conceptualist view of abstract objects or to any other such general view. But this 'weakened' version carries no conviction in its own right, and it obviously has no relevance to how the VCP has in practice been used. In practice the principle has always been associated with a conceptualist view of abstract objects, and it is only in that context that it has any attraction.

Both in Russell's version and in Wang's version and in Dummett's version the VCP is supposed to limit the legitimate quantifications to some lesser totalities than are recognized in ordinary and classical mathematics. In some other authors this limitation is partially rescinded. For example, it may be allowed that one can properly quantify at once over all sets of natural numbers, or indeed over all the sets of any predicative system whatever, *provided that* these very general quantifications are not permitted to occur in the axioms of set abstraction.[38] This is surely sensible. For we do very

---

[38]　Suppose that a set y is defined by a condition of the form

$$\forall x (x \in y \leftrightarrow (—x—)).$$

Then on this account the initial variable 'x' may range over items of all orders, though the bound variables within the defining clause '$(—x—)$' must be restricted to specific orders.

frequently wish to say things about absolutely *all* sets, or *all* real numbers, or *all* ordinals, and so on, without specifying which orders are in question. Some who are sympathetic to the predicativist approach have been ready to accept the use of such general variables in ordinary contexts, while still forbidding them in the formulation of conditions which are supposed to define new sets, for that is where paradoxes would otherwise arise.[39] Should the predicativist allow or disallow such variables?

If his approach is simply pragmatic, namely to find a way of limiting our 'creative definitions' which succeeds in avoiding contradictions, then the more liberal approach has much to be said for it. But in fact predicativists have usually wished to present their position as based on principle rather than mere pragmatics, and they have therefore forbidden *all* quantification that is not suitably restricted. I here add a suggestion about what they must be conceiving as a 'suitable' restriction, namely that the domain should be effectively enumerable. This is true of all the domains that they are willing to consider, for since every object which they admit has a definition we can always enumerate them by a lexicographical ordering of their definitions. It is, moreover, a criterion for legitimate quantification which is quite often satisfied *in the special case of mathematics*. But it is no part of our ordinary understanding of the quantifiers. In many cases, an ordinary quantification does actually have a finite domain (e.g. 'all children born in the year 2006'). But in practice we have no way of enumerating this domain, and we do not regard that as in any way an obstacle to the quantification.

I end with a personal opinion. Predicativism is, I think, the best version of conceptualism that we have yet discovered. Certainly, one can raise objections, and it is not always clear how the conceptualist should respond to them. But that is also true of the main rival theories which I turn to in the next and final chapter, namely realism and nominalism.

---

[39] As an example, I note Feferman (1964). I observe incidentally that in any of Wang's partial systems Σ_α there are such general variables, i.e. the variables of order α, which range over all the entities admitted in that system. But there are no such general variables in Wang's 'total' system, which is the union of all systems Σ_α for permitted ordinals α. I also observe that there is something similar in the extension of the ZF set theory to NBG, which adds 'proper classes'. In NBG one may quantify over all classes – proper or improper – *but* not in the comprehension axioms. Hence NBG is a predicative (and conservative) extension of ZF.

Suggestions for further reading

Those who are curious may like to look at Poincaré (1906), for it was he who really started the idea of predicativity. But basic reading for the early development of this approach is Russell (1908, sections 1–5). The logic developed there is very similar to that finally adopted in *Principia*, vol. I (1910). However, one must add that it is not too easy to extract proper details from what Russell says himself. The most recent exposition of his ramified theory that is known to me is Urquhart (2003), but earlier accounts that are perhaps easier to read (though more debatable) are given in Copi (1971, chapter 3), and in Chihara (1973, chapter 1).[40]

By way of criticism one might look at Ramsey (1925), which is the first place where the distinction between what we now call the *simple* and the *ramified* theory of types is explicitly drawn. (Ramsey accepts the first and rejects the second.) Ramsey was much influenced by the early Wittgenstein (esp. his 1921), and he in turn was probably an important influence on Russell, leading him to change his mind in the second edition of *Principia* (1925). But a more modern criticism is Gödel (1944). It may be debated whether Gödel gives an accurate account of what Russell was trying to do, but in any case his views must be given serious consideration.

Moving on to more modern versions of predicative set theory, I suggest starting with Chihara (1973, chapter 5 and appendix). This gives an outline account of Wang's theories (in Wang, 1962) which is easier to read than Wang himself, though it does slide over several technical details. A rather different approach is to be found in Feferman (1964), which has a résumé of the topic in its part I, and in Feferman (1998). But perhaps the most useful account of the present state of predicative logics is Hazen (1983). (Unfortunately his use of the words 'level' and 'order' is the reverse of mine. See his notes 7 and 9 on pp. 397–8.)

There is a very interesting defence of the VCP in Dummett (1973a, pp. 529–35). This should certainly be read (but – according to me – should not be believed).

---

[40] I remark that Urquhart supposes that *Principia* allows for quantification over propositions, which seems to me to be a mistake on his part, but one that does no real harm to his discussion. (Cf. my 2008.)

Chapter 9

Realism and Nominalism

I said at the beginning of this book that there were two main questions in the philosophy of mathematics, the first being its ontology (what kind of objects is it about?) and the second its epistemology (how do we know it?). I also said that, although these were two distinct questions, one could not in practice keep them separated, for any answer to one of them would be bound to have implications for the other. That remains true. The main focus of this final chapter is the first question, but the second also has a major part to play.

A theme which entered in chapter 2, and has been prominent in the last two or three chapters, is the idea that we human beings *invent* mathematics. The objects concerned – e.g. numbers or functions or sets – are created by us, and we know about them just because they are our own creations. For it is our thinking that determines what laws do and do not apply to these objects. This line of thought will be firmly contradicted in the first part of this chapter, which aims to develop the thoughts of chapter 1, rather than chapter 2. But it will creep back again before the end.

I begin with the 'realist' approach first introduced by Plato, but pursued by many others since. On this view mathematics is concerned to investigate a type of objects – e.g. numbers, or sets – which exist quite independently of human thoughts. In the full Platonist version these objects are also to be regarded as existing quite independently of whatever may be the case in this world that we experience. Apart from Plato himself, whom I shall not discuss further, both Frege and Gödel may be cited as well-known advocates of such a view. Since Frege has been treated already, I begin this chapter with Gödel's ideas, and follow that with the philosophers called 'neo-Fregeans', who present Frege's realism in a way that differs importantly from Frege's own. But there is also a quite different and non-Platonic variety of 'realism' which is due principally to Quine and Putnam, and which has been the spur to much contemporary debate. Part A of the present chapter concerns these various versions of 'realism'.

Part B turns to the rival 'nominalist' account, which might reasonably be said to descend from Aristotle. As noted earlier (p. 34) this has two main versions, one which tries to 'reduce' the objects of mathematics to something simpler, and another which claims that such objects are mere 'fictions' which have no reality. (But it is admitted – as it has to be – that mathematics is still useful, even if it is not true.) The first kind of nominalism can claim the early Russell as an adherent, and the early Wittgenstein, in each case aiming for a logicist conclusion. The second kind is best known as the main thesis of Hartry Field's book *Science without Numbers* (1980). This second version of nominalism will naturally reintroduce the idea that the objects of mathematics are our own creations, but in a very different way. For the claim now is that what we create is not a system of truths but a fairytale world which has no genuine reality.

It may be useful if I say now that it is this last view that I myself subscribe to, so the reader who disagrees with me must think about where he or she would find fault with the arguments that lead up to it.

A. Realism

When we speak of ordinary and familiar objects in the world around us, we use names to refer to particular objects and quantifiers to generalize about them. No one thinks that these familiar objects owe their existence to our ability to name them or to use quantifiers that range over them. So it is natural to assume that the same will apply to the names and quantifiers that figure in any other area of discourse, including mathematical discourse. Its names (for numbers, functions, sets, and so on) will refer to objects that exist quite independently of our ability to name them, or to generalize over them. (I add that the assumption will be that each singular name, such as 'the natural number 3', refers to *just one* such object.)[1]

---

[1] Balaguer, in his (1998, chapters 3–4), has proposed a theory which he calls a variety of Platonism, because it takes our mathematical theories to describe abstract objects, existing independently of us, but which drops this uniqueness assumption. That is, it claims that there are many distinct objects which are all referred to by 'the natural number 3'. This is a position which I shall not discuss, since I see no motivation for it.

As I say, that is the *natural* assumption, and is presumably the one that we all start with. *But* it soon becomes clear that it involves a problem. We can give at least a first attempt at explaining how we know about the objects which our ordinary conversation refers to. But how could we know anything about such things as numbers, if (in Plato's metaphor) they exist 'in a different world from ours'? Benacerraf's article 'Mathematical Truth' (1973) is a classic statement of this dilemma.[2]

Another problem for the realist approach is to say how our supposed knowledge of the objects of mathematics can be in any way related to the applications of mathematics that we constantly make in science and in daily life. For example, how is the number 2 related to the quantifier 'there are 2'? What kind of connection is there between the purely arithmetical claim that 2 + 2 = 4 and such applications as 'if there are 2 coins in my left pocket, and 2 in my right pocket, and I have no other pockets, then there are altogether 4 coins in my pockets'? A Platonic realism may seem to suggest that I know the truth of the purely arithmetical claim by one method (intuition, say), and the truth of what is called its 'application' quite differently. But then it seems that, at least in theory, something might go wrong with the one kind of knowledge (e.g. that the 'perception' that 2 + 2 = 4 is in fact some kind of misperception), while the other remains unscathed (e.g. that 2 coins, and another 2 coins, really do make 4 coins). This divorce between ways of knowing the two different facts evidently creates a problem. I add as an aside that a similar problem affects other approaches which concentrate upon pure mathematics, without any consideration of its applications, e.g. intuitionism. The relation between pure mathematics and its applications will be a central theme of this chapter.

1. Gödel

Gödel's overall position is very similar to Plato's. First, he subscribes to the realist premise that the objects of mathematics exist quite independently of human thought. He also describes them as 'remote from sense experience', which suggests Plato's metaphor 'in another world'. So he very clearly invites the question: how could we know anything about these things,

---

[2] I shall continue to speak of *knowing* about these objects, since that is the simplest way of talking. But the problem is just the same if instead we ask how our beliefs about such objects could be reasonable or reliable or in any way suitably grounded.

if they are as you describe? To answer this, Gödel draws an *analogy* with our ordinary and everyday knowledge of physical objects. To explain this analogy I must here insert a little background.

In the 1920s and 1930s several philosophers attempted to revive a position on ordinary perception which one associates with Berkeley.[3] Berkeley had held that the only things that exist are (i) minds and (ii) ideas, where 'ideas' are to be understood as things which exist in minds. He did not infer that our ordinary beliefs about physical objects are therefore wrong, but that they need to be 'correctly construed', as being really beliefs about ideas and nothing else. The twentieth-century version of this theory was called 'phenomenalism', and it similarly claimed that propositions ostensibly about physical objects must be analysed as propositions which are really about actual and possible experiences (sense-data). This may be regarded as an attempt to *reduce* statements which are apparently about what we do not experience to other statements explicitly about what we do experience. Gödel presumes (with good reason) that this reduction does not succeed.

He compares this situation to Russell's attempt, in his 'no-classes' theory, to reduce talk of classes to talk of something taken to be more familiar, i.e. what Russell calls 'propositional functions'. His view is that this reduction too is unsuccessful, and he comments:

> This whole scheme of the no-class theory is of great interest as one of the few examples, carried out in detail, of the tendency to eliminate assumptions about the existence of objects outside the 'data' and to replace them by constructions on the basis of the data. The result has been in this case essentially negative ... All this is only a verification of the view defended above that logic and mathematics (just as physics) are built up on axioms with a real content which cannot be 'explained away'. (Gödel, 1944, pp. 460–1)

He has already told us that his own position is that:

> The assumption of such objects [as sets] is quite as legitimate as the assumption of physical bodies and there is quite as much reason to believe in their existence. They are in the same sense necessary to obtain a satisfactory system of mathematics as physical bodies are necessary for a satisfactory theory of our sense perceptions. (*ibid*, pp. 456–7)

---

[3] Ayer's (1936) is a well-known example, but the view was common amongst members of the so-called 'Vienna Circle'. Russell put forward a somewhat similar view in his (1914).

One naturally asks just what it is that the (Platonic) existence of these mathematical objects is supposed to explain, and in a later discussion Gödel's answer is more clearly stated:

> Despite their remoteness from sense experience, we do have something like a perception also of the objects of set theory, as is seen from the fact that the axioms force themselves on us as being true. I don't see any reason why we should have less confidence in this kind of perception, i.e. in mathematical intuition, than in sense perception. (Gödel, 1947, pp. 483–4)

This comparison between physical and mathematical objects, and their relation in each case to a suitable kind of 'perception', is evidently quite unconvincing. For the reason why the postulation of ordinary physical objects helps us to explain our ordinary experiences is that these objects are postulated as the *causes* of our experiences. Moreover, we have at least some idea of how this causal process works. For example, in the case of vision it is because light travels from the object perceived to the eye of the perceiver. (It is true that we do not have a *complete* explanation. We cannot say why light which travels from a tomato to me, and impinges on my retina, causing impulses to go from there to my brain, should give me just that kind of experience which I call 'seeing something red'. But at least we know how the explanation starts, if not how it ends.) By contrast, we have not the faintest idea of how there could be a *causal* process, beginning from such abstract objects as numbers are supposed to be, and leading me to a 'mathematical intuition' such as 'every natural number has a successor', or 'induction holds for the natural numbers (but not for other kinds of numbers)'. How could the real existence of the numbers, 'remote from sense experience', be supposed to *explain* why I have such 'intuitions'?

I add that Gödel does not suppose that *all* true mathematical axioms 'force themselves upon us' in quite this way. He notes that some axioms, though perhaps not at all 'intuitive', may yet be strongly recommended by their fruitfulness:

> There might exist axioms so abundant in their verifiable consequences, shedding so much light upon a whole field, and yielding such powerful methods for solving problems ... that, no matter whether or not they are [intuitive], they would have to be accepted at least in the same sense as any well-established physical theory. (Gödel, 1947, p. 477)[4]

---

[4] One notes that Russell (1907), discussed above on p. 241, proposes a similar justification for accepting some thesis as an axiom.

I remark only that this 'second-tier' criterion for accepting some thesis as an axiom must largely depend upon the first. For example, the 'abundance of verifiable consequences' must presumably refer to consequences *already* verifiable from other axioms, and in the end from those established by the original method of direct mathematical intuition. Again, the new 'methods for solving problems' must presumably be constrained by the requirement that the solutions in question do not conflict with the intuitive part of mathematics that we begin from. Gödel does not give any firm criterion for distinguishing those parts of mathematics which are justified by intuitively evident axioms and those which depend upon further axioms of this second tier, and it would not be surprising if this distinction is hard to draw. But in any case it is the problem with the first tier which is basic, and here one can only say that Gödel's parallel with ordinary perception is quite unconvincing. As he has no further explanation to offer, I now turn to a different kind of realist.

2. Neo-Fregeans

Frege's idea was that numbers are 'logical objects', and so we know about them in the same way as we know any truths of logic. But in order to substantiate this claim he was led to believe that *sets* must be 'logical objects', and must be governed by the two principles which form his axiom V, namely:

$$\forall F \exists y (y = \{x{:}Fx\})$$
$$\forall FG((\{x{:}Fx\} = \{x{:}Gx\}) \leftrightarrow \forall x(Fx \leftrightarrow Gx))$$

Since this axiom leads straight to a contradiction, we can hardly follow Frege in this. But perhaps we can stick to a formally analogous consequence that Frege drew for numbers, which is now generally called 'Hume's Principle', namely:[5]

[5] As I have said (p. 119), Frege himself gave no separate statement of the first part of this principle, and neither do the neo-Fregeans, since their rules of inference take it for granted that every name-like symbol does have a denotation. Further, they argue that since the second part introduces an identity-condition for numbers, the concept of a number must be what is called a sortal concept, which they appear to think is relevant to establishing the existence of numbers. But I see no such relevance, and shall pay no attention to their several remarks on the importance of sortal concepts.

$\forall F \exists y (y = Nx{:}Fx)$

$\forall FG((Nx{:}Fx = Nx{:}Gx) \leftrightarrow F{\approx}G)$

(Here '$F{\approx}G$' abbreviates 'there is a relation which correlates the things x such that Fx one-to-one with the things x such that Gx.) Frege showed that Hume's Principle will suffice as the sole basis for all of elementary arithmetic,[6] and we know that it cannot lead to contradiction. So those who are called 'neo-Fregeans' propose to put aside Frege's mistaken views on sets, and to start from what he himself evidently thought of as crucial, namely Hume's Principle about numbers. Their idea is that, if it can be explained how this is known, then it can thereby be explained how all the theory of natural numbers is known, for Frege's work has shown that no more than this is needed as a basis for elementary arithmetic.[7]

Frege himself did not accept this idea, for he himself argued that Hume's Principle would not suffice. Apart from the fact that it might be difficult to regard this principle as a truth *of logic*, he also claimed that there was a problem which it could not solve, which has become known as 'the Julius Caesar problem'. This arises because the principle tells us when the number of the Fs is the same as the number of the Gs, but it does not tell us *which* objects these things called 'numbers' actually are. For example, it does not by itself determine whether Julius Caesar is or is not a number. Frege inferred that even though the principle sufficed as a single axiom for deducing the arithmetic of the finite cardinal numbers, still it does not explain our ordinary understanding of these numbers. For we all believe that Julius Caesar is not a number, but apparently the principle cannot by itself explain this belief.[8] This is a problem which neo-Fregeans have to face, and I shall later come to their response. But first let us look in more detail at their overall programme.

[6] The demonstration is described in full detail in Heck (1993). Earlier versions may be found in Wright (1983) and in several papers by Boolos, reprinted as Part II of his (1998).

[7] The 'neo-Fregean' view is proposed in Crispin Wright (1983), which itself builds on earlier ideas in his unpublished (1969). Bob Hale adds to the case in his (1987). Since then the two together have made a convenient collection of their more recent thoughts on this topic in Hale and Wright (2001). I shall take this as the main work to be considered.

[8] The Julius Caesar problem was one reason that led Frege to give an explicit definition of numbers as sets of a special kind. He introduces sets by his axiom V, which is formally analogous to Hume's Principle. He does not appear to notice that the same problem affects his axiom V, i.e. it does not tell us whether Julius Caesar is or is not a set.

Frege's claim was that all the truths of elementary arithmetic are 'analytic', in the sense that their proofs depend just on logical laws and on definitions. He therefore thought that he had to provide an explicit definition of the operator 'Nx:—x—' from which Hume's Principle could be proved. But the neo-Fregeans do not seek for any further definition than is provided by Hume's Principle itself. As Frege himself had said, they take up the claim that this principle introduces a new concept, i.e. the concept of a number, and they (unlike Frege) go on to claim that it by itself gives a *complete* account of that concept. On their account, Hume's Principle is itself a kind of *definition* of the new concept.[9]

The central case of a definition is an explicit definition, whereby a new expression is introduced by being equated with a string of existing expressions. So the new expression may always be regarded simply as an abbreviation,[10] and it can always be eliminated from any context in which it occurs. What is called a recursive definition does not have this property, and yet it is still reasonably regarded as a definition. For example, where 'x'' denotes the number immediately after x, we may say that addition of finite numbers is *defined* by the two equations

$$x + 0 = x$$
$$x + y' = (x + y)'$$

These equations do not allow us to eliminate the sign '+' from all contexts in which it occurs, but they do determine just what the sign stands for, insofar as we can already and independently prove that there is one and only one arithmetical function that satisfies the equations.[11] Now a recursive definition might be looked at as a special case of the more general notion of an implicit definition, which takes this general form: a new sign is introduced with the stipulation that it is to be understood in such a way that the following sentences containing it are *true*. (And here we list the sentences in question.)[12] There are obviously two conditions that must be satisfied if this stipulation is to succeed in giving the sign a meaning. There must *be* a way of understanding the sign which makes the sentences true, and – at least to within extensional equivalence – only one such way.

---

[9]　For this outline of the neo-Fregean programme, see Hale and Wright (2001, pp. 1–14).

[10]　This may not always be the most useful way of regarding it. For some reservations see e.g. Quine, *Word & Object* (1960, section 39: 'Definition and the double life').

[11]　This is shown by Dedekind's recursion theorem, noted earlier (p. 103).

[12]　Compare Ayer's position on linguistic conventions, as described earlier in chapter 3, section 3.

The first condition is evidently not satisfied if the sentences in question are inconsistent, and the second would not be satisfied if far too little information is given (as with one who says: 'brillig' is to be so understood that 'Twas brillig' is true; and who says no more). Both conditions are satisfied in the case of the pair of recursive equations just noticed, but they surely are not satisfied by Hume's Principle, when that is considered as an 'implicit definition' of the sign '$Nx{:}Fx$'. I take first one and then the other.

Hume's Principle is certainly consistent, for it has a model with a denumerable domain, e.g. a domain consisting of just the finite cardinal numbers, the first infinite cardinal, and nothing else. (To obtain the model we give the signs '$Nx{:}Fx$' their natural interpretation, and we note that since the domain is denumerable no numbers beyond the first infinite cardinal will be needed.) The principle has no models with smaller, i.e. finite, domains. What larger models it may have is a non-trivial question, and one which I propose to sidestep.[13] But I remark that there must be *some* restriction on the principle if it is to hold in arbitrarily large domains. For it would be incompatible with the normal axioms of ZF set theory to suppose that there is such a thing as the number of all the sets that there are, or all the ordinal numbers, or all the cardinal numbers. The simplest question of this sort is whether we should suppose that there is such a thing as the number of all objects taken together, i.e. $Nx{:}x{=}x$. Both Hale and Wright shrink from affirming this. For purposes of the present discussion I shall therefore consider only the *most* restricted form of Hume's Principle, in which it posits only the finite cardinal numbers, for concepts F which have only finitely many instances. It is very likely that some less narrow restrictions would satisfy the neo-Fregean programme, but we need not explore that issue.[14] At any rate the idea of 'restricting' such an abstraction principle to only certain concepts F is in general acceptable to neo-Fregeans. For when Wright is discussing the analogous abstraction principle for sets, i.e. Frege's ill-starred axiom V, he wishes to say that a suitably restricted version of that can be accepted, and would *still* count as having the status

---

[13] There is an important discussion of this question in Fine (2002). Fine also pays attention to another question which I shall disregard, namely why it is that some 'abstraction principles' (such as Frege's axiom V) are inconsistent, and some, while each self-consistent, are not consistent with one another. (This is known as the objection to HP that it 'keeps bad company', since there are so many similar abstraction principles which evidently cannot be accepted. It is debated between Dummett and Wright in Schirn (1998).)

[14] Wright has suggested, e.g. in his (1999, p. 316), that we exclude from Hume's Principle only what Dummett would call 'indefinitely extensible concepts'. I leave this suggestion unexplored.

of a definition. (The restriction he suggests is that there will always be such a thing as the set of all objects x such that Fx *unless* there are 'too many' such objects (Wright, 1997, pp. 298–306).)[15] He could not object, then, to the proposal that Hume's Principle may also be restricted to something that is less daring, and I suggest that for the present discussion we may simply restrict it to the finite case. For that will still allow us to claim that the restricted principle is by itself an adequate basis for all of the arithmetic of the finite cardinal numbers.[16]

Most people will think that this restricted form of Hume's Principle may safely be asserted. One who takes a nominalist position about numbers will at first be inclined to protest, but that is because he will be thinking of the *expected* interpretation of '$Nx:Fx$', i.e. as referring to a *number*. But all that we need so far is that there is *some* way of interpreting '$Nx:Fx$' which makes the principle true, and that there will certainly be, so long as the universe of all objects is not finite. Even one who is nominalist about numbers is quite likely to think that there are at least *some* infinities (perhaps, say, the numerals), and that concession is enough for the present point. But there are those who would deny even this, and in any case the rest of us will surely want to say that, even if the universe is infinite, still it cannot be a matter of *definition* that it is. So, since Hume's Principle can be true only if the universe is infinite, it cannot reasonably be regarded as being just a kind of definition. I think this is a strong objection, though others may not. For example, Wright responds (quite correctly) that Hume's Principle only forces there to be an infinity of *abstract* objects (namely cardinal numbers). It has no implications for how many *concrete* objects there may be.[17] But I would reply that even the number of abstract objects cannot be the subject of a merely stipulative definition.

A proposal due to Horwich (most recently in his 2000) is that implicit definitions should always take a conditional form: '*if* there is something

---

[15] This proposed principle for the existence of sets is known as 'New V'. Some of its implications are explored by Parsons (1997) and by Shapiro and Weir (1999).

[16] I myself advocate a 'free' logic, in which a term such as $Nx:Fx$ may or may not turn out to denote something. In this case, the restriction is most simply given by:

$$\forall F(\exists y(y=Nx:Fx) \leftrightarrow \neg(\exists\infty x:Fx))$$

The quantifier '$\exists\infty x$', meaning 'there are infinitely many things x', may easily be defined in second-level logic. Others may prefer other methods. Various other formulations are given in Heck (1997a).

[17] Wright (1997, pp. 296–7; 1999, pp. 319–20).

which satisfies the condition stipulated, then so-and-so does', where 'so-and-so' is the new expression thereby introduced. To apply this idea in our case, the appropriate stipulation would be: if there is a function from concepts to objects satisfying the conditions imposed on '*N*' by Hume's Principle, then the function 'the number of' does. The idea here is that the stipulation then really can be seen merely as introducing a convenient way of *expressing* some truth that holds independently. One gets essentially the same effect by making the (simpler) stipulation: if numbers exist, then they obey Hume's Principle. Again, this does not claim that anything *does* satisfy the principle, and so it can be true even if the universe is not infinite. As it seems to me, this has quite a strong claim to be an 'analytic' truth, and one which may well be taken as a (partial) definition of what a number is. But, of course, it will not satisfy the neo-Fregeans. For it is obvious that if this is all that we are given by way of a premise then elementary arithmetic will not be forthcoming. Arithmetic makes existential claims, and if numbers are taken to be objects then it claims the existence of infinitely many objects, but no mere *definition* can substantiate that claim.[18]

But let us move on to our second condition on what counts as a 'reasonable' implicit definition. Suppose that we have some independent reason for believing that the universe is infinite. Then we can accept that there is a way of interpreting Hume's Principle (restricted to the finite numbers) that makes it true. But we also said that there should be only one such way, for otherwise the definition will not give any definite meaning to what it introduces, and this too is a condition which is not satisfied here.[19]

Hume's Principle gives a criterion of identity for numbers, but it is obvious that many other things will satisfy that criterion too. The simplest example is probably the numerals, and since the point would apply to the numerals in any notation (whether decimal or binary or whatever), this already gives us many different interpretations that all make the principle true. But

---

[18] Wright sometimes responds (e.g. in his 1990) that Hume's Principle makes no *additional* existence commitments, since its right-hand side, i.e. '*F≈G*', is *already* committed to the existence of numbers, even though it does not explicitly refer to them. This response is evidently untenable. For with equal justice one could claim that the right-hand side of Frege's axiom V, i.e. '∀*x*(*Fx*↔*Gx*)', is *already* committed to Frege's inconsistent assumption about the existence of sets. But that is obviously wrong.

[19] Compare the Frege–Hilbert correspondence on the axioms of geometry (i.e. Frege, 1980). Hilbert wanted to say that his axioms *define* the primitive terms 'point', 'line', and so on, though he also said that there are many different ways of interpreting those terms which make the axioms true. It is not surprising that Frege objected to this.

also, as we all know, numbers may be identified with sets in a variety of ways, and each of these identifications would satisfy Hume's Principle. Generalizing, as Benacerraf did in his 'What Numbers Could Not Be' (1965), *any* set of objects may be taken to be the numbers, provided that they can be seen as arranged in a progression, and the so-called 'Julius Caesar' problem is merely a special case of this. For if we are given any progression which does not already contain Julius Caesar as a member, we can form from it another which does, just by substituting him for (say) its first member. Then, whatever our progression, we interpret '*Nx:Fx*' as denoting whichever member of that progression has as many predecessors in the progression as there are *F*s, and Hume's Principle is satisfied. Thus, if Hume's Principle is *all* that we are given, by way of explanation of what the numbers are, the only conclusion to draw would seem to be the structuralist's conclusion: since arithmetic is the study of numbers, and numbers are explained only as entities that obey Hume's principle, arithmetic must be construed as the study of *all* systems of entities that satisfy that principle, i.e. of all progressions. Needless to say, the neo-Fregeans would not welcome this conclusion.

Their response, attempting to avoid the Julius Caesar problem, is to strengthen Hume's Principle in the hope of ensuring that only numbers will satisfy it. Their basic idea is to say that the criterion of identity provided by Hume's Principle is *essential* to numbers, but not to anything else, and they think that this will make the needed distinction. The thought is that other things that satisfy Hume's Principle have other and independent criteria of identity, e.g.: for numerals, their shape when written; for sets, their membership; and so on. By contrast, the idea is that numbers *have* to be individuated as Hume's Principle says, for there is no other way of doing it.[20] There seem to me to be two important objections to this proposal.

The first is that even if it were true it still would not evade Benacerraf's attack, for even if numbers and only numbers will satisfy the strengthened version of Hume's Principle, still it is not determined *which* numbers are the intended ones. For example, '*Nx:x≠x*' may be interpreted as denoting the number 0, as is presumably desired, but it may just as well be

---

[20] This brief paragraph boils down to a few sentences, the results of the neo-Fregeans' latest and very lengthy treatment of this issue, in Hale and Wright (2001, chapter 14). But I do not think that I am being unfair to them. I observe incidentally that their discussion is preoccupied with the question whether *a person* could be a number, and they do not often mention other candidates.

interpreted as denoting any other number, say 67. (In that case the next Fregean numeral, namely '$Ny{:}y = Nx{:}x{\neq}x$', may be interpreted as denoting the number 68; and so on.) As Benacerraf observed, any progression will contain further progressions within itself, and each of these may also 'play the role of' the numbers, a point which applies to the progression of 'the numbers themselves' (if there is such a thing) just as much as to any other progression. So Hume's principle cannot by itself tell us which is the intended one, for presumably the identity criterion which it provides will apply equally well to all of them.

This problem is discussed at some length in Hale (1987, chapter 8, sections 2–3). His approach relies on Frege's thought that numbers should be identified with classes of concepts, so that the question becomes: *which class is a given number to be identified with?* For the sake of argument, I shall simply accept this approach. (Of course one may object that numbers do not seem to be classes at all, that the existence of the classes in question is – to say the least – controversial, and so on. But this makes almost no difference to the main point.) Hale first claims that, in accordance with the desired interpretation of Hume's Principle, the classes in question would have to have the criterion of one–one correlation as their criterion of identity. He infers that this narrows the field to what we may call the 'Frege classes', i.e. the classes of all the Fregean concepts such that

   the concept is true of nothing
or: the concept is true of just one thing
or: the concept is true of just two things
or: etc.

But he also concedes that from this infinite list we may select any sub-progression whatever, and that in all cases we shall satisfy the requirement that one–one correlation should be the criterion of identity (pp. 208–12). So he then proposes adding a further individuating factor, which makes use of the familiar numerical quantifiers (defined as usual), namely:[21]

$Nx{:}Fx = n$ if and only if $\exists_n x{:}Fx$

This gives the game away. It concedes that what individuates the various numbers is not Hume's Principle, but the ways in which those numbers

---

[21] To avoid some obvious problems, we may regard this as an axiom-schema, for each specific numeral in place of 'n'.

are applied in practice. The criterion of one–one correlation is here playing no essential role, for of course the numerical quantifiers can be defined without making any use of this notion. Naturally, the ideas are connected, for it is quite easy to show that

$$\exists_n x{:}Fx \;\to\; (\exists_n x{:}Gx \leftrightarrow F{\approx}G)$$

But one cannot infer from this that the notion of numerical identity can only be understood via the notion of '$\approx$'.

This shows not only that the idea of one–one correlation is not *sufficient* as an explanation of what numbers are, but also that it is not *necessary*. For we can easily offer other criteria of identity in terms of the various different applications of the numbers, not only in numerical quantifiers but also in ordinal adjectives or in what I call 'numerical comparisons' (such as 'x is 3 times as long as y'). For example, the number n is the same number as the number m if and only if

$\forall F$ (There are n Fs $\leftrightarrow$ There are m Fs)

$\forall R$ (The nth term of the R-series $=$ The mth term of the R-series)

$\forall \phi \; \forall xy$ (x is n times as ϕ as y $\leftrightarrow$ x is m times as ϕ as y)

Clearly, there are yet more uses of the numbers that we might go on to consider, and each of them would suggest a similar criterion of identity.[22] It is also possible to give a quite different criterion, which pays no attention at all to any uses of the numbers, but simply locates them within the number-series itself. For example, only the number n can be the number which has n numbers as predecessors, and hence $n = m$ if and only if m has n predecessors. I add finally that we often assert identities between numbers without having any of these suggested criteria in mind, and apparently without needing them, as when we calculate that the square root of 169 is 13. But how could all this be so, if the notion of there being a one–one correlation between the Fs and the Gs was *essential* to understanding what a number is?

To sum up: Hume's Principle will not do as an implicit definition, (i) because (even when restricted to the finite case) it makes a positive claim

---

[22] I elaborate this thought later, on pp. 286–8. I add that in each case there would be a similar problem generated by the possibility that our universe contains only finitely many objects, and that in each case one might try to meet this as Frege did, i.e. by applying numbers to *themselves*.

about the size of the universe, which no mere definition can do, and (ii) because it does not by itself explain what the numbers are. Taken in a straightforward way, it may well be satisfied (when restricted to the finite case), but if so it is satisfied in many ways, and will not single out one of these ways in particular. Taken in the kind of strengthened way that the neo-Fregeans propose, when attempting to evade the problem of Julius Caesar, it is still the case that if it is satisfied at all then it is satisfied in too many different ways. But, when taken in this way, I see no reason to agree that it is satisfied at all. For there is not just *one* way of giving a criterion of identity for numbers, nor just *one* way of coming to understand what they are (or are supposed to be). So I now abandon the neo-Fregeans, and move on to a rather different kind of realism.

3. Quine and Putnam

This is essentially due to Quine, in various writings from his (1948) on, but it has become known as the Quine/Putnam theory because Putnam has expounded it (in his 1971) at greater length than Quine himself ever did. In broad outline the idea is this. We can know that mathematics is true *because* it is an essential part of all our physical theories, and we have good ground for supposing that they are true (or, anyway, roughly true). Our reason for believing in physical theories is, of course, empirical; a theory which provides satisfying explanations of what we have experienced, and reliable predictions of what we will experience, should for that reason be believed. But all our physical theories make use of mathematics, and therefore they could not be true unless the mathematics that they use is also true. So this is a good reason to believe in the truth of the mathematics, and (it is usually held) a reason sufficiently strong to entitle us to claim *knowledge* of the mathematical truths in question. Knowledge grounded in this way is obviously empirical knowledge. That is the outline of what has come to be called 'the indispensability argument'. I lead up to a more detailed account in a slightly circuitous way.

Rudolph Carnap, who taught Quine, believed in a firm distinction between analytic and synthetic statements, but he also thought that the analytic statements had their special status because they formed part of a 'linguistic framework'. However, he went on to admit that these frameworks themselves might be adjudged useful or non-useful, according to how well they served the purpose of empirical enquiry. In his own words:

The acceptance or rejection of . . . linguistic forms in any branch of science, will finally be decided by their efficiency as instruments, the ratio of the results achieved to the amount of effort and complexity of the efforts required. . . . Let us grant to those who work in any special field of investigation the freedom to use any form of expression which seems useful to them; the work in the field will sooner or later lead to the elimination of those forms which have no useful function. (Carnap, 1950, pp. 256–7)

This fudges the issue. It says that a statement is analytic only within a given 'linguistic framework', and that that framework itself is to be accepted or rejected on empirical grounds. But this means that it is an empirical question whether or not a supposedly 'analytic' statement should be accepted, which is surely not in accordance with what had always been meant by 'analytic', e.g. by Kant, or Frege, or even Ayer.

Quine is more straightforward. He thinks that there is *no* worthwhile distinction between the analytic and the synthetic, and that *all* our beliefs face 'the tribunal of experience'. In this respect mathematics is no different from the physical sciences, and he goes on to draw a moral. It is the orthodox view nowadays that (most) physical theories should be construed 'realistically'. That is, these theories are presented as positing the existence of things which cannot plausibly be regarded as perceptible (e.g. atoms, electrons, neutrinos, quarks, and so on), and we should take such positings at face value. So if the theory is verified in our experience then that is a good reason for supposing that the entities which it posits really do exist. But, as we have said, today's physical theories all make heavy use of mathematics, and mathematics in its turn posits the existence of things (e.g. numbers) which are traditionally taken to be imperceptible. So we should take this too at face value, and accept that if the physical theories are verified then these mathematical entities must also exist. This gives us an empirical reason for supposing that mathematics is true, and that the entities which it posits do really exist, even though they are not thought of as themselves perceptible entities.[23]

A common objection to this theory is that it is only *some* parts of mathematics that could be justified in this way, by their successful application in physical theory (or in daily life), whereas the (pure) mathematician will probably think that all parts of his subject share the same epistemic

---

[23] A relevant comment by Paseau (2007) is that the indispensability argument at best establishes the *truth* of mathematical claims. It does not dictate their *analysis*, but leaves it open whether that analysis might be Platonic, or structuralist, or nominalist, or of any other kind. Quine and Putnam are simply assuming the realist's analysis of these (supposed) truths.

status. Two obvious examples to mention here are the theory of the real numbers (as finally tamed by Cauchy and Weierstrass and Dedekind), and the theory of infinite numbers introduced by Cantor. The first was very largely developed in response to the demands of physical theory: physics *needed* a good theory of the real numbers, and physicists were very happy to take over the mathematicians' best version of this theory. By contrast, physical theory did not in any way require Cantor's development of the theory of infinite numbers, and the higher reaches of this theory still have no practical applications of any significance. In consequence, the indispensability argument that I have just sketched could provide a justification for saying that there really are those things that we call the real numbers, but it could not justify infinite numbers in the same way. But the (pure) mathematician is likely to object that each of these theories deserves the mathematician's attention, and that the same epistemic status (whatever that is) should apply to both.

This objection cuts no ice with proponents of the indispensability argument. They seriously do maintain that if there are only some parts of mathematics that have useful applications in science (or elsewhere), then it is only those parts that we have any reason to think true. Other parts should be regarded simply as fairy-stories. Putnam expresses the point in a friendly way: 'For the present we should regard [sets of very high cardinality] as speculative and daring extensions of the basic mathematical apparatus of science' (1971, p. 56). His thought is that one day we might find applications for this theory, so we may accept that it is worth pursuing, even if there is now no reason to think it true.[24] Quine is rather less friendly: 'Magnitudes in excess of such demands [i.e. the demands of the empirical sciences], e.g. $\beth_\omega$ or inaccessible numbers, I look upon only as mathematical recreation and without ontological rights'.[25] They would say

---

[24] There is a similar claim in Colyvan (2001): 'if there were no use for continuum mathematics anywhere in science . . . then real analysis would just be "mathematical recreation"' (p. 110). But it would still be worth pursuing, in case an application should turn up.

[25] Quine (1986), p. 400. I should perhaps explain that '$\beth$' (pronounced 'beth') is the second letter of the Hebrew alphabet, and the beths are defined thus:

$\beth_0 = \aleph_0$ (= the smallest infinite cardinal)

$\beth_{n+1} = 2^{\beth_n}$

$\beth_\omega$ = the least cardinal greater than all the $\beth_n$, for finite n.

(The natural model for Russell's simple theory of types has cardinal $\beth_\omega$.) Inaccessible cardinals are greater than any that could be reached by the resources available in standard ZF set theory.

the same of any other branch of mathematics that has not found application in any empirically testable area.

I shall ask shortly just how much mathematics could receive the suggested empirical justification, but before I come to this I should like to deal with two other very general objections to this line of argument, one which is due to Charles Parsons and endorsed by Penelope Maddy and another pressed by Elliot Sober. The first objection is that this indispensability argument 'leaves unaccounted for precisely the *obviousness* of elementary mathematics'.[26] This seems to me a misunderstanding, which arises because the proponents of the argument do tend to speak of the applications of mathematics *in science* (and, especially, in physics). This is because they are mainly thinking of applications of the theory of real numbers, which one does not find in everyday life. Of course science applies the natural numbers too, but we do not have to wait until we learn what is called 'science' before we see that the natural numbers have many useful applications. Indeed, one's very first training in school mathematics is a training in how to use natural numbers to solve practical problems. (E.g., if I have 10p altogether, and each toffee costs 2p, how many can I buy?) It is hardly surprising that those who have undergone such training in their early childhood should find many propositions of elementary arithmetic just *obvious*, but that is no objection to the claim that the reason for supposing them to be true is the empirical evidence that they are useful. For that is exactly how such propositions were learnt in the first place.

Next, Sober (1993) has objected to the argument from indispensability that no scientist ever does suppose that a test of his empirical hypothesis is also a test of the mathematics which he employs. For example, he never tries to combine the same empirical hypothesis with rival mathematical theories, so as to compare how well the different combined views stand up to testing. Similarly, when the test disconfirms his hypothesis, he never takes seriously the idea that it may be the mathematics employed that is disconfirmed. (Of course it may be taken to show that the mathematics in question turns out not to be *applicable* to the situation in question, but that is not taken to suggest that the mathematics itself may not be *true*.[27])

---

[26] Parsons (1979/80, p. 101); Maddy (1990, p. 31).

[27] Maddy likes to say (e.g. in her 1996, or 1997, Part II, chapter 6) that scientists often know that the mathematics which they are employing is 'not true', because it idealizes the situation. But it is the empirical situation which is idealized, so as to avoid complications in applying the mathematics to it. It is not the mathematics itself which is 'not true'.

No doubt what Sober says is a fair description of actual scientific practice, but it is no serious challenge to the Quine/Putnam argument. For Quine's theory comes with a 'maxim of minimum mutilation', which claims that we do and should revise our falsified theories in ways as economical as possible, i.e. in ways which give higher priority to the revisions which introduce the least change overall. This at once explains why the mathematics used would be questioned only as a last resort, since the effect of revising it would have all kinds of knock-on effects elsewhere.

I now move to a more detailed question: just *how much* mathematics could be justified on the ground that its applications cannot be dispensed with?

All proponents of the indispensability argument will agree with this first step: we do not need any more than set theory, and the usual ZF set theory is quite good enough (or perhaps one should say ZFC, to make it clear that the axiom of choice is included). One may happily admit that ordinary mathematics speaks of things (e.g. numbers) that are not sets. As we know, the numbers can be 'construed' as sets in various ways, but there are strong philosophical arguments for saying that numbers, as we actually think of them, cannot really *be* sets.[28] But the reply is that, in that case, we can dispense with numbers as ordinarily thought of, for the sets with which they may be identified will do perfectly well instead. The physical sciences do not ask for more than numbers construed as sets, even if that is not how numbers are ordinarily construed. I think that this first step of reduction is uncontroversial.

But how much of ordinary set theory is indispensable? We have already said that its claims about infinite numbers seem to go well beyond anything that physics actually needs. One might say with some plausibility that physics requires there to be such a thing as the number of the natural numbers (i.e. $\aleph_0$), and perhaps that it also requires the existence of the number of the real numbers (i.e. $2^{\aleph_0}$), but it surely has no use for higher infinities than this. Yet this provokes a problem. For how can one stop the same principles as lead us from $\aleph_0$ to $2^{\aleph_0}$ from leading us higher still? Well, one suggestion that is surely worth considering is this: do we really need anything more than *predicative* set theory?

As we have seen, in predicative theories a set is taken to exist when and only when it has a predicative definition, i.e. a definition that conforms to

---

[28] Notably Benacerraf (1965), discussed earlier (p. 161).

the vicious circle principle. It must follow that there cannot be more sets than there are definitions, and therefore not more than denumerably many. All the hereditarily finite sets will qualify, for each of these can (in principle) be given a predicative definition, simply by listing its members. There will also be *some* infinite sets of these, which can be identified with real numbers in the usual way, but the full classical theory of the real numbers is not forthcoming. However, we have noted that a surprisingly large part of the classical theory can in fact be recovered by the predicativist. It is quite plausibly conjectured (by Putnam, 1971, pp. 53–6; by Chihara, 1973, pp. 200–11; and by Feferman, 1998, chapter 14) that *all* the mathematics that is needed in science could be provided by a predicative set theory.[29] (So far as I am aware, no one has tried to put this conjecture to any serious test, but it seems plausible enough to me.)

There is of course no reason why one who accepts the indispensability argument should also subscribe to the conceptualism that is the natural justification for a predicative set theory. So the fact – if it is a fact – that the indispensability argument will only justify a conceptualist mathematics may be regarded as something of an accident. But it provokes an interesting line of thought, which one might wish to take further. The *intuitionist* theory of real numbers is even more restrictive than that which ordinary predicative set theory can provide. But is there any good reason for supposing that science actually *needs* anything more than intuitionistic mathematics? (Of course, the intuitionists themselves are not in the least bit motivated by the thought that they should provide whatever science wants. But perhaps, as it turns out, they do? This question is explored by Fletcher, 2002.)

More drastically still, one might propose that science does not *really need* any theory of real numbers at all. We all know that in practice no physical measurement can be 100 per cent accurate, and so it cannot require the existence of a genuinely irrational number, rather than of some rational number that is close to it (for example, one that coincides for the first 100 decimal places). Discriminations finer than this simply cannot, in practice, be needed. Moreover, physical laws which are very naturally formulated in terms of real numbers can actually be reformulated (but in a more complex manner) in terms simply of rational numbers. The procedure is briefly illustrated in Putnam (1971, pp. 53–6), who comments

---

[29]   Putnam speaks generally of all predicative sets. Chihara more precisely conjectures that the sets of Wang's system Σ_ω would be enough.

that 'A language which quantifies only over *rational* numbers, and which measures distances, masses, forces, etc., only by rational approximations ("the mass of *a* is *m* ± δ") *is*, in principle, strong enough to at least state [Newton's] law of gravitation.' I add that when the law is so stated we can make all the same deductions from it, but much more tediously.[30] The same evidently applies in other cases. At the cost of complicating our reasoning, our physics *could* avoid the real numbers altogether. If so, then there is surely no other empirical reason for wanting *any* infinite sets, and the indispensability argument could be satisfied just by positing the hereditarily finite sets. So we might next ask how many of these are strictly *needed*.

The answer would appear to be that we do not really need any sets at all, but only the natural numbers (or some other entities which can play the role of the natural numbers because they have the same structure, e.g. the infinite series of Arabic numeral types). Our scientific theories do apparently assume the existence of numbers, but they do not usually concern themselves with sets at all, and it has only seemed that sets have a role to play because the mathematicians like to treat real numbers as sets of rationals. But we have just said that real numbers could in principle be dispensed with, so that reason now disappears, and surely there is no other. It is true that the standard logicist constructions also treat rational numbers as sets (namely sets of pairs of natural numbers), but there is no need to do so. The theory of rational numbers can quite easily be reduced to the theory of natural numbers in a much more direct way, which makes no use of sets.[31] So apparently our scientific theories could survive the loss of all kinds of numbers except the natural numbers. But are even these really *needed*?

The most radical answer to this question is that there are absolutely *no* 'mathematical objects' that are strictly indispensable for scientific (or other) purposes. This answer is proposed by Hartry Field in his *Science without Numbers* (1980), and I have argued something similar in the final chapter of my (1979). Field explicitly characterizes his position as a variety of nominalism, but since it is not the usual variety I shall come back to it only after a more general discussion of nominalist approaches to the philosophy of mathematics.

---

[30] The general strategy is given in greater detail by Newton-Smith (1978, pp. 82–4).

[31] I have in mind a reduction in which apparent reference to and quantification over rational numbers is construed as merely a way of abbreviating statements which refer to or quantify over the natural numbers. For a brief account see e.g. Quine (1970, pp. 75–6), or my (1979, pp. 79–80).

B. Nominalism

Traditionally, nominalism is the doctrine that there are no abstract objects. It is called 'nominalism' because it starts from the observation that there are in the language words which appear to be names (*nomina*) of such objects, but it claims that these words do not in fact name anything. The *usual* version of the theory is that sentences containing such words are very often true, because these words are not really names at all, but have another role. The sentences containing them are short for what could be expressed more long-windedly without using the apparent names. (To illustrate with a trivial example: one may say that abstract nouns are introduced 'for brevity' without supposing that the word 'brevity' is here functioning as the name of an abstract object. For (in most contexts) the phrase 'for brevity' is merely an idiomatic variant on 'in order to be brief', and this latter does not even look as if it refers to an abstract object named 'brevity'.) A theory of this kind may be called 'reductive nominalism', for it promises to show how statements which apparently refer to abstract objects may be 'reduced' (without loss of meaning) to other statements which avoid this appearance.

4. Reductive nominalism

One may be a nominalist about some kinds of abstract objects without being a nominalist about all of them. For example, one might feel that numbers, as construed by the Platonist, are incredible, and yet feel no such qualms about properties and relations of a more ordinary kind. (The thought might be that ordinary properties and relations are entities of a higher type than the objects they apply to, and this makes them acceptable, whereas the Platonist's numbers are not to be explained in this way.) Conversely, one might feel that numbers have to be admitted as objects, whereas ordinary properties and relations do not, since in their case it is usually quite easy to suggest a reductive paraphrase. Or, to take a quite different example, one might feel that numbers are highly problematic whereas numerals are entirely straightforward, though numerals (construed as types, rather than tokens) are presumably abstract objects. I shall be concerned here only with nominalism about numbers, and in this section I shall focus mainly on the natural numbers. Can we say that the ordinary arithmetical theory of the natural numbers may be 'reduced' to some alternative theory, in which numerals no longer appear to be functioning as names of objects, and quantification over the numbers no longer appears as an ordinary first-level quantification over objects? Many have thought so.

A reduction that was in effect proposed by Hilbert is just to replace reference to numbers by reference to the numerals which are ordinarily understood as standing for them. For this purpose his preferred numerals were the stroke numerals, i.e. (finite) series of strokes, such as '| | |' for the Arabic numeral '3'. This is part of his explanation of how finitary arithmetic has a genuine content, and can straightforwardly be seen to be true. A more ambitious proposal on the same lines is due to Charles Chihara in his (1973). In this book Chihara advocates a predicative set theory, and we have noted that in such a theory there is for each set an open sentence of the theory which states its defining condition. He therefore proposes to replace all reference to sets by reference to the open sentences that define them. No doubt both numerals and open sentences should be counted as abstract objects, for what is in question is the string of symbols as a type, independently of whether there happen to be any concrete tokens of it. But it is hardly surprising if someone thinks that a string of symbols is a more comprehensible abstract object than is the number or set which it stands for or defines. After all, symbols are clearly human inventions, though this is not obviously true of numbers or sets.

There is an evident objection to this kind of reduction, namely that there are many different styles of numeral (e.g. Greek or Roman or Arabic, decimal or binary), and many different languages in which one can formulate open sentences, and it is obviously arbitrary to choose one rather than another. But I should here add that in Chihara's case the reduction of sets to open sentences was only the first step of his nominalism, for he went on to reduce talk of open sentences (construed as types) to talk of the *possibility* of uttering or writing tokens of them. Presumably he supposes that such possibilities are somehow more comprehensible than are the open sentences themselves, when construed abstractly, though I feel no temptation to follow him in this. Moreover, in his later (1990) he went further still, applying the same idea not just to the sets of a predicative theory but to all the sets recognized in the (impredicative) simple theory of types. Again the plan is that talk about sets of some specified level should be replaced by talk about open sentences of the corresponding level. In this way he aims to obtain a theory that is isomorphic to the simple theory of types, at least in its main outlines, but one which never asserts the actual existence of anything except individuals. Otherwise, it simply claims the *possibility* of constructing the relevant open sentences, in any language that you like.[32]

We could notice some difficulties connected with the idea of a *possible* open sentence: for example, are there as many of them as there are real numbers? But that would be a digression in a discussion which is focused just on the natural numbers, so I shall not explore it. Presumably we can grant that there are denumerably many possible open sentences, which allows for there to be one for each natural number. But which are the relevant open sentences? What is it that makes an open sentence a suitable representative of a natural number?

We meet the same problem with the initial and simpler idea that talk of natural numbers may be paraphrased as just talk about numerals. If we are already given some system of numerals, then such a paraphrase is extremely straightforward. (For example, 'every natural number has a successor' just becomes 'every numeral has a successor', which is likely to be true in all but the most primitive systems of numerals.) But the obvious objection to such a 'reduction' is that it must be arbitrary to specify any one system of numerals, and if no specification is made then we need to be given some general criterion for what is to count as a numeral. The obvious suggestion that a numeral is a sign that stands for a number cannot be used by one who abjures all reference to numbers. He may offer other descriptions of a more formal nature, e.g. that a system of numerals for the natural numbers must form a progression, with a first member, for each member a next, and no more members than these two conditions require. But there are (abstractly speaking) any number of progressions of symbols, most of which are not used at all, and in any case are not used as numerals. What matters, then, is how these symbols are *used*, and this brings us to the thought that what looks like a reference to a number should be paraphrased as an expression which *uses* that number (or numeral) in a non-referring way.

The favoured candidate here has always been the use of numbers (or numerals) in numerical quantifiers, so that what looks like a reference to the number n is replaced by a use of the associated numerical quantifier 'there are n objects x such that ... x ...'. This was Chihara's choice of which (possible) open sentences to employ, but the idea is by no means peculiar

---

[32] Putnam (1967a) contains a somewhat similar proposal to replace reference to abstract objects by appeal to the possibility of constructing concrete objects. But, unlike Chihara, Putnam did not work out this idea in any detail.

to him. Something like it was surely what Aristotle was thinking of when he said that arithmetic should be viewed as a theory of quite ordinary objects, but one that is very general. It is also very natural to say that this is what the Millian theory would come to, when purged of Mill's own talk of the operations of moving things about. As we know, Frege at one stage of his discussion proposed exactly this reduction, but then went on to reject it, because he claimed that we must recognize numbers as objects (1884, pp. 67–9). Those who disagree with him are likely to want to accept the reduction, and certainly it is the cornerstone of Russell's theory of natural numbers. For although he first takes numbers to be certain classes, his 'no-class' theory then eliminates all mention of classes in favour of the 'propositional functions' that define them; and in the case of the numbers these propositional functions just are the numerical quantifiers.

As we have said, Russell's theory runs into two main difficulties: first, it seems that some axiom of infinity is required, in order to ensure that each quantifier 'there are n . . .' is true of something; second, numerical quantifiers apply to (monadic) propositional functions of *all* types (or levels). Chihara aims to avoid the first just by postulating the required infinity of possible open sentences, and this is no doubt reasonable. He does not aim to avoid the second, but evidently thinks that it is of no importance. What appear to be the same numerical quantifiers reduplicate themselves from one type to another, and Chihara merely comments that this is just what to expect of open sentences (p. 74); he does not see it as posing a difficulty. But he should, for it means that there is no way of choosing *which* possible open sentences are to be employed in the reductive paraphrase of the statements of arithmetic. If those of one level will do the job, then so will those of any other.

My own idea that one should recognize 'type-neutral' predicates, as a special kind of predicate that can occur at any level, is aimed at this problem. For if we can recognize type-neutral predicates at all, then it is obvious that the numerical quantifiers will be among them. I have claimed that we can, and the theory is briefly sketched earlier (in section 3 of chapter 5). But I have to admit that this theory introduces ideas which are not familiar, and which do not seem to be as 'clear and distinct' (in the Cartesian sense) as one would like. One cannot have very much confidence in it, and in fact a satisfying theory of the numerical quantifiers proves to be much more difficult to attain than might at first have been expected. Certainly it is a great deal more complex than the ordinary theory of the natural numbers that we all learn in early childhood. So the wholesale 'reduction' of the latter to the former might seem to be a somewhat dubious enterprise. But let us now step back a little, to consider the problem more generally.

As I have said, the attempt to 'reduce' the theory of natural numbers to the theory of numerical quantifiers is certainly the one which has attracted the most attention. But it is not the only reduction worth considering. For example, Wittgenstein's *Tractatus* (1921) contains a different proposal, summed up as 'a number is the index of an operation' (*Tractatus* 6.021). The basic thought here is focused not on 'there is 1 . . .', 'there are 2 . . .', and so on, but rather on the series that begins with 'once', 'twice', 'thrice', understood as applied in this way. Starting from a given object, and operating on it 'twice' is first applying the operation once to the given object, and then applying the operation again to *what results from* the first application. (Thus the instruction 'add 1 to 3 twice' is not obeyed by writing the same equation '3 + 1 = 4' twice over, but by successively writing the two different equations '3 + 1 = 4' and '4 + 1 = 5'.) I would myself prefer to generalize this a bit, on the ground that an 'operation' corresponds to a many-one relation (i.e. the relation between what is operated upon and what results from the operation), and the idea of a numerical index can be applied to all relations, not just those that represent operations. We define what are called the 'powers' of a relation in this way[33]

$$xR^1y \leftrightarrow xRy$$
$$xR^{n+1}y \leftrightarrow \exists z\,(xR^nz \wedge zRy).$$

Using this idea, '7 + 5 = 12' comes out very simply as

$$\forall R\,\forall xy\,(\exists z\,(xR^7z \wedge zR^5y) \leftrightarrow xR^{12}y).$$

Pure logic can obviously provide the proof. (Indeed in this case the logic can be even 'purer' than in the case of the numerical quantifiers, since we do not need to invoke the notion of identity.[34]) But again, when we

---

[33]   If desired, one may add

$$xR^0y \leftrightarrow x = y.$$

[34]   This is *perhaps* the reason why the *Tractatus* prefers this reduction to Russell's, for the *Tractatus* does not allow the introduction of a sign for identity.

pursue in detail the proposal that *all* of arithmetic be reduced in this way, we find exactly the same problems as before. If the definitions are given in the obvious way, then apparently we shall need something like an axiom of infinity; and what I call the 'type-neutrality' of the numbers will again cause problems. For we can consider the powers of a relation of any level whatever, and a logic that will allow us to generalize over all of these at once is not easy to devise. A type-neutral solution such as I have proposed for the numerical quantifiers is easily adapted to this case too, but of course the same objection still applies: the logic proposed is not familiar, and certainly more complex than that of ordinary arithmetic.

Once this line of thought is started, there is no end to it, for in our quite ordinary concerns the natural numbers are applied in *many* ways. We have mentioned so far their use as indices of operations ('double it twice'), or as powers of relations ('cousin twice removed'), and their use as cardinals ('there are two'). But obviously there are others. For example, the natural numbers are used as ordinals ('first', 'second', 'third') in connection with any (finite) series. They are also used in what I call 'numerically definite comparisons' such as 'twice as long' and 'three times as heavy'. One can of course suggest yet further uses (e.g. to state chances), but those I have mentioned will be quite enough to make my point. One can set out to 'reduce' the theory of the numbers themselves to the theory of any one of these uses. In each case one encounters essentially the same problems (infinity and type-neutrality), and if they can be solved in any one case then they can equally be solved in the other cases too. So there is nothing to choose between the various reductions on this score. Moreover, I do not believe that there is any other way of choosing between them either. So we are faced with a further application of Benacerraf's well-known argument in his 'What Numbers Could Not Be' (1965). Assume, for the sake of argument, that the technical problems with each of these proposed reductions can be overcome, so they are each equally possible. Moreover, there is nothing in our ordinary practice that would allow us to choose between them, so they are each equally good. But they cannot *all* be right, for each proposes a different account of what the statements of ordinary arithmetic actually mean. Hence they must *all* be wrong. I find this argument very convincing.[35]

What it shows is that *for philosophical purposes* no such reductive account of the natural numbers will do, for none preserves the meaning

---

[35] A somewhat similar argument may be found in Papineau (1990, esp. section 6).

of the simple arithmetical statements that we began with. As Frege claimed (when commenting on Mill's proposed reduction) the theory of the numbers themselves must be distinguished from the theory of any of their applications (1884, p. 13). But we have earlier argued (p. 263) that it is equally true that the two cannot be completely divorced, i.e. one cannot suppose that our knowledge of the numbers themselves is acquired quite independently of any such applications. So some kind of compromise is required. Moreover, the present section has introduced a complication into this question by stressing that numbers have *many* applications, and it is arbitrary to work with just one of them, e.g. the numerical quantifiers, while ignoring all the others. No doubt, for most people it was the numerical quantifiers that provided their first introduction to the numbers, but there is no necessity in this. One *could* perfectly well begin with the numerical ordinal adjectives ('first', 'second', 'third', and so on), or the numerical powers of relations, or in yet other ways. For none of these applications depends upon any other.[36]

The moral that this suggests is a return to something like the Quinean view: we posit a theory of numbers as self-standing objects, existing in their own right, in order to explain and unify all the various applications. Since this theory is successful, and does do all that it was meant to do, we have good reason to suppose that it is true. After all, scientific theories are accepted because of their success in explaining and unifying the data, and the same idea can surely be applied to mathematical theories too. There is perhaps this difference, that in the scientific case the data are given (ultimately) by observation, whereas in the case of the numbers the data are the so-called applications of the numbers, and a case can be made for saying that these truths may be known *a priori*. But in any case the idea is that justifying a theory by what is called 'inference to the best explanation' is a method of justification that applies in both cases.

The position just outlined, as resulting from the failure of reductive nominalism, is of course not a nominalistic position at all. It urges us to take a realist view of the numbers. But it leads very naturally to a different kind of nominalism, which I shall call 'fictionalism'. The basic idea is that the theory which posits the real existence of 'the numbers themselves' is of course *useful*, but nevertheless what it actually says is *not true*. That is, it is a fiction. But what are true *instead* are those statements in which numbers are applied, and they somehow justify the fiction. This view is presented and

---

[36]   This point is demonstrated at some length in my (1979, chapter 1).

explored by Hartry Field, principally in his book *Science without Numbers* (1980), but also in some later papers. However, I shall begin by offering my own arguments rather than his. This is mainly because I shall begin by continuing the theme of the present section, with its focus just on the natural numbers, whereas Field's book spends little time on this.[37] But it is also because his own opening arguments for the position (in chapter 1 and its appendix) are by themselves quite unconvincing.

These arguments simply assume that we can start with a 'nominalist' or 'physicalist' version of any scientific theory, which does not mention such Platonic entities as the numbers. We then consider adding to this a Platonist theory which does mention such things, and which contains all the vocabulary of the nominalist theory, together with suitable bridge laws connecting the two vocabularies. Let us call this the combined theory. Then what Field tries to argue for is that any proposition that uses only the nominalist vocabulary and that is entailed by the combined theory is also entailed just by the nominalist part of that theory. That is to say, the combined theory is a conservative extension of the nominalist theory. These general arguments of chapter 1 require several assumptions – most obviously, that the combined theory is consistent – and are open to some technical objections.[38] But it is the initial assumption, that there always are such 'nominalistic' versions of any scientific theory, that will strike most people as highly problematic. As a preliminary, I merely note that there is a similarity between Hilbert's position on 'ideal' elements in mathematics and Field's position on 'Platonic' elements in science: in each case the idea is that these are *useful* additions to a more basic theory, but do not have to be *true*, so long as they are *conservative* additions.

5. Fictionalism

In chapter 6, when discussing formalism, I observed that in many areas of mathematics there could not be any serious question over whether the

---

[37] He treats the topic only on pp. 20–3.

[38] I mention some below. I remark that Field himself later became dissatisfied with the Platonist assumptions that these general arguments in his (1980) rely upon, and so has offered a more 'nominalist' version of the argument in his (1992). I remark also that Field aims to defend the nominalist position on *all* abstract objects whatever, but I have already said (on p. 282 above) that my aim is narrower.

relevant axioms are *true*. For the axioms simply define what the subject is (as e.g. in group theory), and the mathematician's problem is to discover what must follow for any system of entities that satisfies them (pp. 181–2). I also remarked that resistance to this approach was found mainly in three areas: the theory of the natural numbers, the theory of the real numbers, and the theory of sets. In these cases one has at least some sympathy with Gödel's claim that the axioms 'force themselves upon us as being true'. I shall therefore consider the rival theory of fictionalism only with respect to these three areas.

(i) Natural numbers[39]
Everyone will agree that ordinary arithmetic is very useful, not just in what is called 'science' but also in many aspects of everyday life. (That is why it is one of the first things that we learn at school.) The Quine/Putnam argument claims that it would not be useful unless it were true, but the fictionalist replies that even false theories may yet be useful. In the philosophy of science this is called an 'instrumentalist' view of theories. The idea is that a scientific theory should be an effective 'instrument' for the derivation of predictions from observational data, but that no more than this is required. And a theory may be a very efficient 'instrument' of this sort even though it is not true, nor even an approximation to the truth. It has sometimes been said that *all* scientific theorizing should be viewed in this way, but that is not a popular view these days.[40] A common view (which I share) is that although a fairy tale may provide very useful predictions, it cannot provide *explanations* for why things happen as they do. In order to do that, a theory must also be true (or, at least, an approximation to the truth). But there are *some* cases of scientific theories which have deliberately been proposed *simply* as instruments of prediction, though this is not common. (A well-known historical example is Ptolemy's theory of planetary motion.)[41]

[39] Two interesting articles that go *most* of the way towards a fictionalist account of this topic are Hodes (1984) and Jeffrey (1996).
[40] The view mostly appeals to those of a strongly verificationist outlook (e.g. Berkeley). An interesting modern exponent is Van Fraassen (1980). It is claimed by Burgess and Rosen (1997, part IA) that one cannot take an instrumentalist view about one part of science without being committed to the same view about all parts. This claim has no foundation. (It is taken over from the earlier Burgess, 1983).
[41] For detailed accounts of the Ptolemaic system see e.g. Neugebauer (1952, Appendix 1) or Dreyer (1953, esp. chapters 7 and 9).

The fictionalist view of ordinary arithmetic is 'instrumentalist' in roughly this sense. That is, it views arithmetic as a calculating device which is not itself true but which does lead us from true non-fictional premises to true non-fictional conclusions. But Hartry Field's approach goes further, in two ways, for in his view arithmetic is not *just* a calculating device. Rather its role is to simplify calculations which could have been done without it, but in a more long-winded way. Connected with this is the idea that it introduces a kind of unity into what would otherwise be unconnected facts. Let us take these points in turn.

The arithmetic thesis that $7 + 5 = 12$ is supposed to unify and explain such facts as these:

(i)   If you have 7 of a thing here, and 5 there, then you have 12 of that thing altogether.

(ii) If you start with the 7th member of any ordering, and go on to the 5th after that, you will end at the 12th.

(iii) If you do something 7 times, and then 5 times more, you will have done it 12 times.

(iv) If one part of a thing is 7 times as heavy as x, and the rest of it 5 times as heavy, then the whole thing is 12 times as heavy as x.

Obviously, many more such examples could be provided. Now the (supposed) fact that $7 + 5 = 12$ is not needed as an *explanation* for such truths as are stated by (i)–(iv), for each can be seen to be true quite independently of this (supposed) fact. If I may simplify a little, they depend only on an explanation of what is meant by such phrases as '7 cows', 'the 7th day', 'knock 7 times', '7 times as heavy'. It may be said that we need some account of the similarity between them, which is shown here by the fact that the same numerals '7', '5', '12' are used in all of them, and that this is due to the fact that they are all applications of the same equation '$7 + 5 = 12$'. But the reply is that this introduces a point which is not needed, for we can surely say instead that expressions of the type 'there are n', 'the nth', 'n times' in each case fall into the same structure, namely that of a progression. That is by itself enough to show why the same numerals are used in each case. There is no need to posit yet another example of that same structure, a progression of abstract objects called the numbers, in order to explain why the initial examples all share it. For this further example adds nothing. You may perhaps say that in a way it simplifies things, for we do find it *easier* to think of a structure as a structure of objects, rather than as a structure of what is signified by certain quantifiers, or adjectives, or adverbs, or whatever. But I still say that this does not explain anything which is not equally explicable without it.

In a similar way, arithmetic simplifies – i.e. makes easier – the task of calculation. Here is an example from Field (1980, p. 22). Suppose that

(i)   there are exactly twenty-one aardvarks;
(ii)   on each aardvark there are exactly three bugs;
(iii) each bug is on exactly one aardvark.

The problem is: how many bugs are there? The *method* is: translate the problem into a problem in pure arithmetic, namely 'what is 21 × 3?'; calculate arithmetically that the answer is '63'; translate back into the language that we began with, concluding

(iv)   there are sixty-three bugs.

Is that not a very convincing account of what we all learnt to do at school? But the point is that this detour through pure arithmetic was not in the strict sense *needed*. For, given standard definitions of the numerical quantifiers 'there are 3', 'there are 21', and 'there are 63', the result *could* have been reached just by applying first-order logic (with identity) to the premises. Of course this proof, if fully written out in primitive notation, would occupy several pages (and would scarcely be surveyable). But, in principle, it is available. This again illustrates the claim that pure arithmetic is *useful*, but is never strictly *needed*.

Field's way of putting this claim is to say that the addition of ordinary arithmetic is a *conservative* addition, which means (as normally understood) that it does not allow us to prove any more results in the language of the numerical quantifiers than could have been proved (no doubt more tediously) without it. This is a very plausible claim. But to investigate it in detail we must first be much more precise about what to count as 'the theory of the numerical quantifiers'. Here Field's view is very different from that which I am drawn to (as given above in chapter 5, section 3). Roughly speaking, his theory is minimal whereas mine is maximal.

For example, does this theory include generalization over the numerical quantifiers? I would say 'yes', but Field would say 'no'. On his view, it is only in the fictional story of the numbers themselves that any kind of quantification over numbers is available. Does the theory also allow us to generalize over all properties of the numerical quantifiers? Again, I would wish to say 'yes', for I think that 'second-level' logic is in fact perfectly clear, and that there are very good reasons for wanting it; but Field would prefer to say 'no', for he would rather avoid second-level logic.[42] These are points on which proponents of essentially the same idea will differ among themselves, but they do affect just what is to be meant by the claim that the arithmetic of the numbers themselves is a 'conservative' addition.

First-level logic has a complete proof procedure, which is to say that in that logic the notions of semantic consequence (symbolized by ⊨) and syntactic consequence (symbolized by ⊢) coincide. Second-level logic is not complete in this way, and so the two notions diverge: whatever proof procedure is chosen there will be formulae which are not provable by that method but which are true in all (permitted) interpretations. So if a second-level logic is adopted as background logic, we have to be clear about what is to count as a 'conservative' addition to some original theory. In each case the idea is that statements of the original theory will be entailed by the axioms of the expanded theory only if they were already entailed by the axioms of the original theory, but we can take 'entailment' here either in its semantical or its syntactical sense. In the first case we are concerned with statements which have to be true if the original axioms are true, and in the second case with those that are provable from the original axioms. In the context of a second-level logic, there will be statements which have to be true (if the axioms are) but which are not provable (from those axioms). In this situation it seems to me that all that is required of a theory that adds the numbers themselves is that the addition be *semantically* conservative, which is enough to ensure that it cannot take us from 'nominalist' truths to 'nominalist' falsehoods. But if the addition allows us to *prove* more (as it may), then that should be regarded as just another way in which the addition could turn out to be useful.[43]

---

[42] My brief explanations of second-level logic in chapter 5 above, especially sections 1 and 2, are evidently favourable to it. I have tried to argue the case more fully in my (1998). See also Shapiro (2005a). Field (1980) does use a second-level logic in his 'nominalistic' construction of the Newtonian theory of gravitation (chapters 3–8), but then (chapter 9) he discusses first-level variants of this theory, and is evidently attracted by them. The preference for first-level theories is stated much more strongly in his (1985).

[43] Shapiro (1983) points out the importance, for Field's programme, of distinguishing between semantic and syntactic conservativeness. But the point had in effect been anticipated by Field himself (1980, p. 104).

The relevance of this point may be seen thus. Suppose first (as I would prefer) that the proposed theory of the numerical quantifiers does allow us to quantify over these quantifiers, does allow us also to quantify over the properties of these quantifiers, and thereby allows us to prove (analogues of) Peano's postulates for these quantifiers. Then the theory will be a *categorical* theory, which means that all its models are isomorphic, and hence that any statement in the language of this theory is either true in all models of the axioms or false in all models.[44] It follows that the addition of any other theory *must* be a conservative extension (in the semantic sense), provided only that the addition is consistent. Suppose, on the other hand, that the original theory of the numerical quantifiers is much more limited: it adds to ordinary first-order logic just the definition of each (finite) numerical quantifier, and adds no more than this. (This is apparently what Field himself envisages in this case, p. 21.) Then again the addition of any other theory must be a conservative extension (in either sense) unless it actually introduces an inconsistency. This is because the language of the original 'nominalistic' theory is now so limited that only very elementary arithmetical truths can be stated in it, and these can all be certified by the first-level theory of identity which is a complete theory. Either way, Field's claim is in this case vindicated: the addition of pure arithmetic to any theory which applies the natural numbers as numerical quantifiers will be a conservative extension.

Of course, I have only argued for this in two particular cases, of which the first was a very ambitious theory of these quantifiers while the second was extremely restricted. Obviously, there are intermediate positions which one might think worth considering.[45] Also, I have not considered

[44] Briefly, the argument is this. When Dedekind discovered the axioms that are now called 'Peano's postulates', he proved that these axioms are categorical. The proof that he gives in his (1888) is very straightforward and easy to understand, but it does presuppose that the logic employed is a second-level logic, with the postulate of mathematical induction understood as quantifying over absolutely all properties of natural numbers, whether or not the vocabulary employed allows us to express those properties. This proof transfers quite straightforwardly to the theory of numerical quantifiers, provided again that we may quantify over absolutely all properties of those quantifiers.

[45] It is obvious that for any particular quantifier '$\exists_n$' you can define its successor '$\exists_{n'}$'. In Field's restricted theory, as I understand it, you cannot even state that these quantifiers are different quantifiers (e.g. that they are not equivalent, or not necessarily equivalent). A position intermediate between mine and his might provide a 'nominalist' way of stating this, and then the question would be whether you could prove it without ascending to the fictional story of 'the numbers themselves'. (Compare p. 28 above, on Aristotle's problem over the infinity of the numbers.)

any of the other ways in which natural numbers may be applied, though we have noted that actually there are very many such ways. I think that we should come to the same conclusion in all cases, namely that the truth of pure arithmetic is not *required* for the explanation of any actual phenomenon. But I do not stay to elaborate that idea here, for much of my (1979) provides a detailed demonstration. Let us now move on from the natural numbers to the real numbers.

(ii) Real numbers
The usual way of stating the Quine/Putnam argument for the indispensability of the numbers is that they are indispensable *for science*, and 'science' here is mainly thought of as physics. It is obvious that in this context it is the real numbers that figure most prominently, so the question is whether we can extend to them the same view as has been argued for the natural numbers. That is, can we say that the theory of the real numbers is certainly useful in science, but need not be viewed as true, because whatever we can do with it we can (in principle) do without it? To put this in Field's terms: is it true that a physical theory can always be formulated in a 'nominalistic' vocabulary, which does not mention the real numbers, and that when it is so formulated the addition of the usual theory of real numbers must always be a conservative extension?

The simplest case to begin with is that of (Euclidean) geometry. When our schoolchildren are introduced to this geometry, it is not long before they learn to speak in terms of the real numbers. For example, they learn that the area of a circle is πr^2, where 'π' is taken to be the name of a real number, and so is 'r', and the theorem is understood in some such way as this: multiplying the number π by the square of the number which measures the length of the radius of a circle gives the measure of the area of the circle, in whatever units were used to measure the length of the radius. Does this not presuppose the existence of the real numbers π and r^2? Well, it is certainly natural to say that, when the theorem is stated in these terms, it does have that presupposition.

But the existence of the real numbers cannot be *necessary* for Euclidean geometry, as can be argued in two ways. Field (1980, p. 25) relies upon the point that a modern axiomatization of geometry, such as is given in Hilbert (1899), does not need to claim the existence of the real numbers anywhere in its axioms. So the basic assumptions say nothing of the real numbers, though in practice real numbers will quite soon be introduced, e.g. by showing that the axioms imply that the points on a finite line are ordered in a way which is isomorphic to the ordering of the real numbers in a finite interval. A line of argument which I prefer goes back to the ancient Greek way of doing geometry, which never introduces real numbers at any point, since the Greeks did not recognize the existence of such numbers. Nevertheless their techniques (with a little improvement) could be used to prove whatever we can prove today, though in a more long-winded fashion. They would instead use the theory of proportion due to Eudoxus. Here is a simple illustration.[46]

What we would say about circles, by introducing a reference to the real number π is said in the Greek fashion by:

(i)   In every circle, the ratio in length of the diameter to the circumference is always the same.
(ii)   In every circle, the ratio in area of the square on its radius to the whole circle is always the same.
(iii) Moreover the ratio in (i) is the same as the ratio in (ii).

(We may if we wish introduce a name for this ratio, and call it π. We may also show, as Archimedes did, how to approximate its value, i.e. as lying between two ratios of natural numbers that are close together.) When I express the claims in this way, it may seem that, even if I am not referring to things called real numbers, I am referring to things called ratios, which is not what a nominalist would desire. This could be avoided just by some rephrasing (e.g. 'the diameter of c_1 is to the circumference of c_1 as the diameter of c_2 is to the circumference of c_2'). But the more important point is that the whole theory is based on a formal definition, as noted on p. 91 above, whereby what we express as:

The ratio of a to b in φ-ness is the same as the ratio of c to d in ψ-ness

is defined simply as:

$$\forall nm((n\cdot a >_\varphi m\cdot b) \leftrightarrow (n\cdot c >_\psi m\cdot d))$$

This quantifies over the natural numbers, and refers to the objects a, b, c, d, but it does not even appear to mention such abstract objects as ratios or real numbers. (And as for the apparent quantification over the natural

---

[46] I have treated the Greek theory of proportion at some length in my (1979), chapters 3–4.

numbers, a suitable analysis of '$n \cdot a$' will show how this can be exchanged for a quantification over the applications of those numbers.)

The use of the real numbers in geometry can be handled either in Field's preferred way or in my preferred way, and so far as the overall project is concerned it does not matter which. But geometry is no doubt a very simple case. Putnam (1971, p. 36) issues a challenge on the Newtonian theory of gravitation, where the basic law may be stated as:

$$F = \frac{g M_a M_b}{d^2}.$$

Here F is the force, g is a universal constant, M_a is the mass of a, M_b is the mass of b, and d is the distance between a and b. On the face of it, *all* these symbols refer to real numbers. Can the law be restated without such a reference? Well, the answer is that it can, and the bulk of Field's (1980) is devoted to establishing this point. As before, my own preference is for a different method, showing once more how these apparent references to real numbers may be replaced by the same Greek theory of proportion. But in either case we agree upon the conclusion. The elimination is possible, and that is what matters.[47,48]

It does not follow that all mention of the real numbers can be eliminated from *every* theory that the physicists have proposed or will propose. Balaguer, in his (1998, chapter 7) has put forward a version of fictionalism which claims that this point need not matter. His idea is that the scientists make use of the ordinary theory of real numbers as a *descriptive tool*, and that for this purpose it does not matter whether the theory is or is not a fiction, for even what is known to be a fiction may still be a useful descriptive tool. (He offers this illustration (p. 140): 'A historical description of the years surrounding the Russian Revolution, for instance, could very easily use talk of the novel *Animal Farm* as a theoretical apparatus, or descriptive aid. We can say something roughly true about Stalin

---

[47] I do not even sketch the elimination in this case, since it is somewhat complex. But I remark that there are *categorical* theories of continuity, which can be applied not only to such abstract things as numbers but also to quantities such as force, mass and distance. So the Newtonian theory can be given a categorical but still 'nominalistic' formulation.

[48] It is sometimes objected to Field, e.g. by Resnik (1985), that *his* kind of nominalism should not admit the existence of space-time points. I see no force in this objection. Field replies to it in his 1984. I also remark that the notion of a point can, if one wishes, be replaced by that of an ever-shrinking series of regions. The idea is due to Whitehead (1919 and 1929), but its proper execution proves tricky, and is best accomplished by Roeper (1997).

by uttering the sentence "Stalin was like the pig Napoleon", even though this sentence is, strictly speaking, false.') It is not easy to believe that a story which is known to be a fiction might be *indispensable* as a descriptive aid, though perhaps it is not impossible.[49] What seems to me (and to Field) to be more important is that the ordinary mathematics is then used to draw consequences from these mathematical descriptions. For if you can draw from a fictional premise a true non-fictional conclusion, and if there is no other way of obtaining that conclusion, then that is a good (but not conclusive) reason for saying that the premise must after all be true, and not a fiction. It is the elimination of pure mathematics from the reasoning, rather than from the initial descriptions, that is crucial to the fictionalist claim. (But presumably the first will imply the second.)

To return, from the few examples given it does not follow that all mention of the real numbers can be eliminated from *every* theory that the physicists have proposed or will propose. Since scientists nowadays have no scruples over presupposing the real numbers, their theories are usually formulated in a way which simply assumes the real numbers right from the start. Nevertheless, it seems to me that the two examples given do create quite a good *prima facie* case, and make it reasonable to hope that sufficient effort could provide versions of other scientific theories which have been freed from this assumption. Besides, there is a very general reason for saying that the existence of such abstract objects cannot really be *needed* in the explanation of why physical objects behave as they do. For who would suppose that, if the real numbers did not exist, then the behaviour of physical objects would be different (e.g. that apples would not fall from trees with the rate of acceleration that we call '32 ft/sec^2')? A good explanation of why some physical object behaves as it does should mention only those factors that affect the behaviour in question. But, despite what Quine suggests, numbers do not act as causes in the way that atoms or electrons do. So, if the behaviour can be explained at all, it must be possible to explain it without assuming that numbers are objects which have an essential role to play.[50] (I am assuming here that scientific explanations are, or should be, causal explanations.)

---

[49] One may once more note how scientists frequently use theories which they know to be idealizations.

[50] Burgess is by no means a fictionalist, and yet even he is prepared to say that 'having a mathematical ontology is a feature imposed on science more by us than by the universe' (1990, p. 14). I add that what we impose need have no reality independent of us.

So far I have mentioned the use of the real numbers only in Euclidean geometry and in Newtonian physics, but of course they have many other uses too. I see no difficulty of principle in accommodating the change from Euclidean to non-Euclidean geometry, or from Newtonian to relativistic physics, for these theories are still broadly of the same type. In each case the basic assumption that legitimizes the application of the real numbers is the assumption that distance, time, and other relevant quantities are both continuous and measurable quantities. But that is also what makes the Greek theory of proportion applicable, and all such quantities can be treated by its means instead.[51] I add that the Greek theory of proportion is already a categorical theory. As before, categoricity is only available against the background of a logic that is of second level at least, but in the present case that is inevitable anyway. For an application of the usual theory of the real numbers must presuppose that the quantity in question is continuous, and continuity can be defined only in a second-level logic. Given this background, the usual proof of the categoricity of the axioms for real numbers (which is due to Cantor) can quite easily be transferred to a categoricity proof for the Greek theory of proportion. Any addition to a theory that is already categorical must be conservative (in the sense of *semantic* entailment), provided only that the addition still leaves a consistent theory. Taking this proviso for granted, it follows that thinking in terms of the real numbers can do no harm, even if no such things actually exist.

I add a final note. The real numbers have *many* applications, and only two have been considered here in any detail, though a few simple extensions of those two have also been mentioned. But there are many further theories within physics that make use of the real numbers, and some that are outside physics altogether (e.g. the theory of probability). The fictionalist position claims that in *all* cases a suitable 'nominalistic' version of the theory can be found, so that we can regard the usual theory as formed by adding to this a conservative theory of the real numbers themselves. For the reasons given earlier, I believe that this is true, though no one could claim that the point has been well verified in a variety of cases.[52]

[51] It is objected that I am here being over-optimistic, and perhaps that is so. But I cannot take the discussion any further here.

[52] In particular, I note that it has been said (e.g. by Malament, 1982) that quantum theory presents special difficulties for this approach, and so too may other theories that lie in the future. I make no comment on this, in the first case because I am not in a position to do so, and in the second case because no one could be in such a position. There is a response to Malament in Balaguer (1998, chapter 5).

Let us sum up. There are varieties of fictionalism. It is common to all of them that numbers of all kinds – natural, rational, real – are held to be fictions when they are construed as Platonic objects, and that is how they seem to be construed in pure mathematics. But there are differing views on how to draw the distinction between uses of numbers that are (or may be) genuinely true, and uses that are taken to be part of the fiction. In general, the truths are taken to be what Frege would call *applications* of the numbers, whereas the fictions are the Platonic theories that he would think of as being applied, but this is only a rough guide. For one thing, consideration of where to draw the distinction will be affected by the choice of a background logic, as we have seen. In general, the stronger the logic the more can be taken as free of fiction, and that is a question on which one may well hesitate to form a definite verdict. But always what is counted as free of fiction is connected with what appears to be the application of a 'purer' theory, and in fact it dictates what the axioms of that 'purer' theory have to be, if it is to fulfil its instrumentalist purpose. The fiction is not just dreamt up out of thin air, but has a definite role to play.

Let that suffice on the topic of numbers as fictions. I now move on to the different subject of set theory.

(iii) Sets

I believe that sets too are mere fictions, though I confess that in this case the arguments are rather different. This is because it is not very clear why such fictions would be useful fictions. For what useful applications does set theory have? From a historical point of view one may fairly say that sets first had an important role to play in mathematics as a background to Cantor's theory of infinite numbers, so I open with a digression on infinite numbers.

Cantor himself did not claim that his infinite numbers, either cardinal or ordinal, are themselves sets. In his way of thinking the crucial claims are (i) that for every set there is something which is its cardinal number, and (ii) that two sets have the same cardinal number if and only if there is a relation which correlates them one-to-one. This, of course, is a version of Hume's Principle (discussed earlier in section 2 of this chapter), but with a quantification over sets replacing the second-level quantification used by Frege and neo-Frege. On the assumption that there are sets, this tells us that there are also cardinal numbers, with an identity-criterion as specified, but it does not tell us what they are. Essentially the same account applies to ordinal numbers; we are given a criterion of identity, and told that there are things that satisfy it, but not what they are. It was a later development to identify these numbers with certain sets, and one that is not found in Cantor himself. We have already noted that in the case of the finite numbers any such identification is arbitrary, and the same obviously applies to the infinite cardinals and the infinite ordinals as well. Consequently, if one is trying to take a realistic attitude to these numbers, one can only say that these basic assumptions are incomplete, for they do not tell us how to resolve the 'Julius Caesar' problem.

If it is already agreed that the finite numbers are best regarded as fictions, then surely the infinite numbers should also be taken as fictions. But there is this difference. I have argued that a style of nominalism which aims to reduce statements about numbers to statements about their applications does not work for the natural numbers, because they have many applications, and it is arbitrary to choose just one of them. But when a number is specified as a *cardinal* number one is in effect referring to a particular application, namely one that answers the question 'how many?'. This use of the natural numbers is performed by the numerical quantifiers 'there are n things x such that —x—', which are easily defined in terms of the ordinary quantifiers and identity, and which introduce no fictional entities. Given a background logic of second level, we can easily go on to define the quantifier 'there are infinitely many things x such that —x—', and the various special cases such as 'there are $\aleph_0$ things x such that —x—'. Again, there is no reason to suppose that these quantifiers introduce any reference to fictional objects. But since this is in practice the only way in which the infinite cardinal numbers are applied – or, in any case, the most central way, on which all others depend – there is no strong objection to the idea of *reducing* the theory of infinite cardinals to the theory of infinite numerical quantifiers. The same applies to the infinite ordinal numbers. The basic ordinal use of the natural numbers is their use as ordinal adjectives, in phrases such as 'the first', 'the second', 'the third', and so on.[53] We can easily define an infinite continuation of this series as 'the ωth', 'the $(\omega + 1)$th', and so on. This again opens up the possibility of *reducing* statements about ordinal numbers to statements which introduce no fictions.

---

[53] Other uses are easily defined in terms of this. Thus a finite series is said to be of length n if and only if its last member is its nth member. An infinite well-ordered series is said to be of length α if and only if it has a βth member for all β less than α, and no other members. (The discrepancy arises because we do not count the initial member of a series as its '0th' member.)

There is a complication here, which I find significant, though others may dismiss it. The finite numerical quantifiers, and ordinal adjectives, are what I call 'type-neutral'. That is, they can be applied at any level in the simple theory of types. One would expect the same to hold of their continuations into the transfinite. But, if that is to be so, one must tread very warily in order to avoid the analogues of familiar paradoxes, such as that which Cantor derived from the assumption that there is a cardinal number of all the cardinal numbers, or that which Burali-Forti derived from the assumption that there is an ordinal number of the series of all ordinal numbers. Since the question soon becomes very complex, I will not explore it here, but I must note that it exists.[54] I add that it may well be argued that it has a significance for our central question 'what is the use of set theory?', as the next paragraph will show. Let us now address that question more directly.

Cantor thought of cardinal numbers, and ordinal numbers, as applying to sets. But if the theory of these numbers can indeed be reduced to the theories of numerical quantifiers, and of ordinal adjectives, then on the face of it sets are not needed in this role. For we do not have to supply any *objects* for the numerical quantifiers to be true of, and the same holds of the ordinal adjectives. However, one must add that, as Cantor himself saw somewhat dimly, and as we now see very clearly, one way of preventing the familiar paradoxes from arising is to restrict attention to sets and not to count all conceivable collections as sets. On the contrary, we put forward a definite conception of what is allowed as a set, namely the iterative conception (discussed in chapter 5, section 4), and we claim that it is only sets that have cardinal numbers or ordinal numbers. In this way we avoid the contradictions which had upset earlier and uncritical uses of this notion. One cannot say that this is the only way of avoiding such contradictions, but it is the way that is most commonly adopted by working mathematicians.

In order to appreciate this use of sets, we do not have to suppose that there really are such things. What is required is a certain *conception* of the nature of a set, for it is this that yields a coherent theory. But we may perfectly well take this whole theory as a piece of fiction. Gödel claimed that the axioms of set theory 'force themselves upon us as being true', but I would gloss this by saying that we see these axioms to be true *to the given conception*, from which it does not follow that they are true of any actual objects.[55]

---

[54] I have said something about this problem in the final section of my (1980), but I cannot pretend to have reached a satisfying solution.

[55] There is a lengthy development of this line of thought in Chihara (1973, chapter 2).

I add as an aside that neither the axiom of infinity nor the axiom of replacement do strike me as being 'forced' by the iterative conception of sets. Their role is rather to make the story of the sets more interesting, to ensure that it becomes what Hilbert called 'the paradise that Cantor has created for us' (1925, p. 191). The contradictions stemmed from the story that Cantor 'created', and they are resolved by filling out his notion of a set into something much more definite than he himself gave us. But this adds to, and clarifies, the story; it does not give us a reason for taking it to be true.

Mathematicians do quite often use the notion of a set to speak of sets of items – e.g. numbers, functions, and so on – which are not themselves thought of as sets. Here the notion employed need not be regarded as falling under the iterative conception at all. (It is close to what Shapiro, in section 1.3 of his (1991), calls the 'logical' conception of a set, as opposed to the 'iterative' conception.) But I would like to make three points about what genuinely are uses of the iterative conception.

First, in its higher reaches, which posit sets of huge cardinalities, set theory is just a fairy story – a delightful fairy story, perhaps, but one with no serious claim to truth. There is behind it a conception of sets which is definite enough to justify a search for how best to extend the existing axioms, i.e. how to make the story more explicit and more detailed, but that does not prevent us from regarding it as just a story. Moreover, it has not found any useful applications either in physics or in other parts of mathematics and it does not seem likely that it will do so. It may be debated quite where these 'higher reaches' start, but I would myself say 'as soon as the axiom of infinity is called upon'. (Recall that I do not count a quantifier such as 'there are $\aleph_0$ things x such that —x—' as part of set theory. These quantifiers may be independently defined and introduce no fictions.)

Second, the axiom of infinity is not called upon in the theory of the hereditarily finite sets, but here there is a different worry. It is now the fashion among set theorists to concentrate upon the *pure* sets, i.e. those that are built up just from the empty set as the one set of lowest rank, and this is very artificial. The rationale is that this is all that is needed for the construction of structures which match the structures of the natural numbers, the signed integers, the rational numbers, and so on. But if we set aside the familiar and arbitrary identifications, I see little use for this match of structures. It may well be true that all the problems that are of interest to the set theorist will arise in this theory which is artificially restricted to the pure sets, but it bears little relation to the way that the notion of a set is used elsewhere – in mathematics, or in physics, or in everyday life. So let us now assume that individuals of various sorts may be included, to be members of the sets of first rank.[56]

Third, even if we confine attention to the sets of finite rank, and so assume no axiom of infinity, we find an oddity. I have said that in many areas of mathematics, and sometimes elsewhere, we find references to sets which have individuals as members, or anyway things which are being treated as individuals (e.g. numbers). Such a use of sets is of course compatible with the iterative conception, but surely does not require it. On occasion one may wish to refer to sets which have these sets as members, i.e. to sets of the second rank, and perhaps even to sets of the third rank. But one can be confident that no one (besides the set theorist) ever refers to sets of the hundredth rank, and yet the theory says that such sets do exist. For example, if a is an ordinary individual (say, an apple), then the theory says that there is a set $\{\{\ldots\{a\}\ldots\}\}$, which is written with one hundred pairs of curly brackets. But presumably no one has ever found a use for such a set, so what is the point of positing its existence?

The answer, I think, is fairly obvious: this comes from extrapolating to all ranks principles that do seem to hold for the more familiar sets of the first few ranks, and such an extrapolation is needed for a non-arbitrary theory. For it must be arbitrary to choose some one finite rank as the maximum of all finite ranks. But now it may be said that the moral to draw from this is that even the theory of the sets of finite rank is something of an idealization, including items posited just in order to yield a smooth theory. And we saw long ago, when discussing Aristotle (chapter 1, pp. 17–20) that while such idealizing theories may be very convenient, that is not a good reason for taking them to be literally true of any real things.

Let that suffice as some reasons for saying that not only numbers but also sets are best regarded as fictions.

6. Concluding remarks

I conclude with a few observations on how fictionalism relates to some other theories that we have considered.

Gödel's version of realism claims that we need to posit the existence of such abstract objects as numbers and sets in order to explain our 'mathematical intuitions'. But the obvious objection is that the supposed existence

---

[56] As noted earlier (p. 145) this casts doubt on the correctness of the axiom of replacement.

of these objects would not actually yield any explanation of why we believe in them, for there would appear to be no way in which they could affect our thinking. However, the fictionalist can offer an alternative explanation of the 'mathematical intuitions' that Gödel was referring to, i.e. of the way that axioms concerning these objects 'force themselves upon us as being true'. On his account, the axioms for the natural numbers seem to most of us to be indubitable because of the way in which they are related to the various applications of the theory. It is these applications, such as the numerical quantifiers, the numerically definite comparisons, the ordinal descriptions, and so on, which do have a good title to be called indubitable. They function as the 'data' for this theory, for they can be grasped quite independently, and the theory is responsible to them. That is why its axioms seem 'forced upon us', i.e. because they are required to fit the data, for that is their *raison d'être*. I add that the fictionalist account of axioms for the real numbers is very similar, for their *raison d'être* is to facilitate working with quantities that are continuous and measurable, so again they have to fit the features of these quantities that can be independently established (or, at least, conjectured). This of course was how the theory of the real numbers began. The position with axioms for sets is somewhat different, as the last section has revealed, and I will not repeat that here. But in all cases the fictionalist's explanation of why most of us believe in these axioms seems to me to be much more credible than Gödel's is.

The realist theory proposed by Quine and Putnam assigns to numbers and sets a different explanatory role: the idea is that they are needed in the explanation of why physical phenomena happen as they do. Again there is an obvious objection, namely that the positing of such unobservable objects as numbers or sets is something very different from the positing of such unobservable objects as atoms or electrons, forces or fields, curved space-times, and so on. For the latter are posited as having genuine causal powers, as interacting both with one another and with more observable things. But such objects as numbers and sets are not taken to have this kind of role, for they are not conceived as 'acting' in any way at all. This makes it entirely reasonable to treat the two kinds of unobservables differently, taking a realistic attitude to the one but viewing the other merely instrumentally, as an aid to thought and calculation which corresponds to nothing in reality. The point is backed up by showing how scientific theories which do apparently assume the existence of these abstract objects may be rephrased without loss of explanatory power in a more nominalist way that lacks this commitment. Admittedly this point has been proved only in a few fairly simple cases, and it is open to doubt whether it holds quite generally. If it does, then the indispensability argument fails altogether. But if there are cases where the 'pure' theory of real numbers cannot be eliminated from a scientific theory that we accept, then one may still dispute the conclusion that Quine draws. For it is still possible to adopt a purely instrumental interpretation of number theory, even if we cannot at present see how to dispense with that 'instrument'. And this position is appealing, because one cannot believe that physical phenomena do really depend upon the existence of such inert abstract objects. They help us to think about the phenomena, but they do not help the phenomena to occur.[57]

I conclude that both of these varieties of realism should be rejected, and that the fictionalist alternative is more attractive than either. I add that the neo-Fregean variety has also been found unsatisfactory, but in this case too there is a connection with the fictionalist alternative. For a main difficulty for the neo-Fregeans is that Hume's Principle cannot by itself tell us *which* objects the numbers are. Their best answer to this question seems to be that proposed by Hale, i.e. one that shows how the numbers which Hume's Principle aims to introduce are related to applications of these numbers which are independently definable. This, for the fictionalist, is an important sign that it is such applications which come first; it is these that are fundamental, and the story of the numbers themselves can only be understood if it is based upon them. But whereas the neo-Fregeans think that this story is true, the fictionalist sees no need for this assumption.

Fictionalism is not the only approach which denies the truth of arithmetical statements, when Platonically construed. Both reductive nominalists and structuralists offer to keep their truth, but only by reconstruing them as not really about Platonic objects after all. The one offers an account of their meaning which 'reduces' them to generalizations over statements which apply the numbers, and the other paraphrases them as being really statements about the structures which we think of as exemplified by the numbers. I have argued against both, but have some sympathy with each. I take structuralism first.

There may be many reasons for thinking that there cannot really be such things as numbers, but the structuralist emphasizes this one. The numbers,

---

[57] Wigner (1967) claimed that it would be some kind of 'miracle' if our sciences worked so well while employing a mathematics that is false. Leng (2005) replies that this is equally a 'miracle' if the mathematics is true. For the objects of mathematics cannot be causes, and so cannot affect what happens.

whether natural or rational or real or complex, exemplify certain kinds of structures, and from the point of view of pure arithmetic they do not seem to have any other properties than those that the structure assigns to them. For example, it is essential to the number 2 that it comes after the number 1 and before the number 3, but there is nothing else about it that ties it down to being one object rather than another. This leads to a comparison of the fictionalist and the structuralist, which could be put in this way. Both agree that what is important about the various kinds of number are the structures that they exemplify. The structuralist infers that the statements of arithmetic are really about those structures themselves, and not about any particular objects that exemplify them, so they do not mean what they appear to mean. By contrast, the fictionalist says that these statements do mean what they appear to mean, do presuppose the existence of these special objects called numbers, and are therefore false. For there are no such objects; they are merely fictions. But he agrees that what makes the fiction *useful* is just the structure that these alleged objects are assumed to have. For the point of the fictional story is to facilitate our thinking about its various applications, and it can do this just because they all share the same structure. So on each view it is the structure that is really important.

The relationship between the fictionalist and the reductive nominalist is rather different. Both agree that the story of 'the numbers themselves' is not true, and that the statements in which numbers are applied very often are true. Moreover the relations between these latter statements can be independently established, and in practice both think of them as established in ways which the logicist would approve of, namely by (a) an analysis of the numerical notions in question, and (b) the application of logic. This claim is not a central part of either theory, and could be dropped, so long as there is some other explanation of how the relations between such application-statements may be independently established. But both are drawn to the logicist's answer to this question, and may wish to hang on to this answer even when it becomes clear that both the analyses and the logic involved are too complex to be the route to knowledge that is actually followed by most people. So they may wish to claim that these relationships *can* be known *a priori*, even if they usually are not. I should have liked to end the book by saying something about these ideas, but unfortunately space does not permit.[58]

---

[58] I have said *something* on this question in the final part of my (2009a), but it is not quite what I would say today. (The piece was written in the first half of 2003.)

Instead, I end with a brief comment on fictionalism and conceptualism. If numbers are fictions then of course it is we who create those fictions, and in this way numbers might be characterized as our 'mental mathematical constructions'. But still this is not a version of conceptualism as ordinarily understood. For the story that we create is one that represents numbers as things that exist quite independently of human minds, and the story is a fiction because numbers do not actually exist at all. This is evidently rather different from the conceptualist theory that numbers *do* exist, but owe their existence to our mental activity. Nevertheless there is some similarity even here.

Suggestions for further reading

Frege should by now be familiar. On Gödel, one must read his (1944) and (1947). There is quite a nice attack on his position in chapter 2 of Chihara (1973), and a useful discussion of mathematical intuition in Parsons (1979/80). On the Neo-Fregeans, I suggest starting with the introduction to Hale and Wright (2001) (but you may stop on p. 23).[59] This outlines the overall programme, and the difficulties that it faces, and describes the various essays on particular aspects which then follow. In the light of this introduction you may wish to pursue some of these essays. I also suggest Hale's earlier discussion of the Julius Caesar problem in chapter 8 of his (1987), and possibly the debate between Boolos (1997) and Wright (1999) on whether Hume's Principle should be counted as analytic. By way of criticism I suggest the admirably straightforward discussion in Field (1984a).[60]

On the indispensability argument due to Quine and Putnam, one should start with Quine (1948). There is a conspectus of all the main themes in Quine's philosophy, and a clarification of how they fit together with one another, in Resnik (2005); this may be useful. But, for a detailed discussion of the indispensability argument, one should read Putnam (1971), which is a short booklet, quite easy to read, and need not be abbreviated. This

[59] Pages 23–7 give a very quick outline of their ideas on real numbers. If you wish to pursue this area, I suggest their more recent (2005, part 2, i.e. pp. 186–200).
[60] There is an over-long reply to Field's criticism in Wright (1988). I suggest that the main thrust of this reply comes in section V, i.e. pp. 448–52, and that the rest may reasonably be omitted.

both proposes the argument and looks at various problems for it. The obvious criticism is Field (1980), but you might reasonably confine attention to his first five chapters, i.e. pp. 1–46, and not try to tackle the detail of his 'nominalistic' account of Newtonian physics. There are many other discussions of this indispensability argument. The fullest is probably Colyvan (2001), which is a full-length book on the topic.

On reductive nominalism, Russell should by now be familiar, but you might like to look at the different reduction which is very hastily sketched in Wittgenstein (1921), propositions 6–6.031. I have given some detail on how several different reductions are equally possible in my (1979), with sections 1–3 of chapter 1 confining attention to the natural numbers, and some morals drawn on pp. 275–92 of chapter 4. Then pp. 293–7 go on to draw a fictionalist conclusion for the natural numbers, before widening the discussion to take in other varieties of numbers. As for objections to fictionalism, an important point of detail is raised in Shapiro (1983), and a curious line of objection which my text does not discuss at all is to be found in Hale (1990).[61] There are many other objections that have been proposed, but in my view none of them have any real force,[62] and you will do best to think things through for yourself. You will no doubt find chapters 8 and 9 of Shapiro (2000) a helpful aid.

---

[61] The argument is first found in Wright (1988, pp. 462–8). It recurs in Hale and Wright (1994). Together with the response in Field (1993), it is discussed in section 4 of chapter 6 of Colyvan (2001).

[62] For example, there is a long string of objections in Chihara (1990, chapter 8, section 5), but it seems to me that they do not get anywhere near the heart of the fictionalist's position. (And I remark *ad hominem* that they come oddly from one who himself once espoused a fictionalist position about ZF set theory, i.e. in his 1973, chapter 2.)

References

Aczel, P. (1988), *Non-Well-Founded Sets*, Center for the Study of Language and Information, Lecture Notes, no. 14.

Annas, J. (1976), *Aristotle's Metaphysics, books M and N*, Oxford: Clarendon Press.

Aristotle, *The Complete Works*, 2 vols, ed. J. Barnes, Princeton, NJ: Princeton University Press, 1984.

Ayer, A.J. (1936), *Language, Truth & Logic*, London: Gollancz; 2nd edn 1946. Chapter 4 is reprinted in Benacerraf and Putnam (1983).

Balaguer, M. (1998), *Platonism and Anti-Platonism in Mathematics*, Oxford: Oxford University Press.

Barrow, I. (1664), *The Usefulness of Mathematical Learning*, republished London: Frank Cass & Co, 1970.

Bell, J.L. (1999), *The Art of the Intelligible: An Elementary Survey of Mathematics in its Conceptual Development*, Dordrecht: Kluwer.

Benacerraf, P. (1965), 'What Numbers Could Not Be', *Philosophical Review* 74: 47–73; reprinted in Benacerraf and Putnam (1983).

Benacerraf, P. (1973), 'Mathematical Truth', *Journal of Philosophy* 70: 661–79; reprinted in Benacerraf and Putnam (1983), and in Hart (1996).

Benacerraf, P. (1996), 'Recantation', *Philosophia Mathematica* 4: 184–9.

Benacerraf, P. and Putnam, H. (eds.) (1983), *Philosophy of Mathematics: Selected Readings*, 2nd edn, Cambridge: Cambridge University Press.

Bennett, J. (1966), *Kant's Analytic*, Cambridge: Cambridge University Press.

Berkeley, G. (1710), *Principles of Human Knowledge* (many editions).

Berkeley, G. (1734), *The Analyst*, reprinted in vol. 4 of A.A. Luce and T.E. Jessop (eds.), *The Works of George Berkeley* (9 vols.), London: Nelson, 1948–57.

Billinge, H. (2003), 'Does Bishop Have a Philosophy of Mathematics?', *Philosophia Mathematica* 11: 176–94.

Bishop, E. (1967), *Foundations of Constructive Analysis*, New York: McGraw-Hill.

Boghossian, P. (1996), 'Analyticity Reconsidered', *Nous* 30: 360–91.

Boghossian, P. (2000), 'Knowledge of Logic', in P. Boghossian and C. Peacocke (eds.), *New Essays on the A Priori*, Oxford: Clarendon Press, 229–54.

Bolyai, J. (1832), 'The Science of Absolute Space', published as an appendix to W. Bolyai, *Tentamen Juventutem Studiosam in Elementa Matheseos*, Budapest.

Boolos, G. (1971), 'The Iterative Conception of Sets', *Journal of Philosophy* 68: 215–32; reprinted in Boolos (1998), and in Benacerraf and Putnam (1983).

Boolos, G. (1975), 'On Second-Order Logic', *Journal of Philosophy* 72: 509–27; reprinted in Boolos (1998).

Boolos, G. (1984), 'To Be Is to Be the Value of a Variable (or Some Values of Some Variables)', *Journal of Philosophy* 81: 430–50; reprinted in Boolos (1998).

Boolos, G. (1990), 'The Standard of Equality of Numbers', in G. Boolos (ed.), *Meaning and Method: Essays in Honor of Hilary Putnam*, Cambridge: Cambridge University Press, 261–77; reprinted in his (1998).

Boolos, G. (1997), 'Is Hume's Principle Analytic?', in R.G. Heck (ed.), *Logic, Language, and Thought*, Oxford: Oxford University Press, 245–62; reprinted in his (1998).

Boolos, G. (1998), *Logic, Logic, and Logic*, Cambridge, MA: Harvard University Press.

Borkowski, L. (1958), 'Reduction of Arithmetic to Logic Based on the Theory of Types without the Axiom of Infinity and the Typical Ambiguity of Arithmetical Constants', *Studia Logica* 8: 283–95.

Bostock, D. (1972/3), 'Aristotle, Zeno, and the Potential Infinite', *Proceedings of the Aristotelian Society* 73: 37–51; reprinted in Bostock (2006).

Bostock, D. (1974), *Logic & Arithmetic*, vol. 1, Oxford: Clarendon Press.

Bostock, D. (1979), *Logic & Arithmetic*, vol. 2, Oxford: Clarendon Press.

Bostock, D. (1980), 'A Study of Type-Neutrality', *Journal of Philosophical Logic* 9: 211–96, 363–414.

Bostock, D. (1986), *Plato's Phaedo*, Oxford: Clarendon Press.

Bostock, D. (1990), 'Logic and Empiricism', *Mind* 99: 572–82.

Bostock, D. (1997), *Intermediate Logic*, Oxford: Clarendon Press.

Bostock, D. (1998), 'On Motivating Higher-Order Logic', in T. Smiley (ed.), *Philosophical Logic*, Oxford: Oxford University Press, 29–43.

Bostock, D. (2006), *Space, Time, Matter, and Form: Essays on Aristotle's Physics*, Oxford: Clarendon Press.

Bostock, D. (2008), 'Russell on *the* in the Plural', in N. Griffin and D. Jaquette (eds.), *Russell vs Meinong: The Legacy of 'On Denoting'*, London and New York: Routledge, 113–43.

Bostock, D. (2009a), 'Empiricism in the Philosophy of Mathematics', in A. Irvine (ed.), *Handbook of the Philosophy of Mathematics* (vol. 4 of D. Gabbay, P. Thagard, J. Woods (eds.), *Handbook of the Philosophy of Science*), Amsterdam: Elsevier, 155–227.

Bostock, D. (2009b), 'Aristotle's Philosophy of Mathematics', in C. Shields (ed.), *The Oxford Handbook on Aristotle*, Oxford: Oxford University Press.

Brouwer, L.E.J. (1905), *Leven, Kunst en Mystiek*, Delft: Waltman; English translation by van Stigt in *Notre Dame Journal of Formal Logic* 37 (1996), xx.

Brouwer, L.E.J. (1907), *Over de Grondslagen der Wiskunde*, Amsterdam: Maas & van Suchtelen; English translation in Brouwer (1975).

Brouwer, L.E.J. (1912), *Intuitionisme en Formalisme*, Amsterdam: Clausen; English translation by A. Dresden, as 'Intuitionism and Formalism', *Bulletin of the American Mathematical Society* 20 (1913); reprinted in Benacerraf and Putnam (1983).

Brouwer, L.E.J. (1948), 'Consciousness, Philosophy, and Mathematics', in Benacerraf and Putnam (1983), 90–6.

Brouwer, L.E.J. (1975), *Collected Works I*, ed. A. Heyting, Amsterdam: North Holland.

Burgess, J.P. (1983), 'Why I Am Not a Nominalist', *Notre Dame Journal of Formal Logic* 24: 93–105.

Burgess, J.P. (1990), 'Epistemology and Nominalism', in Irvine (1990), 1–15.

Burgess, J.P. and Rosen, G. (1997), *A Subject with no Object*, Oxford: Clarendon Press.

Burnyeat, M.F. (2000), 'Plato on Why Mathematics is Good for the Soul', *Proceedings of the British Academy* 103: 1–81.

Cantor, G. (1895, 1897), *Contributions to the Founding of the Theory of Transfinite Numbers*, translated by P.E.B. Jourdain, New York: Dover, 1955.

Carnap, R. (1947), *Meaning & Necessity*, Chicago: University of Chicago Press (2nd edn. 1956).

Carnap, R. (1950), 'Empiricism, Semantics, and Ontology', *Revue Internationale de Philosophie* 4: 20–40; reprinted in the 2nd edn of his (1947), and in Benacerraf and Putnam (1983).

Cauchy, A.L. (1821), *Cours d'Analyse Algebraique*, in his *Oeuvres* vol. 2, Gauthier-Villars, 1897–9.

Chihara, C.S. (1973), *Ontology and the Vicious Circle Principle*, Ithaca: Cornell University Press.

Chihara, C.S. (1990), *Constructibility and Mathematical Existence*, Oxford: Clarendon Press.

Chihara, C.S. (2004), *A Structural Account of Mathematics*, Oxford: Clarendon Press.

Chihara, C.S. (2005), 'Nominalism', in Shapiro (2005b).

Church, A. (1936), 'A Note on the *Entscheidungsproblem*', *Journal of Symbolic Logic* 1: 40–1, 101–2.

Church, A. (1976), 'Comparison of Russell's Resolution of the Semantic Antinomies with that of Tarski', *Journal of Symbolic Logic* 41: 747–60.

Cohen, P.J. (1966), *Set Theory and the Continuum Hypothesis*, New York: Benjamin.

Colyvan, M. (2001), *The Indispensability of Mathematics*, Oxford: Oxford University Press.

Copi, I.M. (1971), *The Theory of Logical Types*, London: Routledge & Kegan Paul.

Cross, R.C. and Woozley, A.D. (1964), *Plato's Republic*, London: Macmillan.

Curry, H.B. (1951), *Outlines of a Formalist Philosophy of Mathematics*, Amsterdam: North Holland.

Davidson, D. (1967), 'Truth and Meaning', *Synthese* 17: 304–23; reprinted in his (1984).

Davidson, D. (1984), *Inquiries into Truth and Interpretation*, Oxford: Clarendon Press.

Dedekind, R. (1872), *Stetigkeit und irrationale Zahlen*, Brunswick: Vieweg; translated as *Continuity and Irrational Numbers* in his (1963).

Dedekind, R. (1888), *Was sind und was sollen die Zahlen?*, Brunswick: Vieweg; translated as *The Nature and Meaning of Numbers* in his (1963).

Dedekind, R. (1963), *Essays on the Theory of Numbers*, translated by W.W. Beman, New York: Dover.

Demopoulos, W. (ed.) (1995), *Frege's Philosophy of Mathematics*, Cambridge, MA: Harvard University Press.

Demopoulos, W. and Clark, P. (2005), 'The Logicism of Frege, Dedekind, and Russell', in Shapiro (2005b), 129–65.

Descartes, R. (1628), *Rules for the Direction of the Mind*, translated in Descartes (1967) and (1985).

Descartes, R. (1637), *Discourse on Method*, translated in Descartes (1967) and (1985).

Descartes, R. (1641), *Meditations on First Philosophy*, translated in Descartes (1967) and (1985).

Descartes, R. (1644), *The Principles of Philosophy*, partly translated in Descartes (1967), and fully translated in Descartes (1985).

Descartes, R. (1967), *The Philosophical Works of Descartes*, translated by E.S. Haldane and G.R.T. Ross, vol. I, Cambridge: Cambridge University Press.

Descartes, R. (1985), *The Philosophical Writings of Descartes*, translated by J. Cottingham, R. Stoothoff, and D. Murdoch, Cambridge: Cambridge University Press.

Detlefsen, M. (1986), *Hilbert's Program*, Dordrecht: Reidel.

Diels, H. revised W. Kranz (1961), *Die Fragmente der Vorsokratiker*, 10th edn, Berlin: Weidmannsche Verlagsbuchhandlung.

Dreyer, J.L.E. (1953), *A History of Astronomy*, 2nd edn, New York: Dover.

Dummett, M. (1959), 'Wittgenstein's Philosophy of Mathematics', *Philosophical Review* 68: 324–48. Reprinted in Benacerraf and Putnam (1983), and in his (1978).

Dummett, M. (1963), 'The Philosophical Significance of Gödel's Theorem', *Ratio* 5: 140–55; reprinted in his (1978).

Dummett, M. (1973a), *Frege: Philosophy of Language*, London: Duckworth.

Dummett, M. (1973b), 'The Philosophical Basis of Intuitionist Logic', in *Logic Colloquium '73*; reprinted in his (1978), and in Benacerraf and Putnam (1983), and in Hart (1996).

Dummett, M. (1977), *Elements of Intuitionism*, Oxford: Clarendon Press (2nd edn, 2000).

Dummett, M. (1978), *Truth and Other Enigmas*, London: Duckworth.

Dummett, M. (1991), *Frege: Philosophy of Mathematics*, London: Duckworth.

Enderton, H.B. (1977), *Elements of Set Theory*, New York and London: Academic Press.

Feferman, S. (1964), 'Systems of Predicative Analysis', *Journal of Symbolic Logic* 29: 1–30; reprinted in Hintikka (1969).

Feferman, S. (1998), *In the Light of Logic*, Oxford: Oxford University Press.

Feferman, S. and Hellman, G. (1995), 'Predicative Foundations of Arithmetic', *Journal of Philosophical Logic* 24: 1–17.

Field, H. (1980), *Science without Numbers*, Princeton: Princeton University Press.

Field, H. (1984a), 'Critical Notice of Crispin Wright (1983)', *Canadian Journal of Philosophy* 14: 637–62; reprinted as 'Platonism for Cheap?' in his (1989).

Field, H. (1984b), 'Can We Dispense with Space-time?', in P. Asquith and P. Kitcher (eds.), *PSA 1984: Proceedings of the 1984 Biennial Meeting of the Philosophy of Science Association*, vol. 2, 33–90; reprinted in his 1989.

Field, H. (1985), 'On Conservativeness and Incompleteness', *Journal of Philosophy* 82: 239–60; reprinted in his (1989).

Field, H. (1989), *Realism, Mathematics and Modality*, Oxford: Blackwell.

Field, H. (1992), 'A Nominalistic Proof of the Conservativeness of Set Theory', *Journal of Philosophical Logic* 21: 111–23.

Field, H. (1993), 'The Conceptual Contingency of Mathematical Objects', *Mind* 102: 285–99.

Fine, K. (2002), *The Limits of Abstraction*, Oxford: Clarendon Press.

Fletcher, P. (2002), 'A Constructivist Perspective on Physics', *Philosophia Mathematica* 10: 26–42.

Forbes, G. (1994), *Modern Logic*, Oxford: Clarendon Press.

Fraenkel, A.A. (1922), 'Zu den Grundlagen der Cantor-Zermeloschen Mengenlehre', *Mathematische Annalen* 86: 230–7.

Fraenkel, A.A., Bar-Hillel, Y. and Levy, A. (1973), *Foundations of Set Theory*, 2nd edn, Amsterdam: North Holland.

Frege, G. (1879), *Begriffsschrift*, Halle: Louis Nebert; translated in van Heijenoort (1967), 1–82.

Frege, G. (1884), *Grundlagen der Arithmetik*, Breslau: Koebner; translated by J.L. Austin as *Foundations of Arithmetic*, Oxford: Blackwell, 1959. (The positive part, i.e. sections 55–91 and 106–9, is also given a different translation, by M.S. Mahoney, in Benacerraf and Putnam, 1983.)

Frege, G. (1892), 'On Concept and Object', translated by P.T. Geach in P.T. Geach and M. Black (eds.), *Frege: Philosophical Writings*, Oxford: Blackwell, 1960.

Frege, G. (1893), *Grundgesetze der Arithmetik* vol. 1, Jena: H. Pohle; partially translated by M. Furth as *The Basic Laws of Arithmetic*, Berkeley and Los Angeles: University of California Press, 1964.

Frege, G. (1902), Letter to Russell, in van Heijenoort (1967), 126–8.

Frege, G. (1903), *Grundgesetze der Arithmetik* vol. 2, Jena: H. Pohle; partially translated in P. Geach and M. Black, *Frege: Philosophical Writings*, Oxford: Blackwell, 1960.

Frege, G. (1976), *Wissenschaftlicher Briefwechsel*, eds. G. Gabriel, H. Hermes, F. Kambartel, C. Thiel, A. Veraart, Hamburg: Felix Meiner Verlag.

Frege, G. (1980), *Philosophical and Mathematical Correspondence*, ed. B. McGuinness, translated by H. Kaal, Oxford: Blackwell. (This is a selection from Frege, 1976.)

Geach, P.T. (1965), 'Assertion', *Philosophical Review* 74: 449–65.

Gentzen, G. (1936), 'Die Widerspruchsfreiheit der reinen Zahlentheorie', *Mathematische Annalen* 112: 493–565.

Giaquinto, M. (2002), *The Search for Certainty*, Oxford: Clarendon Press.

Gillies, D.A. (1982), *Frege, Dedekind, and Peano on the Foundations of Arithmetic*, Assen: Van Gorcum.

Gödel, K. (1930), 'Die Vollständigkeit der Axiome des logischen Funktionenkalkuls', *Monatschefte für Mathematik und Physik* 37: 349–60; translated in van Heijenoort (1967), 582–91.

Gödel, K. (1931a), 'Über formal unentscheidbare Sätze der *Principia Mathematica* und verwandter Systeme', *Monatschefte für Mathematik und Physik* 38: 173–98; translated in van Heijenoort (1967), 596–617.

Gödel, K. (1931b), 'Über Vollständigkeit und Widerspruchsfreiheit', *Ergebnisse eines mathematischen Kolloquiums* 3: 12–13; translated in van Heijenoort (1967), 616–17.

Gödel, K. (1944), 'Russell's Mathematical Logic', in P.A. Schilpp (ed.), *The Philosophy of Bertrand Russell*, Evanston, IL: Northwestern University Press; reprinted in Benacerraf and Putnam (1983), 447–69.

Gödel, K. (1947), 'What Is Cantor's Continuum Problem?', *The American Mathematical Monthly* 54: 515–25; revised and expanded in Benacerraf and Putnam (1983), 470–85.

Grover, D.L. (1974), review of Copi (1972), *Philosophical Review* 83: 281–3.

Hale, B. (1987), *Abstract Objects*, Oxford: Blackwell.

Hale, B. (1990), 'Nominalism', in Irvine (1990), 121–44.

Hale, B. and Wright, C. (1994), 'Reductio ad Surdum?', *Mind* 103: 169–84.

Hale, B. and Wright, C. (2001), *The Reason's Proper Study*, Oxford: Clarendon Press.

Hale, B. and Wright, C. (2005), 'Logicism in the Twenty-First Century', in Shapiro (2005b), 166–202.

Hart, W.D. (ed.) (1996), *The Philosophy of Mathematics*, Oxford: Oxford University Press.

Hazen, A. (1983), 'Predicative Logics', in D. Gabbay and F. Guenthner (eds.), *Handbook of Philosophical Logic, vol. 1*, Dordrecht: Reidel, 331–407.

Heath, T.L. (1912), *The Works of Archimedes, with supplement*, reprinted New York: Dover, n.d.

Heath, T.L. (1925), *The Thirteen Books of Euclid's Elements*, 2nd edn, Cambridge: Cambridge University Press; reprinted New York: Dover, 1956.

Heck, R.G. (1993), 'The Development of Arithmetic in Frege's *Grundgesetze*', *Journal of Symbolic Logic* 58: 579–601; reprinted in Demopoulos (1995).

Heck, R.G. (1995), 'Definition by Induction in Frege's *Grundgesetze*', in Demopoulos (1995), 295–333.

Heck, R.G. (1997a), 'The Julius Caesar Objection', in Heck (ed.) (1997b), 273–308.

Heck, R.G. (ed.) (1997b), *Language, Thought, and Logic: Essays in Honour of Michael Dummett*, Oxford: Oxford University Press.

Hellman, G. (1989), *Mathematics without Numbers*, Oxford: Oxford University Press.

Hellman, G. (1990), 'Modal Structural Mathematics', in Irvine (1990), 307–30.

Hellman, G. (1998), 'Beyond Definitionism – But Not Too Far Beyond', in Schirn (1998), 215–26.

Heyting, A. (1930), 'Die formalen Regeln der intuitionistichen Logik', Stizungsberichte der Preussischen Akademie der Wissenschaften, 42–71, 158–69; English version in his (1956), chapter 7.

Heyting, A. (1956), *Intuitionism: an Introduction*, Amsterdam: North Holland (2nd edn, 1966).

Hilbert, D. (1899), *Foundations of Geometry*, translated by E. Townsend, La Salle, IL: Open Court, 1959.

Hilbert, D. (1925), 'On the Infinite', *Mathematische Annalen* 95: 161–90; translated in Benacerraf and Putnam (1983) and in van Heijenoort (1967).

Hilbert, D. (1927), 'The Foundations of Mathematics', *Abhandlungen aus dem mathematischen Seminar der Hamburgischen Universität* 6: 65–85; translated in van Heijenoort (1967).

Hintikka, J. (1967), 'Kant on Mathematical Method', *Monist* 51: 325–75; reprinted in Posy (1992).

Hintikka, J. (ed.) (1969), *The Philosophy of Mathematics*, Oxford: Oxford University Press.

Hodes, H. (1984), 'Logicism and the Ontological Commitments of Arithmetic', *Journal of Philosophy* 81: 123–49.

Horwich, P. (2000), 'Stipulation, Meaning, and Apriority' in P. Boghossian and C. Peacocke (eds.), *New Essays on the A Priori*, Oxford: Clarendon Press, 150–69.

Hughes, G.E. and Cresswell, M.J. (1968), *An Introduction to Modal Logic*, London: Methuen.

Hume, D. (1739), *A Treatise of Human Nature*, ed. L.A. Selby-Bigge, revised P.H. Nidditch, Oxford: Clarendon Press, 1978.

Hume, D. (1748), *Enquiry into Human Understanding*, ed. L.A. Selby-Bigge, revised P.H. Nidditch, Oxford: Clarendon Press, 1975.

Hussey, E. (1983), *Aristotle's Physics, books III and IV*, Oxford: Clarendon Press.

Hylton, P. (1990), *Russell, Idealism and the Emergence of Analytic Philosophy*, Oxford: Clarendon Press.

Irvine, A.D. (ed.) (1990), *Physicalism in Mathematics*, Dordrecht: Kluwer.

Jeffrey, R. (1996), 'Logicism 2000: A Mini-Manifesto', in Morton and Stich (1996), 160–4.

Kant, I. (1781, 1787), *Critique of Pure Reason*, translated by N. Kemp Smith, London: Macmillan, 2nd edn, 1933.

Kant, I. (1783), *Prolegomena to any Future Metaphysics*, translated by P.G. Lucas, Manchester: Manchester University Press, 1953.

Kenny, A. (1981), *Descartes: Philosophical Letters*, Oxford: Blackwell.

Kenny, A. (2005), *Medieval Philosophy*, Oxford: Clarendon Press.

Kitcher, P. (1975), 'Kant and the Foundations of Mathematics', *Philosophical Review* 84: 23–50; reprinted in Posy (1992).

Kitcher, P. (1980), 'Arithmetic for the Millian', *Philosophical Studies* 37: 215–36.

Kitcher, P. (1984), *The Nature of Mathematical Knowledge*, Oxford: Oxford University Press.

Kleene, S.C. (1952), *Introduction to Metamathematics*, Amsterdam: North Holland.

Kline, M. (1972), *Mathematical Thought from Ancient to Modern Times*, New York: Oxford University Press.

Kneale, W. and Kneale, M. (1962), *The Development of Logic*, Oxford: Clarendon Press.

Kneebone, G.T. (1963), *Mathematical Logic and the Foundations of Mathematics*, London: Van Nostrand.

Kreisel, G. (1967), 'Informal Rigour and Completeness Proofs', in I. Lakatos (ed.), *Problems in the Philosophy of Mathematics*, Amsterdam: North Holland; reprinted with some omissions in Hintikka (1969).

Kripke, S. (1965), 'Semantic Analysis of Intuitionistic Logic I', in J.N. Crossley and M.A.E. Dummett (eds.), *Formal Systems and Recursive Functions: Proceedings of the 8th Logic Colloquium*, Amsterdam: North Holland.

Kripke, S. (1972), *Naming & Necessity*; reprinted as a book, Oxford: Blackwell, 1980.

Kripke, S. (1975), 'Outline of a Theory of Truth', *Journal of Philosophy* 72: 690–716.

Kronecker, L. (1899), *Werke*, ed. K. Hensel, Leipzig: Teubner.

Kuratowski, C. (1920), 'Sur la notion de l'ordre dans la théorie des ensembles', *Fundamenta Mathematicae* 2: 161–71.

Landini, G. (1996a), 'Will the Real *Principia Mathematica* Please Stand Up?' in R. Monk and A. Palmer (eds.), *Bertrand Russell and the Origins of Analytical Philosophy*, Bristol: Thoemmes Press, 287–330.

Landini, G. (1996b), 'The Definability of the Natural Numbers in the Second Edition of *Principia Mathematica*', *Journal of Philosophical Logic* 25: 597–614.

Lear, J. (1979/80), 'Aristotelian Infinity', *Proceedings of the Aristotelian Society* 80: 187–210.

Lear, J. (1982), 'Aristotle's Philosophy of Mathematics', *Philosophical Review* 91: 161–92.

Leibniz, G.W. (1765), *New Essays on Human Understanding*, translated by P. Remnant and J. Bennett, Cambridge: Cambridge University Press, 1981.

Leng, M. (2005), 'Platonism and Anti-Platonism: Why Worry?', *International Studies in the Philosophy of Science*, 19: 65–84.

Lewis, D. (1986), *On the Plurality of Worlds*, Oxford: Blackwell.

Lobachevsky, N.I. (1830), 'On the Foundations of Geometry', *Kazan Journal*, 1829–30.

Locke, J. (1706), *Essay Concerning Human Understanding*, 5th edn, ed. P.H. Nidditch, Oxford: Clarendon Press, 1975.

MacBride, F. (2005), 'Structuralism Reconsidered', in Shapiro (2005b), 563–89.

Mackie, J.L. (1976), *Problems from Locke*, Oxford: Clarendon Press.

Maddy, P. (1990), *Realism in Mathematics*, Oxford: Clarendon Press.

Maddy, P. (1996), 'The Legacy of "Mathematical Truth"', in Morton and Stich (1996), 60–72.

Maddy, P. (1997), *Naturalism in Mathematics*, Oxford: Clarendon Press.

Malament, D. (1982), 'Review of Field (1980)', *Journal of Philosophy* 19: 523–34.

Malcolm, J. (1962), 'The Line and the Cave', *Phronesis* 7: 38–45.

Mancosu, P. (1996), *Philosophy of Mathematics and Mathematical Practice in the Seventeenth Century*, Oxford: Oxford University Press.

McCarty, D.C. (2005), 'Intuitionism in Mathematics', in Shapiro (2005b), 356–86.

Mendelson, E. (1964), *Introduction to Mathematical Logic*, Princeton: Van Nostrand. (4th edn, New York: Chapman & Hall, 1997.)

Mill, J.S. (1843), *System of Logic*, many editions.

Mirimanoff, D. (1917), 'Les antinomies de Russell et de Burali-Forti et le problème fondamental de la théorie des ensembles', *L'Enseignement Mathématique* 19: 37–52.

Morton, A. and Stich, S.P. (eds.) (1996), *Benacerraf and his Critics*, Oxford: Blackwell.

Mueller, I. (1970), 'Aristotle on Geometrical Objects', *Archiv für Geschichte der Philosophie*, 52: 156–71.

Mueller, I. (1992), 'Mathematical Method and Philosophical Truth', in R. Kraut (ed.), *The Cambridge Companion to Plato*, Cambridge: Cambridge University Press, 170–99.

Myhill, J. (1960), 'Some Remarks on the Notion of Proof', *Journal of Philosophy* 57: 461–71.

Myhill, J. (1974), 'The Undefinability of the Natural Numbers in the Ramified *Principia*', in G. Nakhnikian (ed.), *Bertrand Russell's Philosophy*, London: Duckworth, 19–27.

Myhill, J. (1979), 'A Refutation of an Unjustified Attack on the Axiom of Reducibility', in G. Roberts (ed.), *The Bertrand Russell Memorial Volume*, London: Allen & Unwin, 81–90.

Neugebauer, O. (1952), *The Exact Sciences in Antiquity*, reprinted New York: Harper Torchbooks, 1962.

Newton-Smith, W.H. (1978), 'The Underdetermination of Theory by Data', *Aristotelian Society Supplementary Volume* 52: 71–91.

Papineau, D. (1990), 'Knowledge of Mathematical Objects', in Irvine (1990), 155–82.

Parsons, C. (1969), 'Kant's Philosophy of Arithmetic', in S. Morganbesser, P. Suppes and M. White (eds.), *Philosophy, Science & Method: Essays in Honor of Ernest Nagel*, New York: St Martin's Press; reprinted in Parsons (1983) and in Posy (1992).

Parsons, C. (1977), 'What is the iterative conception of set?', in E. Butts and J. Hintikka (eds.), *Proceedings of the 5th International Congress of Logic, Methodology and Philosophy of Science*, Dordrecht: Reidel, 335–67; reprinted in his (1983), and in Benacerraf and Putnam (1983).

Parsons, C. (1979/80), 'Mathematical Intuition', *Proceedings of the Aristotelian Society* 80: 145–68.

Parsons, C. (1983), *Mathematics in Philosophy*, Ithaca: Cornell University Press.

Parsons, C. (1990), 'The Structuralist View of Mathematical Objects', *Synthese* 843: 303–46; reprinted in Hart (1996).

Parsons, C. (1992), 'The Transcendental Aesthetic', in P. Guyer (ed.), *The Cambridge Companion to Kant*, Cambridge: Cambridge University Press, 62–100.

Parsons, C. (1997), 'Wright on Abstraction and Set Theory', in Heck (1997b), 263–71.

Paseau, A. (2007), 'Scientific Platonism', in M. Leng, A. Paseau and M. Potter (eds.), *Mathematical Knowledge*, Oxford: Oxford University Press.

Peacocke, C. (1993), 'How Are A Priori Truths Possible?', *European Journal of Philosophy* 1: 175–99.

Peano, G. (1889), *Arithmetices Principia*, Turin; partially translated in van Heijenoort (1967), 83–97.

Plato, *Collected Dialogues*, ed. E. Hamilton and H. Cairns, New York: Pantheon Books, 1961.

Poincaré, H. (1906), 'Les Mathématiques et la Logique', *Revue de Metaphysique et de Morale* 14: 17–34.

Poincaré, H. (1908), *Science et Methode*, Paris: Flammarion; translated by F. Maitland as *Science and Method*, New York: Dover, n.d.

Posy, C.J. (1984), 'Kant's Mathematical Realism', *Monist* 67: 115–34; reprinted in Posy (1992).

Posy, C.J. (2005), 'Intuitionism and Philosophy', in Shapiro (2005b), 318–55.

Posy, C.J. (ed.) (1992), *Kant's Philosophy of Mathematics*, Dordrecht/Boston/London: Kluwer.

Potter, M. (2000), *Reason's Nearest Kin: Philosophies of Arithmetic from Kant to Carnap*, Oxford: Oxford University Press.

Potter, M. (2004), *Set Theory and its Philosophy*, Oxford: Oxford University Press.

Prawitz, D. (2005), 'Logical Consequence from a Constructivist Point of View', in Shapiro (2005b), 671–95.

Prior, A.N. (1960), 'The Runabout Inference-Ticket', *Analysis* 21: 38–9; reprinted in P.F. Strawson (ed.), *Philosophical Logic*, Oxford: Oxford University Press, 1967.

Prior, A.N. (1961), 'On a Family of Paradoxes', *Notre Dame Journal of Formal Logic* 2: 16–32.

Putnam, H. (1967a), 'Mathematics without Foundations', *Journal of Philosophy* 64: 5–22; reprinted in Benacerraf and Putnam (1983), 295–311, and in Hart (1996).

Putnam, H. (1967b), 'The Thesis that Mathematics is Logic', in R. Schoenman (ed.), *Bertrand Russell, Philosopher of the Century*, London: Allen & Unwin; reprinted in Putnam (1979).

Putnam, H. (1971), *Philosophical Logic*, New York: Harper & Row; reprinted in his (1979).

Putnam, H. (1979), *Mathematics, Matter, and Method: Philosophical Papers*, vol. 1, 2nd edn, Cambridge: Cambridge University Press.

Putnam, H. (1980), 'Models and Reality', *Journal of Symbolic Logic* 45: 464–82; reprinted in his *Realism and Reason: Philosophical Papers, vol. 3*, and in Benacerraf and Putnam (1983).

Quine, W.V. (1936), 'Truth by Convention', in O.H. Lee (ed.), *Philosophical Essays for A.N. Whitehead*, New York: Longmans Green & Co; reprinted in Benacerraf and Putnam (1983) and in Quine (1976).

Quine, W.V. (1937), 'New Foundations for Mathematical Logic', *American Mathematical Monthly* 44: 70–80; reprinted in his (1953).

Quine, W.V. (1948), 'On What There Is', *Review of Metaphysics* 2: 21–38; reprinted in his (1953).

Quine, W.V. (1950), *Methods of Logic*, New York: Holt; revised editions 1959, 1972.

Quine, W.V. (1951a), 'Two Dogmas of Empiricism', *Philosophical Review* 60: 20–43, reprinted in his (1953), and in Hart (1996).

Quine, W.V. (1951b), *Mathematical Logic*, revised edition, Cambridge, MA: Harvard University Press.

Quine, W.V. (1953), *From a Logical Point of View*, 2nd edn, Cambridge, MA: Harvard University Press, 1961.

Quine, W.V. (1955), 'On Frege's Way Out', *Mind* 64: 145–59.

Quine, W.V. (1960), *Word & Object*, Cambridge, MA: MIT Press.

Quine, W.V. (1963), *Set Theory and its Logic*, Cambridge, MA: Harvard University Press.

Quine, W.V. (1970), *The Philosophy of Logic*, Englewood Cliffs, NJ: Prentice-Hall.

Quine, W.V. (1976), *The Ways of Paradox*, Cambridge, MA: Harvard University Press.

Quine, W.V. (1986), 'Response to my Critics', in L.E. Hahn and P.A. Schilpp (eds.), *The Philosophy of W.V. Quine*, La Salle, IL: Open Court.

Ramsey, F.P. (1925), 'The Foundations of Mathematics', *Proceedings of the London Mathematical Society* 25: 338–84; reprinted in his (1931).

Ramsey, F.P. (1931), *The Foundations of Mathematics and Other Essays*, ed. R.B. Braithwaite, London: Routledge & Kegan Paul (2nd edn by D.H. Mellor, 1978.)

Rayo, A. (2002), 'Frege's Unofficial Arithmetic', *Journal of Symbolic Logic* 67: 1623–38.

Rayo, R. and Yablo, S. (2001), 'Nominalism through De-Nominalisation', *Nous* 35: 74–92.

Reck, E.H. and Price, M.P. (2000), 'Structures and Structuralism in Contemporary Philosophy of Mathematics', *Synthese* 125: 341–87.

Resnik, M.D. (1980), *Frege and the Philosophy of Mathematics*, Ithaca: Cornell University Press.

Resnik, M.D. (1985), 'How Nominalist Is Hartry Field's Nominalism?', *Philosophical Studies* 47: 163–81.

Resnik, M.D. (1997), *Mathematics as a Science of Patterns*, Oxford: Clarendon Press.

Resnik, M.D. (2005), 'Quine and the Web of Belief', in Shapiro (2005b), 412–36.

Riemann, B. (1854, 1867), in his *Gesammelte mathematische Werke*, reprinted New York: Dover, 1953.

Robinson, A. (1965), 'Formalism 64', in Y. Bar-Hillel (ed.), *Logic, Methodology, and Philosophy of Science: Proceedings of the 1964 International Congress*, Amsterdam: North-Holland, 228–46.

Robinson, A. (1966), *Non-Standard Analysis*, Amsterdam: North Holland.

Robinson, A. (1967), 'The Metaphysics of the Calculus', in I. Lakatos (ed.), *Problems in the Philosophy of Mathematics*, Amsterdam: North Holland, 28–40; reprinted in Hintikka (1969).

Robinson, R. (1953), *Plato's Earlier Dialectic*, Oxford: Clarendon Press.

Roeper, P. (1997), 'Region-Based Topology', *Journal of Philosophical Logic* 26: 251–309.

Ross, W.D. (1924), *Aristotle's Metaphysics*, Oxford: Clarendon Press.

Rosser, J.B. (1936), 'Extensions of some theorems of Gödel and Church', *Journal of Symbolic Logic* 1: 87–91.

Russell, B. (1901), 'The Logic of Relations', *Revue des Mathématiques* 7: 115–48; translated in Russell (1956).

Russell, B. (1902), Letter to Frege, in van Heijenoort (1967), 124–5.

Russell, B. (1903), *The Principles of Mathematics*, London: George Allen & Unwin.

Russell, B. (1905), 'On Denoting', *Mind* 14: 479–93; reprinted in Russell (1956) and (1973).

Russell, B. (1906a), 'On Some Difficulties in the Theory of Transfinite Numbers and Order Types', *Proceedings of the London Mathematical Society* 4: 29–53; reprinted in Russell (1973).

Russell, B. (1906b), 'On the Substitutional Theory of Classes and Relations', first published in Russell (1973).

Russell, B. (1906c), 'On Insolubilia and their Solution by Symbolic Logic', first published in English in Russell (1973).

Russell, B. (1907), 'The Regressive Method of Discovering the Premises of Mathematics', first published in Russell (1973).

Russell, B. (1908), 'Mathematical Logic as Based on the Theory of Types', *American Journal of Mathematics* 30: 222–62; reprinted in Russell (1956).

Russell, B. (1910), with A.N. Whitehead, *Principia Mathematica*, vol. 1, London: Cambridge University Press.

Russell, B. (1912), *Problems of Philosophy*, Oxford: Oxford University Press.

Russell, B. (1914), *Our Knowledge of the External World*, London: Open Court; 2nd edn with revisions London: George Allen & Unwin, 1926.

Russell, B. (1918), 'The Philosophy of Logical Atomism', *The Monist* 28: 495–527 and 29: 33–63, 190–222, 344–80; reprinted in Russell (1956).

Russell, B. (1919), *Introduction to Mathematical Philosophy*, London: George Allen & Unwin; reprinted New York: Dover, 1993.

Russell, B. (1925), with A.N. Whitehead, *Principia Mathematica*, vol. 1, 2nd edn, London: Cambridge University Press.

Russell, B. (1956), *Logic & Knowledge*, ed. R.C. Marsh, London: George Allen & Unwin.

Russell, B. (1959), *My Philosophical Development*, London: George Allen & Unwin.

Russell, B. (1967), *Autobiography*, London: George Allen & Unwin.

Russell, B. (1973), *Essays in Analysis*, ed. D. Lackey, London: George Allen & Unwin.

Sainsbury, R.M. (1979), *Russell*, London: Routledge & Kegan Paul.

Sainsbury, R.M. (1995), *Paradoxes*, 2nd edn, Cambridge: Cambridge University Press.

Schirn, M. (ed.) (1998), *The Philosophy of Mathematics Today*, Oxford: Clarendon Press.

Scott, D. *et al.* (1981), *Notes on the Formalization of Logic*, Oxford: Sub-faculty of Philosophy.

Shapiro, S. (1983), 'Conservativeness and Incompleteness', *Journal of Philosophy* 80: 521–31; reprinted in Hart (1996).

Shapiro, S. (1991), *Foundations without Foundationalism: A Case for Second-Order Logic*, Oxford: Oxford University Press.

Shapiro, S. (1997), *Philosophy of Mathematics: Structure and Ontology*, New York: Oxford University Press.

Shapiro, S. (2000), *Thinking about Mathematics*, Oxford: Oxford University Press.

Shapiro, S. (2005a), 'Higher-Order Logic', in Shapiro (2005b), 751–80.

Shapiro, S. (ed.) (2005b), *The Oxford Handbook of Philosophy of Mathematics and Logic*, Oxford: Oxford University Press.

Shapiro, S. and Weir, A. (1999), 'New V, ZF, and Abstraction', *Philosophia Mathematica* 7: 293–321.

Skolem, T. (1923), 'The Foundations of Elementary Arithmetic, Established by Means of the Recursive Mode of Thought', translated in van Heijenoort (1967), 302–33.

Skorupski, J. (1989), *John Stuart Mill*, London: Routledge.

Sober, E. (1993), 'Mathematics and Indispensability', *Philosophical Review* 102: 35–57.

Spector, C. (1955), 'Recursive Well-orderings', *Journal of Symbolic Logic* 20: 151–63.

Strawson, P.F. (1959), *Individuals*, London: Methuen.

Stroud, B. (1965), 'Wittgenstein & Logical Necessity', *Philosophical Review* 74: 504–18.

Takeuti, G. (1987), *Proof Theory*, 2nd edn, Amsterdam: North Holland.

Tennant, N. (1987), *Anti-Realism and Logic*, Oxford: Oxford University Press.

Urquhart, A. (2003), 'The Theory of Types', in N. Griffin (ed.), *The Cambridge Companion to Bertrand Russell*, Cambridge: Cambridge University Press.

Uzquiano, G. (2005), review of Potter (2004), *Philosophia Mathematica* 13: 208–12.

van Fraassen, B.C. (1980), *The Scientific Image*, Oxford: Oxford University Press.

van Heijenoort, J. (1967), *From Frege to Gödel*, Cambridge, MA: Harvard University Press.

van Stigt, W.P. (1990), *Brouwer's Intuitionism*, Amsterdam: North Holland.

van Stigt, W.P. (1998), 'Brouwer's Intuitionist Programme', in P. Mancosu (ed.), *From Brouwer to Hilbert*, Oxford: Oxford University Press, 1–22.

von Neumann, J. (1925), 'Eine Axiomatisierung der Mengenlehre', *Journal für reine und angewandte Mathematik* 154: 219–40; translated in van Heijenoort (1967).

Wagner, S. (1987), 'The Rationalist Conception of Logic', *Notre Dame Journal of Formal Logic* 28: 3–35.

Wang, H. (1954), 'The Formalization of Mathematics', *Journal of Symbolic Logic* 19: 241–66; reprinted in his (1962).

Wang, H. (1959), 'Ordinal Numbers and Predicative Set Theory', *Zeitschrift für Mathematische Logik und Grundlagen der Mathematik* 5: 216–39; reprinted in his (1962).

Wang, H. (1962), *A Survey of Mathematical Logic*, Amsterdam: North Holland.

Wedberg, A. (1955), *Plato's Philosophy of Mathematics*, Stockholm: Almqvist & Wiksell.

Weierstrass, K. (1857, 1872), *Mathematische Werke*, Mayer & Müller, 1894–1927.

Weyl, H. (1918), *Das Kontinuum*, Leipzig: Veit.

Whitehead, A.N. (1919), *An Enquiry into the Principles of Nature*, Cambridge: Cambridge University Press.

Whitehead, A.N. (1929), *Process and Reality*, Cambridge: Cambridge University Press.

Wigner, E. (1967), 'The Unreasonable Effectiveness of Mathematics in the Natural Sciences', in his *Symmetries and Reflections*, Indiana: Indiana University Press.

Wittgenstein, L. (1921), *Tractatus Logico-Philosophicus*, translated by D.F. Pears and B.F. McGuiness, London: Routledge & Kegan Paul, 1961.

Wittgenstein, L. (1953), *Philosophical Investigations*, translated by G.E.M. Anscombe, Oxford: Blackwell.

Wittgenstein, L. (1964), *Remarks on the Foundations of Mathematics*, translated by G.E.M. Anscombe, Oxford: Blackwell; reprinted Cambridge, MA: MIT Press, 1978.

Wolff, C. (1739), *A Treatise of Algebra*, translated by J. Hanna, London: Bettesworth & Hitch.

Wright, C. (1980), *Wittgenstein on the Foundations of Mathematics*, London: Duckworth.

Wright, C. (1983), *Frege's Conception of Numbers as Objects*, Aberdeen: Aberdeen University Press.

Wright, C. (1988), 'Why Numbers Can Believably Be', *Revue Internationale de Philosophie* 42: 425–73.

Wright, C. (1990), 'Field and Fregean Platonism', in Irvine (1990), 73–93; reprinted in Hale and Wright (2001).

Wright, C. (1997), 'On the Philosophical Significance of Frege's Theorem', in R. Heck (ed.), *Language, Thought, and Logic: Essays in Honour of Michael Dummett*, Oxford: Oxford University Press, 201–44; reprinted in Hale and Wright (2001).

Wright, C. (1999), 'Is Hume's Principle Analytic?', *Notre Dame Journal of Formal Logic* 40: 6–30; reprinted in Hale and Wright (2001).

Zermelo, E. (1908), 'Untersuchungen über die Grundlagen der Mengenlehre', *Mathematische Annalen* 65: 261–81; translated in van Heijenoort (1967).

Zermelo, E. (1930), 'Über Grenzzahlen und Mengenbereiche', *Fundamenta Mathematica* 16: 29–47.

Index

Printed in the United States
By Bookmasters